Chemical Engineering Essentials 1

Chemical Engineering Essentials 1

Comprehensive Chemical Engineering

Edited by
Raj Kumar Arya
George D. Verros
J. Paulo Davim

iSTE

WILEY

First published 2025 in Great Britain and the United States by ISTE Ltd and John Wiley & Sons, Inc.

ISTE Ltd
27-37 St George's Road
London SW19 4EU
UK

www.iste.co.uk

John Wiley & Sons, Inc.
111 River Street
Hoboken, NJ 07030
USA

www.wiley.com

Library of Congress Control Number: 2024951138

British Library Cataloguing-in-Publication Data
A CIP record for this book is available from the British Library
ISBN 978-1-78630-984-6

Contents

Preface

The field of chemical engineering has evolved significantly over the decades, expanding its horizons to encompass a range of interdisciplinary applications and cutting-edge advancements. *Chemical Engineering Essentials 1* is designed as a comprehensive resource for both students and professionals, providing fundamental insights along with in-depth discussions on advanced topics. With the rapid advancements in technology and growing concerns over sustainability and safety, chemical engineering is at a transformative juncture, poised to offer sustainable and innovative solutions across industries. This handbook seeks to serve as a definitive guide for the current generation of engineers, providing both foundational knowledge and modern approaches required to navigate and excel in this dynamic field.

Volume 1 is organized into three sections, as is Volume 2, each addressing critical aspects of chemical engineering.

Part 1 of this volume: Fundamental Principles lays the groundwork, beginning with an overview of the field and covering essential topics such as material and energy balances, thermodynamics and phase equilibrium. These fundamental principles form the bedrock upon which more specialized knowledge is built, equipping readers with a strong theoretical base.

Part 2 of this volume: Fluid Mechanics and Transport Phenomena delves into the physics and behavior of fluid systems and heat transfer. Chapters on fluid flow, heat conduction, convection, radiation and mass transfer provide essential understanding for designing and analyzing chemical processes.

Part 3 of this volume: Separation Processes covers the vital area of chemical separations, with discussions on extraction, distillation, filtration and emerging membrane processes such as pervaporation. Mathematical modeling techniques for binary and multicomponent distillation are also covered to provide a deeper insight into process design and optimization.

Part 1 of Volume 2: Reaction Engineering introduces readers to various reaction mechanisms and reactor designs, with a special focus on applications within the pharmaceutical industry and the concept of reactive chromatography. The section concludes with the mathematical modeling of batch reactors and nonisothermal continuous stirred-tank reactors (CSTRs), complemented by simulation studies using MATLAB and ASPEN PLUS.

Part 2 of Volume 2: Material Properties and Advanced Applications focuses on the diverse materials used in chemical engineering, alongside advanced applications such as hydrogen production, vinyl chloride production, process intensification and the integration of nanotechnology in the field. These topics underscore the importance of selecting the right materials and processes to achieve efficiency and innovation in chemical engineering applications.

Part 3 of Volume 2: Sustainability and Safety highlights the importance of green chemistry and sustainable practices in chemical engineering. This section also covers waste minimization, resource recovery and safety management, including hazard identification and risk assessment – critical components of responsible engineering practice in today's world.

This handbook represents the collective effort of numerous contributors, whose expertise and dedication have shaped this comprehensive guide. We extend our deepest gratitude to each author who has contributed invaluable insights and research to make this work possible. Our heartfelt thanks also go to our families, whose unwavering support, love and encouragement made this journey possible.

Special acknowledgment goes to the team at ISTE. Their efforts and guidance were instrumental in bringing this book to completion.

It is our sincere hope that the two volumes of *Chemical Engineering Essentials* will serve as a valuable resource, offering knowledge, inspiration and practical tools for all of those engaged in the field of chemical engineering.

Raj Kumar ARYA
George D. VERROS
J. Paulo DAVIM
January 2025

PART 1

Fundamental Principles

1

Overview of Chemical Engineering

1.1. Introduction

Chemical engineers have been enhancing our overall welfare for over a century. Chemical engineers have contributed to several advancements that impact our daily lives, ranging from the creation of more efficient computer chips to improvements in recycling, illness treatment, water purification and energy generation.

Chemical engineering (CE), also referred to as process engineering, is a branch of engineering that uses mathematics, economics, physical sciences and biological sciences to create and modify chemicals, energy and materials. Chemical Engineering was originally developed as a combination of mechanical engineering and applied chemistry, with a specific focus on its application in the petrochemical and heavy chemical industries.

Chemical engineers convert laboratory-developed procedures into practical applications for the commercial manufacturing of products, and then strive to uphold and enhance such processes. They depend on the fundamental principles of engineering, namely mathematics, physics and chemistry. Biology is assuming a progressively significant role.

Chemical engineers primarily conceptualize and develop procedures related to chemical production on a large scale. Chemical engineers primarily focus on designing and resolving issues related to the manufacturing of various substances such as chemicals, fuels, foods, pharmaceuticals and biologicals. They are primarily hired by industrial facilities of significant size to optimize efficiency and enhance the quality of products, while minimizing expenses.

Chapter written by Devyani THAPLIYAL, Kshitij TEWARI, Chitresh Kumar BHARGAVA, Avinash CHANDRA, Pramita SEN, Rahul KUMAR, Amit K. THAKUR, George D. VERROS and Raj Kumar ARYA.

Chemical engineering encompasses the synthesis and fabrication of goods using chemical reactions and procedures. This includes the creation of machinery, systems and procedures for the purification of basic substances and the blending, combining and treatment of chemical compounds.

Chemical engineers enhance food processing processes and develop more efficient methods for fertilizer production, with the aim of augmenting both the amount and quality of food resources. In addition, they fabricate artificial fibers that enhance the comfort and water resistance of our garments; they devise techniques for the large-scale production of pharmaceuticals, thereby reducing their cost; and they invent safer and more efficient processes for refining petroleum products, thereby increasing the productivity and cost-effectiveness of energy and chemical resources.

Additionally, chemical engineers devise strategies to address environmental issues, such as pollution control and remediation. Indeed, they engage in the manipulation of chemicals, which are employed in the production or enhancement of virtually all the objects within your surroundings.

1.2. History of chemical engineering

Chemical engineering is a field that emerged from the practice of "industrial chemistry" in the late 19th century, as stated by Wikipedia (2024a, 2024b). Prior to the Industrial Revolution in the 18th century, the production of industrial chemicals and consumer items such as soap primarily relied on batch processing methods. Batch processing involves manual labor, when workers combine predetermined quantities of materials in a container and then subject the mixture to heating, cooling or pressurization for a specific duration. Subsequently, the product can be separated, refined and examined in order to obtain a marketable product.

Currently, batch processes are still used for more valuable products, such as pharmaceutical intermediates, specialty and formulated products such as perfumes and paints, or in food production such as pure maple syrups. Despite being slower and less efficient in terms of labor and equipment usage, batch methods are still profitable in these cases.

Chemical engineering approaches have enabled the production of higher volumes of chemicals through continuous "assembly line" chemical processes in industrial process development. The Industrial Revolution marked the transition from batch processing to a more continuous processing approach. The Industrial Revolution resulted in an unparalleled increase in demand, in terms of both quantity and quality, for large quantities of chemicals such as soda ash. This entailed two objectives: firstly, expanding the scale of the activity and enhancing operational efficiency, and secondly, exploring viable alternatives to batch processing, such as continuous operation. The practice of industrial chemistry dates back to the 1800s, and its academic study in British universities commenced with the publishing of the influential book "Chemical Technology" in 1848 by Friedrich Ludwig Knapp, Edmund Ronalds and Thomas Richardson.

According to Wikipedia (2024a, 2024b), James F. Donnelly is credited with referencing an 1839 mention of chemical engineering in the context of sulfuric acid manufacturing. However, it was noted in the same piece that George E. Davis, an English consultant, was attributed with originating the word. In addition, Davis made an attempt to establish a Society of Chemical Engineering. However, it ended up being called the Society of Chemical Industry in 1881, with Davis serving as its inaugural secretary.

The term "chemical engineering", which refers to the application of mechanical equipment in the chemical industry, gained widespread usage in England after 1850. The inaugural effort to establish a Society of Chemical Engineers in London took place in 1880. Consequently, the Society of Chemical Industry was established in 1881. The American Institute of Chemical Engineers (AIChE) was established in 1908, while the U.K. Institution of Chemical Engineers (IChemE) was established in 1922. The term "chemical engineer" was widely used in Great Britain and the United States by 1910.

1.3. Chemical engineering profession

Commonly known as process engineering, chemical engineering is a discipline of engineering that uses physical and biological sciences, mathematics and economics to produce and modify chemicals, energy and materials, as stated by AIChE (2024).

Traditionally, it encompasses the fields of heat transfer, mass transfer, momentum transfer, kinetics and reaction engineering, chemical thermodynamics, control systems and dynamic simulation, separation processes and unit operations.

Chemical engineering, traditionally used in the petrochemical and heavy chemical sectors, has experienced significant advancements and expanded its applications to many domains such as climate change, environmental systems, biomedical science, novel materials and complex systems. Chemical engineers have a significant impact on the manufacturing of nearly every item produced on a large scale in industry.

Common tasks include:

– ensuring adherence to health, safety and environmental regulations;

– engaging in research to enhance manufacturing processes;

– creating and strategizing the arrangement of equipment implementing safety protocols for handling hazardous compounds;

– supervising and enhancing the efficiency of manufacturing procedures. Calculating the expenses associated with production.

Chemical engineers employed in corporate and managerial settings frequently make visits to research and production sites. Engaging with individuals and fostering teamwork are essential for the achievement of chemical engineering projects.

Chemical engineers commonly operate in manufacturing plants, research laboratories or pilot plant facilities. They operate in the vicinity of expansive manufacturing machinery that is located both indoors and outdoors. Hence, it is frequently necessary for them to don personal protective gear such as hard hats, goggles and steel-toe shoes. Chemical engineering is used in several industries and sectors, and is primarily employed in industrial facilities that focus on large-scale production, aiming to optimize efficiency and the excellence of the final product, while minimizing expenses. Chemical engineering is used by several industries including aerospace, automotive, biomedical, electronic, environmental, medical and military sectors to enhance and advance their technical goods. These products encompass a wide range of applications.

Advanced materials have exceptional strength, durability and bonding properties for use in automobiles. Materials that are compatible with living organisms can be used for implants and prosthesis. Optoelectronic devices are designed to use light for various purposes. One important aspect of these devices is the use of films, which play a crucial role in their functionality. These films are specifically engineered to enhance the performance and efficiency of optoelectronic devices.

1.4. Features and challenges in chemical engineering

According to Garnier (2014) and Hipple (2017), numerous captivating and productive difficulties in the field of chemical engineering arise from the amalgamation of chemical engineering with chemistry, physics and biology, which necessitates a redefining of the control volume. This list does not encompass all of the forthcoming difficulties in the field of chemical engineering and the areas where chemical engineering abilities will be applicable:

– Energy resources and use: in this context, it is necessary to disregard any "agendas" that may be primarily driven by political constituents, such as corn-based ethanol. Energy is necessary for the provision of food, shelter and transportation. In order to accommodate the growing global population and the increasing demand for Western standards of living in what are currently referred to as "Third World" countries, it is imperative that we adopt more efficient sources of energy, unless we choose to revert to a completely agrarian society and lower our current standard of living. A recent breakthrough has occurred in the understanding of underground fluid flow and mechanics, commonly known as "fracking". This breakthrough enables the recovery of hydrocarbons that were previously trapped in underground rock formations. It involves injecting high-pressure fluids to break apart the rock formations. Enhanced oil recovery incorporates advanced surface chemical techniques into the conventional process of drilling wells into liquid reservoirs underground. Energy conservation will hold equal significance. It is impractical to believe that inhabitants in developing countries will not aspire to the same luxuries that are commonly enjoyed by those in more developed nations. If this is to occur, there will be a need for greater energy resources and improved energy efficiency. In recent decades, there has been a significant improvement in energy efficiency in various industries and goods, which can be attributed, at least in part, to the increase in energy prices. While there has been a recent shift in this trend, once these energy saving measures are implemented, it is improbable that they will be reversed. As a component of this endeavor, there will be a gradual advancement in the conversion of solar energy. While solar energy is abundant worldwide, its energy density is far lower than that of petroleum fuels. Efforts to enhance the efficiency of solar power conversion, for both thermal and electrical energies, have made consistent advancements. However, these improvements are still insufficient to sustain a solar-based economy. Chemical engineers will have a crucial role in the progress of this technology, particularly in terms of improving catalysts, enhancing collection material efficiency and advancing energy distribution technologies. Affordable energy is crucial for enhancing the quality of life for the majority of individuals in underdeveloped countries. Given the well-established fact of anthropogenic greenhouse gasses leading to gradual global warming, a major task is to generate energy that has a minimal impact on the environment. Chemical engineers are tasked with the job of verifying and ensuring that energy balances and thermodynamics are optimized in the most economically feasible manner. The utilization of renewable sources and the application of green chemistry in chemical production is an expansion of the task. It is the duty of chemical engineers to identify processes and reactions with favorable thermodynamics and energy balances, and subsequently enhance these processes through collaboration with economists, environmental scientists and society as a whole. The efficient storage of solar energy, which includes energy derived from wind and ocean currents, in order to facilitate its distribution during periods of high human demand, continues to be a significant challenge. Hence, it is crucial to prioritize the advancement of reversible processes that enable efficient energy storage and usage, while also exhibiting prompt initiation and termination properties. Efficiently releasing significant amounts of electrical energy is crucial for meeting societal demands. However, it is equally important to recognize the immense advantages of capturing and storing solar energy in a manner that imitates natural photosynthesis. This involves storing solar energy in chemical bonds, rather than as heat or electronic charge separation. If the artificial photosynthetic reaction, which harnesses solar energy, uses carbon dioxide, then two significant goals would be accomplished with a single technological advancement. It is important to note that although the reaction between carbon monoxide and oxygen releases a large amount of heat, the opposite reaction, which is the breaking apart of carbon dioxide into carbon monoxide and oxygen due to heat, can occur at the temperatures achievable in a solar furnace. The outstanding technological challenges involve the creation of sophisticated refractory materials capable of enduring the necessary temperatures to facilitate the reaction, effective heat exchange and the efficient separation of the resulting products. The process of dissolving carbon monoxide in a solution of alkali metal in water to produce alkali metal forms appears to be a highly favorable method.

– Water: the issue of providing drinking water to a growing worldwide population is compounded by the fact that only a tiny percentage of water resources are readily available for consumption. Therefore, it is necessary to explore alternative sources of non-fresh water for drinking purposes, as well as implement effective water recycling and reuse strategies. The provision of safe drinking water, together with food and energy, is sometimes referred to as the "great nexus" of engineering challenges in the 21st century.

– Green chemicals: it is important to fully use renewable feedstock and make the most of all its components. Due to the relatively low energy density of biomass compared to fossil carbon sources, it is necessary to carefully reassess the energy efficiency of biomass processing. This includes the creation of smaller mobile processing units that can be transported to regions where biomass is accessible during specific seasons. It is important to consider potential social and community advantages when doing a re-evaluation. An essential element in optimizing the utilization of biomass will involve the advancement of novel chemical routes that enable more efficient exploitation of the compositions of polysaccharides and lignins. The manipulation of cell differentiation and tissue formation in higher plants by certain insects in the families *Hemiptera* and *Hymenoptera*, resulting in the formation of galls and related protective structures, is a topic that deserves thorough multidisciplinary investigation. Although numerous beneficial enzymes are already manufactured, extracted and used in large-scale commercial applications, their catalytic rates are often constrained by heat instability, denaturation caused by surfactants and deviations in pH from the neutral range. Chemical engineers have conventionally employed heat, pressure and pH to expedite chemical reactions. However, the investigation of extremophile organisms and their enzymes, which have evidently developed to endure extreme temperatures, pressures and pH levels found in deep ocean vents and volcanic pools, seems to be at an early stage.

– Materials: chemical engineering will play a significant role in two long-term elements of this topic. Plastics recycling is the first and most frequently discussed method. The process of segregating polymers into readily recyclable components is highly expensive and inefficient, yielding materials that are seldom as functional as the initial raw ingredients. The economic justification for the separation process of materials such as polyethylene, polystyrene, ABS, nylon, etc., is seldom found, and instead, is supported by taxpayer financing. During this process, the characteristics of the polymer usually deteriorate, mostly due to a decrease in molecular weight, rendering them unsuitable for their original intended use. There are numerous instances of products in the store that bear a label indicating that they include a maximum of 10% recycled materials. Adding more would render the properties that the user is paying for unattainable. Several long-term events will occur, all of which will necessitate substantial involvement from chemical engineering. One possible solution is the implementation of separation technologies that can effectively and cost-efficiently separate different types of discarded plastics. This would enable a greater utilization of these plastics in their original products. The second objective is minimizing plastic consumption by implementing packaging and system modification. For instance, there are situations where a sizable plastic container can be transformed into a "shrink wrap" style packaging, resulting in a significant reduction in the amount of plastic used. It might also be feasible to completely eliminate conventional packaging by using different kinds of coatings. Furthermore, cost-effective pyrolysis methods (heating materials to high temperatures without oxygen, in contrast to combustion) could potentially turn polymers back into their original monomers, such as polystyrene to styrene, polyethylene to ethylene, polypropylene to propylene, etc. By generating conventional hydrocarbon monomers, it becomes simpler to separate them using typical chemical engineering techniques such as distillation. Additionally, these monomers can be used in various applications, not limited to the plastic they were derived from. Efforts are now being made in all of these areas, although the second option is likely to be the most resilient and sustainable in the long run. In addition, we must include "nanomaterials" in this category, as they are increasingly being incorporated into consumer items to offer distinctive features. The possible influence of 10–9 particle size materials and their distribution within the environment is currently being debated and analyzed. Nevertheless, these materials of significant size have a profound and beneficial effect on numerous material characteristics. Chemists and chemical engineers will encounter significant hurdles in the surface chemistry, contact with other materials and affordable manufacture of such materials. The performance of any separation process will vary dramatically when the particles to be separated fall within this particular size range. Within the previously discussed periodic table, there exists a category of substances referred to as "rare earth" metals. Illustrative instances encompass metallic elements such as dysprosium and neodymium. These elements are often used in modest amounts to enhance the characteristics of primary metals. They are infrequent in their incidence and costly to produce. A significant chemical engineering obstacle is the reduction of costs associated with the recovery and manufacturing of rare materials, as well as the exploration of alternative materials and product designs that do not rely on such scarce resources.

– Medical, biomedical and biochemical applications: artificial organs, which are now commonly used, are familiar to all of us. Artificial hearts and kidney dialysis machines can be understood as pumps and filters that serve as chemical engineering substitutes for the original organs in the human body. The principles we have covered, such as fluid dynamics, fluid characteristics, polymers and filtration, form the foundation for the development of these synthetic organs. As our comprehension of human body functionality advances, it is probable that we will witness the development of more artificial and replacement organs and body parts, with chemical engineering fundamentals playing a crucial role in their design. Examples may encompass artificial lungs, prosthetic joints and bodily fluid filtration. Chemical engineering principles, such as fluid flow, friction, filtration fundamentals, porosity, pressure drop and others, are essential components of these products. Currently, numerous educational institutions have integrated the chemical engineering department with this particular field of study, resulting in its rebranding as either biochemical engineering or biomedical engineering. An artificial heart is a man-made reproduction of the natural human heart. However, from a chemical engineering perspective, it can be described as a pump. All the factors involved in pump design, such as friction, energy requirements, flow rates, valve restriction and control, are equally applicable to these devices as they are to traditional pumps. However, it is important to note that the limitations on available energy, friction and pressure drop are significantly more restrictive in these devices. The materials used in the production of these devices must possess compatibility with human tissue to prevent rejection. This issue of incompatibility is significantly more complex than the previously described concerns of corrosion. It involves not only the degradation of materials, but also the potential risk of human fatalities. There are numerous prescription medications that are preferable to be taken in modest, continuous doses, rather than large doses once or twice a day. The practice of enclosing and gradually releasing various categories of medications using a dermal patch is increasingly prevalent. The process of encapsulating a drug and gradually releasing it is achieved by comprehending the rates at which molecules pass through the skin and aligning this absorption rate with an encapsulation method that releases the drug at an equivalent rate. Creation of a systematic methodology involves modeling and governing the behavior and functionality of the human body and brain processes using engineering principles. Use of simulation and control techniques to address several levels of biological systems, including DNA, RNA, cells, tissues, organs and the human body, in order to enhance the quality of life for those with genetic and related illnesses. Use of minimally intrusive sensors for the purpose of monitoring and regulating blood pressure, blood lipid content and heart rate. Nanotechnology enables precise targeting in the field of oncology and medicine delivery. Advancements in biotechnologies and enhanced biomaterials have the purpose of regenerating organs.

– Biochemical engineering is the integration of chemical engineering and biology to efficiently manufacture valuable medical products and systems on a larger scale. From a chemical engineering perspective, there are several distinct obstacles involved in this endeavor. Upon revisiting our analysis of membrane materials, it becomes evident that biological molecules, including viruses and proteins, possess particle sizes ranging from 0.01 to 10 μm. These sizes are considerably lower than the typical particle sizes generated in traditional inorganic and organic chemical reactions. This complicates the separation and retrieval of active, desirable molecules, making it necessary to employ more costly and advanced recovery methods. Typically, the processing of biological entities begins with solutions that are very diluted. The confluence of this component and the particle size presents a significant obstacle in achieving liquid–solid separation in this field, particularly in meeting the stringent pharmaceutical and FDA purity standards. The majority of biological materials are able to exist and carry out their functions under normal body and environmental settings. Therefore, it is impractical to carry out the majority of reactions at elevated temperatures, which could potentially enhance reaction speeds. This fact may also impose constraints on other unit operation notions for consideration. The efficacy of numerous biopharmaceuticals and medications is such that the quantity of substance required for a substantial market share may be comparable to that of a small-scale pilot plant in the conventional chemical or petrochemical sector. While certain scale-up guidelines may not be applicable, the importance of efficient production and high yields of highly valued minerals remains significant. The challenges involved in developing the medicine into an acceptable form for human consumption, such as its dissolving rate in the stomach and gastric tract, as well as the necessary inert components, are significant. The concept of stereo specificity is typically not taken into account in traditional chemical engineering and processing. Nevertheless, it holds great significance in the process of designing and producing physiologically active compounds. The carbon atom possesses a distinctive geometric structure, resembling a pyramid, due to its four bond connections, with the "C" molecule positioned at the middle. If the four molecules connected to this carbon are distinct, there is a chance that the carbon will exhibit chirality, indicating that it lacks symmetry in terms of its geometry. One of the peripheral groups will seem to be receding into the paper when drawn, while the other will protrude outward. The differentiation between these molecules is referred to as "left-hand" and "right-hand" rotations (or d-, l-; denoting dextro or levo rotation, respectively). Biologically relevant or active compounds, such as medicines, are crucial because our human body specifically recognizes and interacts with l- (levo) rotated molecules. These two molecules, which have undergone a chiral rotation, are referred to as enantiomers. This means that they cannot be overlapped with each other, regardless of the physical rotation of the molecule. The human body can only identify and metabolize molecules that are in the "levo" or "l-" configuration. Therefore, whenever there is an optical isomer present, it indicates the existence of an identical molecule that needs to be separated before the material can be used. The lack of absorption of the opposing isomer forms the foundation for many artificial sweeteners available in the market that possess a "right-handed" structure. We have the sensation of the molecule's taste (sweetness), but our bodies do not assimilate it and it is excreted as part of regular bodily fluids. The distinction between molecules with left-handed and right-handed orientations is referred to as various "enantiomers". Separating enantiomers may not always be essential, depending on the potential adverse effects of the other optical isomer. However, when separation is required, it significantly influences decisions in chemical engineering processes. The main point is that while having distinct "optical" characteristics, these molecules share the same physical properties, such as boiling or melting points. Conventional and less expensive methods such as distillation cannot be used to separate these enantiomers. Minor variations in solubility properties can be used in crystallization methods, chromatography and specialized ion-exchange resins. Furthermore, apart from the aforementioned fundamental obstacles, we are also confronted with the obstacle of sluggish response rates. Biological processes generally exhibit slower kinetic speeds in comparison to conventional chemical reactions. Consequently, the capacity to increase temperatures during processes such as tissue culture growth is severely restricted due to temperature sensitivity restrictions. Turbulence is typically necessary in biological reactions, just as it is in other chemical reactions. However, the heightened susceptibility to mechanical forces may need the use of specialized configurations. Additionally, it is worth mentioning that chemical engineers, due to their extensive knowledge of safety, can make valuable contributions to safety assessments of biochemical processes and products. The potential dissemination of hazardous, biochemically strong substances with minute particle dimensions, such as viruses, is a constant source of worry, and the specialized knowledge of chemical engineering in gas management and filtration can be employed to address this issue.

– Reaction engineering: the utilization of a blend of organic, inorganic and biochemical catalysis techniques aims to lower the energy required for chemical reactions, enhance the specificity of the reactions, decrease energy consumption, facilitate the separation of by-products and substitute toxic organic solvents and reagents containing scarce elements with reactions conducted in water-based or bio-based solvents, following environmentally friendly chemical principles. Using photosynthesis to transform solar energy and carbon dioxide into glucose, ligno-cellulosic polymers and their intermediates by the use of enzyme catalysts and/or aqueous systems. Comprehend and enhance the movement of substances, the transfer of energy, the degree of reaction and the ability to choose specific reactions in the field of medicine. Applications encompass the targeted eradication of cancer cells, bacteria, fungi and viruses (infection), as well as the control of immunological responses. Predictive reaction engineering involves manipulating the rate of reactant and product removal in accordance with the kinetics of the reaction in order to reduce the occurrence of side reactions. This approach aims to facilitate separation processes by making them easier and more efficient.

– Unit operations and transport phenomena: enhanced separation techniques with more selectivity, specificity and reduced energy consumption for gas–gas and liquid–liquid systems. Efficient and resistant reverse osmosis and membrane separations with a high flow rate and the ability to prevent fouling. Enhanced isolation of heat-sensitive compounds with comparable boiling points by fractional distillation or alternative methods. Improved techniques for the pumping and transportation of mixtures of solid particles in liquids, particularly when the solid concentration is high.

– Multiscale engineering: establishing connections between the nano-, micro- and mesoscales and the macroscale in both materials and processes will be essential for addressing the bulk of difficulties mentioned. Improved molecular dynamic simulations are crucial for the advancement of nanotechnology through molecular engineering. Materials should be used that are capable of being recycled into similar products, or if not feasible, into a series of products with decreasing value, ultimately resulting in entirely biodegradable end-products. Enhance the production of materials and composites by the comprehensive comprehension of component structures, ranging from atomic scale to macroscopic qualities, using low-energy techniques. Efforts should be focused on replacing energy-intensive concrete and metals used in commodity applications.

Advancements in chemical engineering have typically been gradual and step-by-step. Chemical engineering originated from the fusion of mechanical engineering and applied chemistry, and has evolved into a comprehensive and expansive field that consistently pursues novel challenges. One area that is now focusing on addressing these difficulties is the development of advanced technologies to efficiently use matter and energy in order to create innovative goods, such as organs, energy storage systems and molecularly designed composites.

A closely associated field is process optimization, which aims to ensure that both current and future products are produced in the most efficient and sustainable manner, with regard to energy consumption and waste generation. Another area of difficulty involves constructing new facilities and adapting existing ones in a manner that ensures they are authorized by society to function and use the technologies that are essential for maintaining satisfactory living conditions.

1.5. Conclusions

In recent decades, the field of chemical engineering has made rapid progress and found applications in several industries such as biomedicine, environmental systems and innovative materials. Currently, research in chemical engineering is thriving in various areas such as nanomaterials, microfluidics, nanoreactors, biomedicine and bioinformatics, artificial intelligence (AI), etc. Currently, ongoing research is mostly centered on materials science and the creation of new processes. However, there is also a strong emphasis on reducing waste and emissions by recycling and reusing, as well as using environmentally friendly methods to generate useful chemicals. These environmentally friendly procedures prioritize not only the efficient utilization of raw materials and energy, but also adhere to society's environmental regulations, such as the circular economy concept.

The primary objective in creating an industrial process is to effectively meet consumer expectations while simultaneously minimizing the negative impact on the environment. Nevertheless, the primary responsibility of the chemical engineer remained unchanged: to achieve the most efficient design of industrial processes by extrapolating data obtained from laboratory tests.

Chemical engineers worldwide employ conventional techniques such as applied mathematics and numerical analysis, transport phenomena, physical chemistry principles such as reaction kinetics and surface phenomena, engineering thermodynamics and control theory to achieve their objectives.

The recent progress in various interconnected disciplines, such as AI and nanotechnology, has prompted a reassessment of the established concepts of chemical engineering. This reassessment focuses specifically on the design of complicated processes at the nanoscale.

Challenges in the field of chemical engineering applications encompass various contemporary aspects of reaction engineering, including photocatalysis and nanoscale reactions. Additionally, there are biomedical challenges related to extending human life and improving the quality of living through the production of artificial organs on a large scale.

Energy problems, such as the storage of solar energy and the reduction of CO_2 emissions, are also significant. Furthermore, there is a focus on developing more efficient and sustainable chemical industries, such as the better utilization of biomass. Lastly, there is a need for the advancement of novel intelligent materials, such as functional coatings.

The aforementioned issues in both research and applications necessitate a reassessment of the theory and practice of chemical engineering that has been created in recent decades. This book aims to connect theoretical concepts with practical applications, as well as integrate current research with industrial use, in response to the latest advancements in the field.

1.6. References

Garnier, G. (2014). Grand challenges in chemical engineering. *Frontiers in Chemistry*, 2, 17. doi: 10.3389/fchem.2014.00017.

Hipple, J. (2017). *Chemical Engineering for Non-Chemical Engineers*, 1st edition. Wiley-AIChE, New York.

List of websites

ACS (2024). Chemical engineering [Online]. Available at: https://www.acs.org/careers/chemical-sciences/areas/chemical-engineering.html [Accessed on 4 April 2024].

AIChE (2024). What do chemical engineers do? [Online]. Available at: https://www.aiche.org/k-12/what-do-chemical-engineers-do [Accessed 4 April 2024].

Wikipedia (2024a). Chemical engineering [Online]. https://en.wikipedia.org/wiki/Chemical_engineering [Accessed 4 April 2024].

Wikipedia (2024b). History of chemical engineering [Online]. Available at: https://en.wikipedia.org/wiki/History_of_chemical_engineering [Accessed 4 April 2024].

2

Fundamentals of Material and Energy Balances

This chapter introduces the fundamental concepts and terminologies essential to chemical engineering, and lays the groundwork for understanding material and energy balances. It begins with clear definitions and an overview of unit systems and conversions across different measurement frameworks to ensure precision in engineering calculations. A systematic methodology for performing material and energy balances in diverse chemical processes is presented, with an emphasis on a logical and structured approach. To enhance comprehension, this chapter includes illustrative examples that demonstrate the practical application of these principles in real-world scenarios. This comprehensive guide serves as a critical resource for mastering the fundamentals of chemical engineering processes. In this chapter, concise views of major chemical process calculations books (Himmelblau and Riggs 2012; Felder et al. 2020) are presented for a quick recap. For detailed analysis, readers are advised to go through these primary books.

2.1. Introduction

Material and energy balances are fundamental concepts in chemical engineering and related fields, providing the basis for the analysis and design of chemical processes. These balances ensure the efficient use of raw materials and energy resources, while minimizing waste and environmental impact. By applying the principles of conservation of mass and energy, engineers can predict the process behavior, optimize operations and troubleshoot inefficiencies.

Chapter written by Kshitij TEWARI, Devyani THAPLIYAL, Chitresh Kumar BHARGAVA, Avinash CHANDRA, Pramita SEN, Rahul KUMAR, Amit K. THAKUR, George D. VERROS and Raj Kumar ARYA.

Material balance focuses on tracking the flow and distribution of substances in a process, ensuring that input, output, accumulation, and losses are accounted for. This approach is essential in industries such as pharmaceuticals, food processing and petrochemicals, where precise control of compositions and quantities is critical.

The energy balance complements the material balance by evaluating the energy transformations that occur within a system. It considers the energy input through heat, work and chemical reactions and compares it with the energy output and accumulation. Understanding energy balances is crucial for improving energy efficiency and sustainability in process industries.

This chapter explores the basic principles of material and energy balances, covering steady-state and unsteady-state processes, reactive and non-reactive systems, and open and closed systems. Real-world examples and problem-solving techniques will demonstrate their application in designing, optimizing and scaling up industrial processes. Mastering these concepts is indispensable for engineers who aim to develop innovative, efficient and environmentally friendly solutions.

2.2. Process term definitions

Understanding essential process terms is crucial for analyzing, designing and optimizing chemical and industrial systems. These terms describe the physical phenomena, equipment and operations that form the basis of process engineering. Below is a detailed discussion of these terms and their significance in material and energy balances (Himmelblau and Riggs 2012; Felder et al. 2020).

Gas–liquid contact and absorption

Gas–liquid contact involves the interaction of a gas mixture with a liquid solvent, enabling the transfer of one or more gas components into the liquid phase. This process commonly occurs in absorption columns, where a countercurrent flow arrangement maximizes contact. The liquid solvent, introduced at the top of the column, flows downward, while the gas enters from the bottom and rises. This setup ensures efficient mass transfer, with the gas exiting the column depleted of the absorbed components. Absorption is widely used in industries for gas purification, scrubbing and chemical synthesis.

Adiabatic processes

Adiabatic processes are characterized by the absence of heat exchange between the system and its surroundings, meaning that all energy changes are confined to internal transformations. Such processes are critical in thermodynamic studies, particularly in systems such as compressors, turbines and insulated reactors, where simplifications based on this assumption enable easier energy calculations and system design.

Adsorption and its applications

Adsorption refers to the adhesion of molecules (adsorbates) from a fluid phase – gas or liquid – onto the surface of a solid material (adsorbent). This surface phenomenon is central to applications such as gas purification, separation processes and catalysis. Unlike absorption, which involves bulk phase transfer, adsorption is surface specific, making it effective for targeting specific contaminants or reactants in chemical processes.

Barometer

A barometer is a fundamental instrument for measuring atmospheric pressure. Its accurate readings are essential for weather forecasting, altitude measurement and various engineering applications where pressure influences material behavior or process efficiency.

Boilers and the boiling point

Boilers are critical components in energy-intensive industries. These devices convert feedwater into steam by transferring heat from combustion gases through tubes, enabling the steam to power turbines, heaters or industrial equipment. Closely related to this is the concept of the boiling point, the temperature at which a liquid simultaneously exists as a vapor under equilibrium conditions. This property is pressure dependent and vital for understanding phase transitions in processes such as distillation and evaporation.

Distillation and key product streams

Distillation is a complex separation technique that leverages differences in the volatility of components to purify mixtures. In a distillation column, vapor rises while liquid descends, allowing mass transfer between the phases. The lighter, more volatile components concentrate in the vapor phase and exit at the top as the overhead product, while heavier components accumulate at the bottom as the bottom product. This interplay is of fundamental importance to industries such as petrochemicals, pharmaceuticals and beverages.

Thermodynamic properties: bubble point, dew point and vapor pressure

The bubble point and dew point represent critical temperatures for phase changes in mixtures. The bubble point is the temperature at which vapor first forms in a liquid mixture under constant pressure, while the dew point is the temperature at which vapor begins to condense into a liquid. Vapor pressure, the equilibrium pressure exerted by the vapor of a liquid, is pivotal in predicting phase behavior and is determined using models such as the Antoine equation.

Calibration and catalysis

Calibration ensures that instruments provide accurate readings by comparing them with known standards. This process is integral to maintaining precision in industrial operations. Catalysis, on the other hand, dramatically speeds up chemical reactions without consuming the catalyst itself. Catalysts play a central role in refining, polymerization and emission control, making processes more efficient and sustainable.

Compressors, compressibility factor and condensation

Compressors are mechanical devices that increase gas pressure to enable its transportation or storage in various industrial applications. The compressibility factor (Z), which quantifies deviations of real gases from ideal behavior, is calculated using $Z = PV/nRT$. Condensation involves converting a gas into a liquid by cooling or compressing it, producing separate streams of liquid condensate and uncondensed gases. These processes are fundamental to refrigeration, natural gas processing and chemical synthesis.

Crystallization and decanters

Crystallization separates solutes from a liquid solution by forming solid crystals through cooling or solvent evaporation. Decanters, on the other hand, use gravity to separate immiscible liquids or liquid–solid mixtures. Both are essential in industries such as pharmaceuticals, food processing and chemical manufacturing.

Enthalpy, internal energy and work

Enthalpy ($H=U+PV$) represents the total energy in a system, combining the internal energy (U) with pressure–volume work (PV). The internal energy, which depends on the molecular motion and phase, changes with temperature and state transitions. In thermodynamics, work involves the transfer of energy due to action of force, excluding heat transfer. These concepts form the basis for analyzing energy flows in industrial processes.

Heat, heat exchangers and drying

Heat transfer is fundamental to process engineering, representing energy flow due to temperature differences. Heat exchangers facilitate this transfer between two fluid streams, enhancing thermal efficiency. Drying is another critical process that removes liquid from solids by heating or exposing them to hot gases, which is extensively used in the food, chemical and pharmaceutical industries.

Advanced separation techniques: filtration, extraction and flash vaporization

Filtration separates solid particles from a liquid or gas using a porous medium, while extraction isolates a solute using a carefully chosen solvent. Flash vaporization, the rapid evaporation of a high-pressure liquid upon pressure reduction, is used in processes such as petroleum refining to separate components based on volatility.

Stripping, scrubbing and membranes

Stripping removes dissolved gases from liquids using counterflowing gas, while scrubbing eliminates undesirable gas components through absorption or chemical reactions. As selective barriers, membranes enable the separation of species based on size or chemical affinity, making them crucial for desalination, gas separation and medical applications.

2.3. Engineering calculations – introduction

Dimension

A dimension represents a measurable property such as length, mass, time or temperature, etc. These properties are of fundamental importance for engineering calculations because they form the basis for defining physical quantities. Dimensions can be directly measured or derived from mathematical operations such as multiplication or division of base quantities (Himmelblau and Riggs 2012; Pritchard and Mitchell 2016).

Conversion of units

Unit conversion is a critical calculation skill in engineering to express quantities in different units. This involves multiplying a given value by a conversion factor to convert it into the desired unit. For example, converting one year to seconds involves a systematic multiplication process:

$$1\text{ year} = 365\,\text{days} \times \frac{24\,\text{hr}}{1\,\text{day}} \times \frac{60\,\text{min}}{1\,\text{hr}} \times \frac{60\,\text{s}}{1\,\text{min}} = 31{,}536{,}000\,\text{s}$$

This structured approach ensures accuracy in converting units across different systems.

Systems of units

Systems of units provide a framework for quantifying physical properties using base units for fundamental dimensions such as mass, length, time, temperature, electric current and light intensity.

For example:

– Base units: kilogram (kg), meter (m), second (s), Kelvin (K), ampere (A) and candela (cd).

Derived units are obtained by combining base units.

For example:

– Velocity (v) is calculated as the ratio of length (L) to time (T), with units expressed as m/s.

– Area (A) is determined as the product of two lengths, resulting in units of m^2.

Derived units and examples

Derived units, also called compound units, are created by multiplying or dividing base or multiple units. It is also defined as the equivalent of compound units.

Energy in ergs:

– 1 erg = 1 $(g.cm)/s^2$.

Force in pound-force:

– 1 $lb_f = 32.174\ (lb_m.ft)/s^2$.

Such calculations help to define physical relationships and ensure consistent use of units across engineering disciplines.

2.4. International system of units (SI)

The units of measurement are as follows: length is measured in meters (m), mass is measured in kilograms (kg), time is measured in seconds (s), electric current is measured in amperes (A) and luminous intensity is measured in candelas (cd).

Prefixes: mega (M) = 10^6, kilo (k) = 10^3, centi (c) = 10^{-2}, milli (m) = 10^{-3}, micro (μ) = 10^{-6}, nano (n) = 10^{-9}.

CGS units are a system of measurements in which mass is expressed in grams and length in centimeters (cm). Various quantities and their corresponding unit in various systems are given in Table 2.1.

Quantity	SI	BG	EE
Force	N	lb_f	lb_f
Mass	kg	Slug	lb_m
Length	m	ft	ft
Time	s	s	s
Temperature	K	R	R

Table 2.1. *Fundamental quantities and their units in various systems*

1 L = 10^{-3} m^3 = 1,000 cm^3.

1 N = 1 (kg.m)/s^2 = 10^5 (g.cm)/s^2 = 10^5 dyne.

1 Pascal = 1 (N)/sm^2.

1 J = 1 N.m = 1 (kg.m^2)/s^2.

1 cal = 4.184 J.

1 slug = 1 (lbf.S^2)/ft.

g_c = 32.2 (ft.lb_m)/$lb_f.s^2$.

1 slug = 32.2 lbm.

R = F + 459.67.

g = 32.2 ft/s^2.

K = C + 273.15.

Mass	$1\ kg = 2.205\ lb_m = 6.852 \times 10^{-2}$ slug
Length	$1m = 3.281\ ft = 39.37\ in = 6.214 \times 10^{-4}$ mile
Temperature	$°F = 1.8°C + 32$ $K = 1.8\ R$
Pressure	$1\ N/m^2 = 2.961 \times 10^{-4}$ in Hg $= 2.089\ x10^{-2}\ lb/ft^2$ (psf) $= 1.450 \times 10^{-4}\ lb/in^2$ (psi) $1\ N = 0.2248\ lb_f$
Force	$1\ lb_f = 32.174\ (lb_m.ft)/s$
Area	$1\ m^2 = 10.76\ ft^2$
Acceleration	$1\ m/s^2 = 3.281\ ft/s^2$
Density	$1\ kg/m^3 = 6.243 \times 10^{-3}\ lb_m/ft^3 = 1.94 \times 10^{-3}\ slug/ft^3$
Energy	$1\ J = 9.478 \times 10^{-4}\ BTU = 0.7376\ ft.lb_f$

Significant figures are the digits of a number that contribute to its precision, reflecting the accuracy of a measured or calculated value. These include all digits starting from the first non-zero digit on the left to the last digit on the right, provided there is a decimal point.

If there is no decimal point, significant figures extend only to the last non-zero digit.

The number of significant figures indicates the precision of a value, with a higher count signifying greater accuracy. In arithmetic operations, the treatment of significant figures varies.

– Multiplication or division: the result should retain the same number of significant figures as the factor with the fewest significant figures.

– Addition or subtraction: the precision of the result is determined by the least precise position of the significant figures relative to the decimal point.

When rounding, if the digit to be dropped is exactly 5, the preceding digit is adjusted to ensure that the final number ends with an even digit. This practice minimizes rounding biases in calculations (Pritchard and Mitchell 2016).

2.5. Analysis of processes and variables affecting density

– Density: this refers to the mass of a substance divided by its volume, and is typically measured in units such as kilograms per cubic meter (kg/m^3), grams per cubic centimeter (g/cm^3) or pounds per cubic foot (lbm/ft^3).

– Specific volume: this refers to the amount of space that a given mass of a substance occupies. The symbol γ represents the reciprocal of the variable ρ.

$$\gamma = 1/\rho$$

The densities of pure solids and liquids are virtually unaffected by changes in pressure and exhibit very minor variations with temperature.

– Specific gravity: this refers to the density of a substance compared to the density of a reference substance. Specific gravity (SG) is calculated by dividing the density (ρ) by the reference density (ρ_{ref}).

$$SG = \rho/\rho_{ref}$$

$$\rho_{H2O/4°C} = 1{,}000\ g/cm^3 = 1{,}000\ kg/m^3 = 62.43\ lb_f/ft^3$$

The density of water at 4°C is 1,000 g/cm^3, which is equivalent to 1,000 kg/m^3 or 62.43 lbf/ft^3.

– Flow rate: it refers to the speed at which a substance is moved through a process line. The rate at which a process stream moves can be described as either the mass flow rate or the volumetric flow rate.

The equation $\rho = m/V$ represents the relationship between density (ρ), mass (m) and volume (V).

2.5.1. *Measurement of flow rate (using a rotameter and an orifice meter)*

A rotameter is a vertical tube with a tapered shape that contains a float. The height of the float in the tube increases as the flow rate increases. An orifice meter is a device that obstructs the flow channel and has a tiny hole through which the fluid flows. The fluid pressure decreases as it flows from the upstream side of the orifice to the downstream side. The pressure drop can be quantified using various instruments, such as a differential manometer. It is important to note that the pressure drop increases proportionally with the flow rate.

– **Atomic weight**: this term refers to the mass of a single atom of an element, measured on a scale where a carbon-12 atom has a mass of exactly 12 units. This standardized measurement allows for a consistent comparison of atomic masses across different elements.

– **Molecular weight of carbon (C)**: the molecular weight of a compound is determined by summing the atomic weights of all the atoms constituting a molecule of that substance. For example, the molecular weight of carbon (C) would be calculated based on its atomic composition.

– **Gram-mole**: this is also known as a g-mole or mol in SI units, which refers to the quantity of a species whose mass in grams is equal to its molecular weight. When referring to a species as an element, it is more accurate to use the term "gram-atom" instead of "gram-mole".

The molecular weight of a substance, denoted as M, corresponds to M kg/kmol, M g/mol and M lbm/lbmole of the substance.

A single gram-mole of a substance consists of approximately 6.02×10^{23} molecules of that substance.

– **Mass fraction**: this is denoted as x_A, which is defined as the ratio of the mass of component A to the overall mass. It can be expressed in the units of kg A/kg total, g A/g total or lbm A/lbm total.

$$x_A = \text{mass of A/total mass}$$

– **Mole fraction**: this is denoted as y_A, which is defined as the ratio of the number of moles of component A to the total number of moles in a system. It can be expressed in kilomoles or pounds of component A per kilomole or pound mole, respectively.

$$y_A = \text{moles of A/total moles} = \text{kmol A/kmol} = \text{mol A/mol} = \text{lb mol A/lb mole}$$

– **Average molecular weight**: this is also known as the mean molecular weight of a mixture, which is calculated by dividing the mass of a sample of the mixture (m_f) by the total number of moles of all species (n_f) in the sample. Let y_i be the mole fraction of the ith component in the mixture and n_i be the molecular weight of this component. Average molecular weight = sum(y_i n_i).

– **Concentration**: this term refers to the amount of a component per unit volume in a mixture or solution. Molar concentration specifically denotes the number of moles of a component that is present in a given volume of the mixture. Molarity, a measure of concentration, represents the number of moles of solute per liter of solution.

– **Parts per million**: the terms parts per million (ppm) and parts per billion (ppb) are used to quantify the amounts of trace substances in gas or liquid mixtures. Mass ratios, typically used for liquids, or mole ratios, typically used for gases, indicate the number of parts (in grams or moles) of a certain species present per million or billion parts (in grams or moles) of the mixture.

If y_i is the fraction of component i, then

$ppm_i = y_i \times 10^6$;

$ppb_i = y_i \times 10^9$;

American billion = 10^9;

English billion = 10^{12}.

– **Fluid pressure**: this is defined as the force applied per unit area in a fluid. It is mathematically expressed as P = F/A, where P is the pressure, F is the force and A is the area. Common units for pressure include Newtons per square meter (N/m^2), dynes per square centimeter ($dynes/cm^2$) or pounds-force per square inch (lbf/in^2 or psi). In the context of static fluids, hydrostatic pressure is calculated as the force exerted on a surface, divided by the area of that surface.

– **Atmospheric pressure**: this refers to the force exerted by the Earth's atmosphere on a given area at its surface. The standard atmospheric pressure at sea level is defined as 1 atmosphere, which is equal to 760.0 millimeters of mercury (mm Hg). This standard serves as a reference point for many pressure-related calculations and measurements in various scientific and engineering fields.

– **Absolute pressure**: this is defined as the total pressure measured relative to a perfect vacuum (zero pressure). In contrast, gauge pressure is the pressure measured relative to the local atmospheric pressure.

A gauge pressure of zero signifies that the absolute pressure of the fluid is equal to the atmospheric pressure:

$$P_{absolute} = P_{gauge} + P_{atmospheric}$$

2.5.2. *Measurement of fluid pressure*

– The elastic element approach involves the use of Bourdon tubes, bellows or diaphragms.

– The liquid column method is a technique used to measure pressure using manometers.

– The electrical approach involves the use of strain gauges, piezoresistive transducers and piezoelectric transducers.

A Bourdon gauge is a device consisting of a hollow tube that is sealed at one end and bent into a C-shaped configuration. The exposed end of the tube is in contact with the fluid whose pressure is to be measured. As the pressure rises, the tube has a tendency to become straight, resulting in the rotation of a pointer that is connected to the tube.

A differential manometer is used to quantify the disparity in pressure between two specific locations within a process line.

A sealed end manometer contains a near perfect vacuum at one end. A portion of the fluid that is exposed converts to vapor and disperses into the surrounding empty space, therefore preventing the formation of a completely empty region known as a perfect vacuum. When the open end of the sealed manometer is in contact with the atmosphere, with a pressure of P_{atm}, the instrument operates as a barometer.

– **Temperature scale**: there are two primary temperature scales that are commonly used. These scales are defined based on the freezing point (T_f) and boiling point (T_b) of water at a pressure of 1 atmosphere.

The Celsius or Centigrade scale assigns T_f a value of 0°C and T_b a value of 100°C. Absolute zero, theoretically the coldest temperature achievable in nature, is defined on this scale as –273.15°C.

The Fahrenheit scale assigns a temperature of 32 °F to T_f and a value of 212 °F to T_b. Absolute zero is defined as the temperature of –459.67 °F.

The Kelvin and Rankine scales are defined such that the value of absolute zero is 0, and the magnitude of each degree is equal to that of a Celsius degree (in the case of the Kelvin scale) or a Fahrenheit degree (in the case of the Rankine scale).

$$T\ (K) = T\ (^{o}C) + 273.15.$$

$$T\ (^{o}R) = T\ (^{o}F) + 459.67.$$

$$T\ (^{o}R) = 1.8T\ (K).$$

$$T\ (^{o}F) = 1.8T\ (^{\circ}C) + 32.$$

An interval of 1°C or Kelvin therefore contains 1.8 F or Rankine (Himmelblau and Riggs 2012; Felder et al. 2020).

2.6. Classification of material balances in process fundamentals

Batch process

A batch process involves introducing all the feed material into a vessel at the beginning of the operation, where it remains until the procedure is complete. During the process, no mass is transferred in or out of the system. This method is commonly used for producing small quantities of products or when precise control over reaction time and conditions is critical.

Batch processes are preferred in industries such as pharmaceuticals, specialty chemicals and food production, where product quality and flexibility are essential.

Continuous process

In a continuous process, materials flow steadily into and out of the system throughout the duration of the operation. Inputs and outputs occur simultaneously, allowing for uninterrupted production. This method is ideal for large-scale manufacturing, such as in petrochemical and refining industries, where consistent operations and efficiency are paramount. Continuous processes typically operate under steady-state conditions to maximize throughput and minimize downtime.

Semi-batch process

A semi-batch process combines elements of both batch and continuous operations. In this method, some materials are introduced or removed intermittently during the process. For example, in a pressurized system, gases may be released into the atmosphere, or multiple liquids may be blended in a tank without immediate withdrawal.

Semi-batch processes are particularly useful in reactions where one reactant needs to be gradually introduced or where intermediate control over the reaction conditions is necessary.

2.6.1. *State of processes*

Steady-state process

A steady-state process occurs when all process variables – such as temperature, pressure, flow rates and composition – remain constant over time, with only negligible fluctuations around their mean values. Steady-state operation is typical of continuous processes and is advantageous for maintaining predictable and stable production conditions, reducing variability in output quality.

Transient or unsteady-state process

In a transient (or unsteady-state) process, one or more process variables change over time. Batch and semi-batch processes are inherently transient, as the operating conditions evolve during the process. Continuous processes can also experience transient behavior, particularly during start-up, shutdown or when adjusting to new operating conditions. Over time, many transient processes stabilize and transition to a steady state.

2.6.2. *Applications and selection criteria*

– Batch processing: ideal for small-scale production, high-value products or when frequent changes in product type are required.

– Continuous processing: suitable for large-scale, high-volume production over extended periods of time with consistent output requirements.

– Semi-batch processing: used in scenarios that require precise control over reactant addition or gradual changes in operating conditions.

2.7. General balance equation

A conservation balance ensures the maintenance of a specific quantity – such as total mass, mass of a specific species, energy or momentum – within a defined system. This system can range from a single process unit to an entire industrial operation. The fundamental balance equation is expressed as:

$$\text{Input} + \text{Generation} - \text{Output} - \text{Consumption} = \text{Accumulation}$$

Each term in this equation is defined as follows:

– Input: the quantity entering the system through its boundaries.

– Generation: the quantity produced within the system, often due to chemical reactions or energy transformations.

– Output: the quantity exiting the system through its boundaries.

– Consumption: the quantity consumed or used within the system, often due to reactions or losses.

– Accumulation: the buildup or storage of the quantity within the system over time.

This equation forms the basis for analyzing material, energy and momentum balances in engineering systems.

2.7.1. *Types of balances*

Differential balances

Differential balances focus on the instantaneous state of a system, representing rates of change over time. Each term in the balance equation corresponds to a rate, such as the rate of input, generation or consumption. The terms are expressed in units of the conserved quantity (e.g. mass, energy) per unit time (e.g. grams per second or joules per second). Differential balances are typically applied to continuous processes, where dynamic conditions and time-dependent variations need to be analyzed.

Integral balances

Integral balances examine the cumulative change over a specific time interval. Each term in the balance equation represents a total quantity, such as the total input or total consumption, measured in units such as grams, liters or joules. Integral balances are commonly used for batch processes where the focus is on the changes between defined starting and ending points.

2.7.2. *Applications*

Understanding and applying conservation balances is essential for the following:

– Process design: ensuring systems meet production and safety goals.

– System optimization: identifying inefficiencies and optimizing resource utilization.

– Dynamic analysis: investigating transient conditions using differential balances.

– Batch analysis: evaluating time-dependent processes using integral balances.

By combining both differential and integral approaches, engineers can comprehensively analyze and design systems to maintain conservation principles effectively

Thumb rules for material balance

In conservation balances, specific conditions simplify the terms in the balance equation. For systems at equilibrium where the total mass is conserved, the values for generation and consumption are set to zero, as mass cannot be created or destroyed (except in nuclear reactions). Similarly, for non-reactive species – substances that are neither reactants nor products – the generation and consumption terms are also set to zero. Additionally, when a system operates under steady-state conditions, the accumulation term is always zero, regardless of the type of balance being analyzed. These simplifications streamline calculations and provide clarity when analyzing different systems.

Equilibrium in a continuous steady-state process

Accumulation = 0

Input + generation = output + consumption.

If the balance is on a non-reactive species or on total mass:

Generation = 0

Consumption = 0

Input = output.

Integral balance on batch processes

Since no matter enters or leaves through the system boundaries, therefore:

Generation = Accumulation

Quantity of product produced = buildup (accumulation)

= final amount – initial amount

Accumulation = final output – initial input .... (1) (by definition)

= generation – consumption ....(2)

Input + generation – output – consumption = accumulation

From (1) and (2):

Input + generation = output + consumption.

2.8. Flow charts and process balancing

Flow charts are essential tools in process analysis, visually representing the sequence and components of a system. They use boxes or symbols to depict process units such as reactors, mixers and separation units, with lines and arrows illustrating the input and output streams. Key information, including values and units for known variables, is recorded at specific points on the chart. For unknown variables, algebraic symbols are assigned, and their names and corresponding units are also documented on the chart. When volumetric flow rates are given, it is often practical to convert or calculate mass or molar flow rates, as balances are not typically expressed in volumetric terms.

2.8.1. *Flow chart scaling and basis of calculation*

A process is balanced when material balances for all system components are satisfied. Scaling in a flow chart adjusts all stream amounts or flow rates proportionally while maintaining stream compositions. This scaling can either increase (scale-up) or decrease (scale-down) stream quantities, depending on the requirements. The adjustments ensure that the process analysis remains consistent and adaptable to different scales of operation.

2.8.2. *Guidelines for balancing non-reactive processes*

For non-reactive systems, the maximum number of independent equations obtainable is equal to the number of chemical species present in the input and output streams. To simplify the balancing process, it is advisable to begin by writing equations with the least number of unknowns. Solving these equations first provides a structured approach to addressing more complex balances later, ensuring efficiency and accuracy in process calculations.

2.9. Degree-of-freedom analysis

Before starting the calculations, it is crucial to construct a well-designed and appropriately labeled flow chart to assess whether the available information is sufficient to solve the problem. The degree of freedom (DOF) is determined by the difference between the number of unknowns and the number of independent equations available to describe the system:

$$\text{DOF} = \text{Number of unknowns} - \text{Number of independent equations.}$$

2.9.1. *Case analysis*

– If the DOF is zero, the system is solvable, meaning that the number of unknowns matches the number of independent equations.

– If the DOF is greater than zero, there are more unknowns than independent equations, indicating that additional variable values are required to solve the system. To resolve this, at least as many extra variable values as the DOF must be provided. If no additional information is provided, the problem may be underdefined or may have an infinite number of possible solutions.

– If the DOF is less than zero, it means that there are either too many equations or conflicting relations in the flow chart, which makes the problem unsolvable. It is not advisable to attempt to solve the problem unless the relationships and unknowns are appropriately defined and aligned.

2.9.2. *Sources of equations to establish relationships between unknown variables*

– Material balance: in non-reactive processes, material balances can be written for a limited number of independent molecular species. These balances help establish relationships between the input and output materials in the system.

– Energy balance: this equation relates material flows and temperature changes at the inlet and outlet of the system. It can be used, whether the quantity of energy transferred between the system and its surroundings is known or needs to be calculated.

– Process specifications: these provide constraints and parameters based on the design or operational specifications of the process.

– Physical properties and laws: the behavior of materials under certain conditions is governed by physical laws and properties, which provide further relationships for calculating unknowns.

– Physical constraints: constraints based on system limitations or operational bounds can also provide equations for solving unknowns.

– Stoichiometric relations: in reactive processes, stoichiometric relations define the proportions of reactants and products that are essential for balancing equations and determining unknown quantities.

2.10. Procedure for calculating material balances in single unit processes

(a) Select the quantity or flow rate of a process stream to be used as the basis for the calculation.

(b) Create a flow chart and input the values of all known variables, including the fundamental principles of computation. Next, annotate unidentified stream variables on the chart.

(c) Determine the problem statement in an equation that expresses what needs to be determined using the given variables.

(d) To standardize the units for a stream with mixed mass and mole measurements, convert all quantities to a single basis.

(e) Perform the analysis of degrees of freedom.

(f) If the number of unknowns is equal to the number of equations, meaning that the system has zero degrees of freedom, arrange the equations in an efficient sequence and identify the variables that you will solve for by circling them.

(g) Determine the solutions to the equations.

(h) Determine the quantities specified in the problem statement if they have not yet been determined.

(i) To reach the final result, scale the balancing process by the ratio of the given stream quantity of flow rate (n_g) to either the chosen basis value (n_c) or the estimated value for this stream.

2.11. Recycling and bypassing

Reactions do not achieve 100% completion. Therefore, we must compensate for all of the reactants used in a process, not just the portion that reacts. Additionally, any unreacted A that is present in the final product is considered waste.

If we isolate all of the unreacted reactant from the stream containing the product and reintroduce the unreacted reactant into the reactor, then this unreacted material can be reused rather than wasted. There is a cost associated with the separation and recycling system, but we may offset this expense by reducing the amount of fresh reactant we need to purchase and selling the purified product at a higher price.

Reasons for recycling:

1. Recovery of the catalyst.
2. Dilution of a process stream.
3. Control of the process variable.
4. Circulation of the working fluid – e.g. refrigeration cycle.

Bypass: a portion of the input to a process unit is redirected around the unit and mixed with the output stream from the unit.

2.12. Limiting reactant

It is the reactant that would be completely consumed if the reaction went to completion, while the other reactants are known as excess reactants.

Let n_A be the number of moles of A in feed.

Let $(n_A)_{stoich}$ be the number of moles of A required to react with the limiting reactant.

$$\text{Fraction excess of A} = \frac{n_{A\,feed} - n_{A\,stoich}}{n_{A\,stoich}}.$$

Fractional conversion (f) = moles reacted/moles fed.

Yield (Y)

$$Y = \frac{moles\ of\ the\ desired\ product\ formed}{moles\ that\ would\ have\ formed\ if\ there\ were\ no\ side\ reactions\ and\ the\ limiting\ reactant\ had\ reacted\ completely}$$

Selectivity (S)

$$S = \frac{moles\ of\ the\ desired\ product\ formed}{moles\ of\ the\ undesired\ product\ formed}$$

S is defined as the ratio of the moles of the desired product formed to the moles of the unwanted product formed.

In a batch or continuous steady-state reactor, if a set of reactions occur, V_{ij} represents the stoichiometric coefficient of substance i in reaction j. A negative value indicates reactants, while a positive value indicates products.

$$n_i = n_{io} + \sum_j v_{(ij)} \varepsilon_{(ij)}.$$

2.13. Product separation and recycling

$$Overall\ conversion = \frac{Reactant\ input\ to\ the\ process\ -\ Reactant\ output\ from\ the\ process}{Reactant\ input\ to\ the\ process}$$

$$Single\ pass\ conversion = \frac{Reactant\ input\ to\ the\ reactor\ -\ Reactant\ output\ from\ the\ reactor}{Reactant\ input\ to\ the\ reactor}$$

2.14. Combustion

This is the process of burning. The prompt and efficient response of fuel with oxygen is of greater significance than any other category of industrial chemical reactions. The importance of these reactions stems from the vast amounts of energy they generate. The primary fuels used in power plant combustion furnaces are typically coal (composed of carbon, hydrogen, sulfur (S) and other non-combustible substances), fuel oil (consisting primarily of higher molecular weight hydrocarbons (HCs) with some sulfur), gaseous fuels (such as natural gas, which is primarily methane) or LPG (propane/butane).

Carbon fuel combustion results in the formation of CO_2, CO, H_2O and SO_2. NO is also formed at temperatures exceeding 1,800°C. Partial or incomplete combustion is a combustion reaction that results in the formation of carbon monoxide (CO).

$C + O_2 \rightarrow CO_2$ complete combustion.

$C_3H_8 + 5O_2 \rightarrow 3CO_2 + 4H_2O$ complete combustion of propane.

$C_3H_8 + (7/2)\ O2 \rightarrow 3CO + 4H_2O$ partial combustion of propane.

$CS_2 + 3O_2 \rightarrow CO_2 + 2SO_2$ complete combustion of CS_2.

Theoretical oxygen

The term "theoretical oxygen" refers to the amount of oxygen that would be required for a complete combustion reaction. The quantity of O_2 required for the complete combustion of all the fuel supplied to the reactor, assuming that all carbon in the fuel is converted to CO_2 and all hydrogen is converted to H_2O, can be determined by calculating the moles of molar flow rate of O_2.

Theoretical air

The term "theoretical air" refers to the ideal or perfect composition of air, assuming it consists only of its main components such as nitrogen, oxygen and trace amounts of other gases.

The quantity of air that contains the theoretical oxygen is given by:

$$\%\ Excess\ air = \frac{\text{moles air}_{\text{fed}} - \text{moles air}_{\text{theoretical}}}{\text{moles air}_{\text{theoretical}}} \times 100$$

2.15. Single-phase systems

2.15.1. *Liquid and solid densities*

$$\frac{1}{\rho} = \sum_{i=1}^{n} \frac{x_i}{\rho_i}$$

$$\rho = \sum_{i=1}^{n} x_i\, \rho_i$$

Volume additivity is valid for straight-chain HCs of nearly equal molecular weight, such as n-pentane, n-hexane and n-heptane.

2.15.2. *Ideal gases*

$PV = nRT.$

$V/n = RT/P.$

$\rho = P/RT$, mole/L for an ideal gas.

$\rho = P\bar{M}/RT$, g/L for an ideal gas.

$\bar{M}$ = average molecular weight.

SCMH (standard cubic meters per hour) is the flow rate of STP (273 K and 760 mmHg).

2.15.3. *Ideal gas mixture*

$$\frac{p_A}{P} = \frac{n_A}{n} = y_A.$$

$$PV = nRT.$$

$$p_A V = n_A RT.$$

$$p_A = y_A P.$$

The partial pressure of a component in an ideal gas mixture is the mole fraction of that component multiplied by the total pressure.

$$y_A + y_B + y_C + \ldots\ldots\ldots = 1.$$

$$p_A + p_B + p_C + \ldots\ldots\ldots = (y_A + y_B + \ldots\ldots)P = P.$$

– **Dalton's law**: the partial pressures of the components of an ideal gas mixture add up to the total pressure.

$$PV_A = n_A RT.$$

$$PV = nRT.$$

$$V_A/V = n_A/n = y_A.$$

$$V_A = y_A V.$$

$V_A + V_B + V_C + \ldots\ldots\ldots = V$ Amagat's law.

The volume fraction of a substance in an ideal gas mixture is equal to the mole fraction of that substance.

2.16. Clausius–Clapeyron equation

$$\ln P^* = -\frac{\widehat{\Delta Hv}}{RT} + B.$$

B is a constant that varies with a substance.

P* = vapor pressure of a pure substance.

$\Delta\widehat{Hv}$ = latent heat of vaporization (energy required to vaporize one mole of liquid).

Antoine equation:

$$\text{Log } P^* = A - \frac{B}{T+C}$$

Gibbs' phase rule:

DF = 2 + C - π for no reaction.

DF = degree of freedom.

π = number of phases in a system at equilibrium.

C = number of chemical species.

DF as the number of intensive variables that can be specified independently for a system at equilibrium is called the DOF of the system.

Extensive viable: the variables that describe the condition of a process system, which depend on the size of the system. For example: mass and volume.

Intensive variable: intensive variables include temperature (T), pressure (P), density (ρ), specific volume ($\hat{V}$), as well as mass and mole fractions.

Raoult's law: this is a principle that applies to a single condensable species:

$$p_i = y_i P = p_i^*(T).$$

If a gas at temperature T and pressure P contains a saturated vapor with a mole fraction of y_i, and if this vapor is the only species that would condense if the temperature was slightly decreased, then the partial pressure of the vapor in the gas is equal to the pure component vapor pressure p_i*(T) at the system temperature.

– **Superheated vapor**: a vapor that is present in a gas in an amount below its saturation point.

In this case, $p_i = y_i p < p_i^*(T)$.

– **Dew point**: the dew point of a gas is the temperature at which a single superheated vapor, when cooled under constant pressure, becomes saturated:

$$p_i = y_i p < p_i^*(T_{dp}).$$

– **Degrees of superheat**: the difference between the temperature and the dew point of a gas is called the degrees of superheat of gas.

– **Relative saturation/relative humidity**: consider a gas at temperature T and pressure P that contains a vapor with a partial pressure of pi and a vapor pressure of $p_i(T)$.

$$S_r\ (H_r) = (p_i/p_i^*(T)) * 100.$$

A relative humidity of 40% indicates that the amount of water vapor in the air is 40% of the maximum amount it can hold at the current temperature (Himmelblau and Riggs 2012; Felder et al. 2020).

Molal saturation/molal humidity:

$$S_m(h_m) = \frac{p_i}{P - p_i} = \frac{moles\ of\ vapor}{moles\ of\ vapor\ -\ free\ dry\ gas}$$

Absolute saturation/absolute humidity:

$$S_a(h_a) = \frac{p_i M_i}{(P - p_i) M_{dry}} = \frac{mass\ of\ vapor}{mass\ of\ dry\ gas}$$

– Mi = molecular weight of the vapor.

– M_{dry} = average molecular weight of the dry gas.

– Percentage saturation/percentage humidity:

$$Sp\ (hp) = \frac{Sm}{Sm^*} \times 100 = \frac{p_i/(P - p_i)}{p_i^*(P - p_i^*)} \times 100\ \%$$

Dew point pressure:

$$P_{dp} = \frac{1}{\frac{y_A}{p_A^*(T)} + \frac{y_B}{p_B^*(T)} + \cdots \ldots\ldots\ldots\ldots}$$

Bubble point pressure:

$$P = x_A p_A{}^*{}_{(Tbp)} + x_B p_B{}^*{}_{(Tbp)}$$

$$y_i = \frac{p_i}{P_{bp}} = \frac{x_i p_i^*(T)}{P_{bp}}$$

2.17. Mass balance problem (Felder et al. 2020)

Let us consider a continuous process for the production of ethylene oxide (C_2H_4O) from ethylene (C_2H_4) and oxygen (O_2).

A chemical plant produces ethylene oxide through the oxidation of ethylene. The process involves a reactor, a separator and a recycle stream. The following information is given:

1. Fresh feed: 1,000 kg/h of ethylene (C_2H_4) and 900 kg/h of oxygen (O_2).
2. Reactor conversion: 80% of ethylene.
3. Side reaction: 10% of reacted ethylene forms CO_2 and H_2O.
4. Separator efficiency: 95% recovery of ethylene oxide.
5. Recycle stream: 90% of unreacted ethylene and oxygen are recycled.
6. Inert gas (N_2) enters with the oxygen stream at 50 kg/h.

Calculate: a) the production rate of ethylene oxide; b) the composition of the reactor effluent; c) the composition of the recycle stream; and d) the overall mass balance of the system.

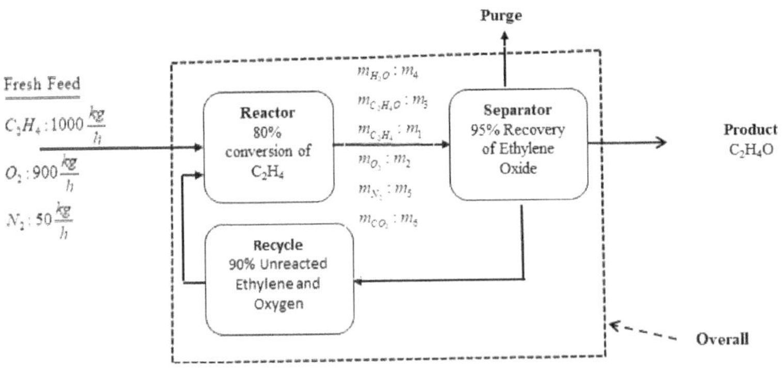

Figure 2.1. *Flow chart of ethylene oxide production*

Solution:

1. Define the system and its boundaries:

Our system includes the reactor, the separator and the recycle stream.

2. Write out the relevant reactions.

Main reaction: $C_2H_4 + \frac{1}{2} O_2 \rightarrow C_2H_4O$

Side reaction: $C_2H_4 + 3\ O_2 \rightarrow 2\ CO_2 + 2\ H_2O$

3. Balance around the reactor.

Ethylene balance

$$\underbrace{1000\frac{kg}{h} + 90\%\ \text{of}\ m_1}_{\text{Feed to reactor}} - \underbrace{80\%\left(1000\frac{kg}{h} + 90\%\ \text{of}\ m_1\right)}_{\text{Conversion in the reactor}} = \underbrace{m_1}_{\text{converted}\ C_2H_4}$$

$$\Rightarrow 1000 + 0.9\,m_1 - 800 - 0.72m_1 = m_1$$
$$\Rightarrow m_1 = 243.9024$$

$$\text{Ethylene fed to reactor} = 1000 + 0.9\,m_1 = 1000 + 0.9 \times 243.9024$$
$$= 1219.5122\frac{kg}{h}$$

$$\text{Total ethylene converted} = 80\%\left(1000 + 0.9\,\text{m}_1\right) = 80\%\left(1000 + 0.9 \times 243.9024\right)$$
$$= 975.60\frac{\text{kg}}{\text{h}}$$

$$\text{Ethylene used in main reaction} = 90\%\ \text{ethylene converted}$$
$$= \frac{90}{100} \times 975.60\frac{kg}{h} = 878.039\frac{kg}{h}$$

$$\text{Ethylene used in side reaction} = 10\%\ \text{ethylene converted}$$
$$= \frac{10}{100} \times 975.60\frac{kg}{h} = 97.56\frac{kg}{h}$$

Oxygen balance around the reactor

$$\underbrace{O_2}_{\text{consumed } O_2} = O_2 \text{ consumed in main reaction} + O_2 \text{ consumed in side reaction}$$

$$= 878.039\frac{kg}{h} C_2H_4 \text{ consumed} \times \frac{0.5 \times 32\, kg\ O_2}{28\ kg\, C_2H_4} + 97.56\frac{kg}{h} C_2H_4 \text{ consumed} \times \frac{3 \times 32\, kg\ O_2}{28\ kg\, C_2H_4}$$

$$= 501.7365 + 334.4914$$

$$= 836.22\frac{kg}{h}$$

Oxygen balance

$$\underbrace{900\frac{kg}{h} + 90\% \text{ of } m_2}_{O_2 \text{ fed to reactor}} - \underbrace{m_2}_{O_2 \text{ uncoverted}} = \underbrace{O_2}_{\text{consumed } O_2}$$

$$\Rightarrow 900 + \frac{90}{100} \times m_2 - m_2 = 836.22$$

$$\Rightarrow m_2 = 637.72\frac{kg}{h}$$

Ethylene oxide formed: m₃

$$m_{C_2H_4O} = C_2H_4 \text{ consumed in main reaction} \times \frac{44\, kg\ C_2H_4O}{28\ kg\, C_2H_4}$$

$$= 878.039\frac{kg}{h} \times \frac{44\, kg\ C_2H_4O}{28\ kg\, C_2H_4}$$

$$= 1379.7755\frac{kg}{h}$$

Water produced: m₄

$$m_{H_2O} = C_2H_4 \text{ consumed in side reaction} \times \frac{2 \times 18\, kg\ H_2O}{28\ kg\, C_2H_4}$$

$$= 97.56\frac{kg}{h} \times \frac{2 \times 18\, kg\ H_2O}{28\ kg\, C_2H_4}$$

$$= 125.4343\frac{kg}{h}$$

Nitrogen balance: m_5

$$\text{Nitrogen in} = \text{Nitrogen out} \left(\because \text{ intert material}\right)$$

$$\Rightarrow m_5 = 50\frac{kg}{h}$$

Carbon dioxide balance: m_6

$$m_{CO_2} = C_2H_4 \text{ consumed in side reaction} \times \frac{2\times 44\, kg\ CO_2}{28\ kg\, C_2H_4}$$

$$= 97.56\frac{kg}{h}\times\frac{2\times 44\, kg}{28\ kg} = 306.6171\frac{kg}{h}$$

$$\Rightarrow m_6 \ = 306.6171\frac{kg}{h}$$

Reactor effluents

$$\begin{cases} m_{C_2H_4} = m_1 = 243.9024\dfrac{kg}{h} \\ m_{O_2} = m_2 = 637.72\dfrac{kg}{h} \\ m_{C_2H_4O} = m_3 = 1379.7755\dfrac{kg}{h} \\ m_{H_2O} = m_4 = 125.4343\dfrac{kg}{h} \\ m_{N_2} = m_5 = 50\dfrac{kg}{h} \\ m_{CO_2} = m_6 = 306.6171\dfrac{kg}{h} \end{cases}$$

4. Balance around the separator.

$$C_2H_4O \text{ produced } = 95\%\ of\ C_2H_4O \text{ formed in reactor}$$

$$= \frac{95}{100}\times 1379.7755\frac{kg}{h} = 1310.7867\frac{kg}{h}$$

Recycle stream $= 90\%$ of uncoverted C_2H_4 and O_2

$$= \frac{90}{100} \times 243.9024 \frac{kg}{h} + \frac{90}{100} \times 637.72 \frac{kg}{h}$$

$$= 793.4602 \frac{kg}{h}$$

5. Overall balance.

Total input = Total output

$$\Rightarrow \underbrace{1000}_{C_2H_4 \text{ In}} + \underbrace{900}_{O_2 \text{ In}} + \underbrace{50}_{N_2 \text{ In}} = \underbrace{1310.7867}_{C2H4O \text{ Produced}} + \text{Purge stream}$$

$$\Rightarrow \text{Purge stream} = 639.2132 \frac{kg}{h}$$

2.18. Energy and the concept of energy balance

Kinetic energy is the energy resulting from the overall translational motion of a system relative to a reference frame or from its rotation around an axis. Potential energy, on the contrary, is the energy of a system based on its position within a potential field, such as a gravitational or electromagnetic field. Internal energy is a comprehensive measure of a system's total energy content, excluding macroscopic, kinetic and potential energies. This includes several microscopic forms of energy, including translational energy, which is associated with the random motion of molecules relative to the system's center of mass; rotational energy, which arises from the spinning motion of molecules around their axes; vibrational energy, which results from the oscillation of atoms within molecules; intermolecular forces, which relate to electromagnetic interactions between molecules; and subatomic energy, which is associated with the motion and interactions of atomic and subatomic particles within molecules. This broad definition of internal energy highlights its complex nature, incorporating various forms of molecular and subatomic energies that contribute to a system's overall energy state.

2.18.1. *Energy balances for closed systems*

$$\begin{aligned} &\textit{Final system energy} - \textit{initial system energy} \\ &\qquad = \textit{net energy transferred to the system } (in - out). \end{aligned}$$

Initial system energy $= U_i + E_{ki} + E_{pi}$.

Final system energy = $U_f + E_{kf} + E_{pf}$.

Energy transferred = Q-W.

Q = energy transferred to the system from its surroundings.

W = work done by the system on the surroundings.

$(U_f - U_i) + (E_{kf} - E_{ki}) + (E_{pf} - E_{pi}) = Q - W$.

$\Delta U + \Delta E_k + \Delta E_p = Q - W$.

(a) If there are no alterations in temperature, phase or chemical processes within a closed system and pressure variations are minimal (less than a few atmospheres), then the change in internal energy (ΔU) is equal to zero.

(b) If a system experiences no acceleration, the change in kinetic energy (ΔE_k) is equal to zero. If a system does not experience a decrease in height, the change in potential energy (ΔE_p) is equal to zero.

(c) If the temperature of a system and its surroundings are equal or if the system is completely insulated, the heat transfer (Q) is zero, and this condition is called adiabatic.

(d) If there are no elements in motion, no flow of electricity or no emission of radiation at the system boundary, then the value of W (work) is zero.

2.19. Energy balances on open systems in the steady state

Flow work and shaft work are two important concepts in thermodynamics:

$W = W_s + W_{FL}$.

W_s = shaft work or rate work done by the process fluid on a moving part within the system (e.g. pump rotor).

W_{FL} = the flow work or the rate of work done by the fluid at the system outlet minus the rate of work done on the fluid at the system inlet.

$W_{FL} = P_{out} V_{out} - P_{in} V_{in}$

$H = U + PV$

2.19.1. *Steady-state energy balance of the open system*

$$\Delta H + \Delta E_k + \Delta Ep = Q - W_s$$

$$\Delta H = \sum_{\substack{output\\streams}} m_j H_j - \sum_{\substack{input\\streams}} m_j H_j$$

$$\Delta E_k == \sum_{\substack{output\\stream}} m_j u_j^2/2 - \sum_{\substack{input\\streams}} m_j u_j^2/2$$

$$\Delta E_p = \sum_{\substack{output\\streams}} m_j g z_j - \sum_{\substack{input\\streams}} m_j g z_j$$

For a single input on output

$$\Delta H = \dot{m}\,(H_{out} - H_{in}) = \dot{m}\,\Delta H.$$

2.19.2. *Procedures for energy balance*

(a) Calculate the flow rates of each individual component in the streams by applying the principle of material balance.

(b) Calculate the precise enthalpies of each individual component in the stream.

(c) The last stage involves formulating the suitable expression for the energy balance and then resolving it to obtain the necessary quantity.

2.19.3. *Mechanical energy balance*

Open systems prioritize kinetic and potential energy changes and shaft work over heat flow, internal energy and enthalpy changes.

$$\frac{\Delta p}{p} + \frac{\Delta u^2}{2} + g\Delta z + F = \frac{W_s}{\dot{m}}$$

where F is the frictional loss (N.m/kg).

2.19.4. *Balances on non-reactive processes*

$$Q = \Delta H = \sum_{out} n_i H_i - \sum_{in} n_i H_i$$

2.19.5. *Hypothetical process paths*

(a) Isothermal pressure change: altering pressure while maintaining constant temperature and phase state.

(b) Isobaric temperature change: modifying temperature at constant pressure, without changing the state of matter.

(c) Phase transitions: transformations between solid, liquid and gaseous states at constant temperature and pressure. For example: melting, freezing, vaporization, condensation and sublimation.

(d) Mixing and dissolution: combining two liquids or dissolving a gas or solid into a liquid. It occurs at constant temperature and pressure.

(e) Isothermal and isobaric chemical reaction: chemical transformations taking place under constant temperature and pressure conditions.

2.19.6. *Methodology for calculating energy balance*

(a) Prepare the material balance: complete all necessary material balance calculations for the system.

(b) Select the energy balance equation: choose the appropriate energy balance equation (closed or open system) and eliminate irrelevant or negligible terms for the given process. For the stationary closed system, ΔE_k, $\Delta E_p = 0$ and $W = 0$, if the system volume is constant. If there is no mixing part or impeller, no energy is transferred by electricity or radiation.

For an open system, $\Delta E_p = 0$.

(c) Choose a reference state: establish a reference state (phase, temperature and pressure) for each species in the process.

(d) For a closed system with constant volume, construct a table with columns for the initial and final amounts of each species (m_i or n_i) and specific internal energies relative to the chosen reference states (U_i). For an open system, construct a table with columns for the inlet and outlet stream component flow rates (m_i or n_i) and specific enthalpies relative to the chosen reference states (H_i).

(e) Compute all the required values of U_i or H_i and insert the values into the appropriate places in the table.

(f) Calculate the closed system:

$$\Delta U = \sum_{out} m_j U_j - \sum_{in} m_j U_j$$

The open system:

$$\Delta H = \sum_{out} m_j H_j - \sum_{in} m_j H_j$$

(g) Calculate any work, kinetic energy or potential energy terms that have not been dropped from the energy balance.

(h) Solve the energy balance for any variable that is unknown, often Q.

Closed system: $Q - W = \Delta U + \Delta E_k + \Delta E_p$.

Open system: $Q - W_s = \Delta H + \Delta E_k + \Delta E_p$.

2.19.7. *Variations in pressure under constant temperature conditions*

If the pressure of a solid or liquid undergoes a shift while maintaining a steady temperature, then:

$\Delta U = 0$;

$\Delta H = \Delta U + \Delta (PV)$;

$= V\Delta P$.

U and H are independent of 1 for ideal gases.

$\Delta U \cong 0$ and $\Delta H \cong 0$ for a gas undergoing an isothermal pressure change unless gases at temperatures well below 0°C and well above 1 atm.

2.19.8. *Sensible heat (Felder et al. 2020)*

This is defined as the heat that must be transformed to raise or lower the temperature of a substance or mixture of substances:

$Q = \Delta U$ (closed system).

$Q = \Delta H$ (open system).

$$Cv = \left(\frac{\partial u}{\partial T}\right)_v.$$

$dU = C_v\, dT$.

$$\Delta U = \int_{T1}^{T2} Cv(T)\, dT$$

(for an ideal gas: exact; solid and liquid: good approximation; for a non-ideal gas: valid only if V is constant).

$$Cp(T) = \left(\frac{\partial H}{\partial T}\right)_P$$

$\Delta H = \int_{T1}^{T2} Cp(T)\, dT$ (for an ideal gas: exact; for a non-ideal gas: exact only if P is constant).

$\Delta U = \int_{T1}^{T2} Cv(T)\, dT$ solid or liquid.

For solids and liquids: Cp = Cv.

For ideal gases: $C_p = C_v + R$.

2.20. Estimation of heat capacity

(a) To determine the total change in enthalpy for a mixture of gases or liquids, add up the individual enthalpy changes for each component of the mixture.

(b) When dealing with very low concentrations of solids or gases dissolved in liquids, we can ignore the enthalpy change of the solute.

$$C_{p,mixture} = \sum_{\substack{all\ mix \\ components}} y_i C_{P,i}(T)$$

$$\Delta H = \int_{T1}^{T2} C_{p,mix}(T)\, dT$$

2.20.1. *Energy balances on single-phase systems*

1. Evaluate, $\Delta U = \int_{T1}^{T2} Cv(T)\, dT$; $\Delta H = \int_{T1}^{T2} Cp(T)\, dT$ (correct for pressure changes if necessary).

2. For a closed system at constant volume:

$\Delta U = n\, \Delta \hat{U}$, where n is the amount of the species that is heated or cooled.

For an open or closed system of constant pressure:

$\Delta H = n\, \Delta \hat{H}$.

3. Substitute for ΔU, ΔH in the appropriate energy balance equation to determine the required heat transfer Q.

Phase change operations: they refer to the processes in which a substance undergoes a change in its physical state, for example, from a solid to a liquid or from a liquid to a gas.

(a) The heat of fusion/melting $\Delta \hat{H}_m(T,P)$ refers to the specific enthalpy difference between the solid and liquid states of a substance at a given temperature (T) and pressure (P).

(b) The head of vaporization, denoted as $\Delta \hat{H}_v$ (T,P), represents the specific enthalpy difference between the liquid and vapor states of a substance at a given temperature and pressure.

2.21. Estimation and correlation of latent heat (Felder et al. 2020)

Trouton's rule:

$\Delta \hat{H}_v$ (kJ/mol) = 0.088 T_b (K)………. non-polar liquids;

= 0.109 T_h (K)………. water, low molecular weight alcohols.

Chen's equation:

$$\Delta \hat{H}_v \text{ (kJ/mol)} = \left[\frac{Tb[0.0331 \frac{Tb}{Tc} - 0.0327 + 0.0297 \log Pc}{1.07 - \frac{Tb}{Tc}}\right]$$

where:

T_b = normal boiling point (K);

T_c = critical temperature (K);

P_c = critical pressure (atm).

Standard heat of fusion (Felder et al. 2020):

$\Delta\hat{H}_m$ (kJ/mol) = 0.0092 T_m(K) metallic elements;

= 0.0025 T_m(K) inorganic compounds;

= 0.050 T_m(K) organic compounds.

Clapeyron equation (Felder et al. 2020):

$$\frac{d \ln P^*}{d\frac{1}{T}} = -\frac{\Delta H_r}{R}$$

Watson's correlation (Felder et al. 2020):

$$\Delta H_r(T_2) = \Delta H_r(T_1)\left[\frac{T_c - T_2}{T_c - T_1}\right]^{0.38}$$

Humid volume (V_H):

The humid volume refers to the total volume occupied by 1 kilogram of dry air, along with the accompanying water vapor.

For fully saturated air, the dry- and wet-bulb temperatures are equal.

2.22. Heats of solution and mixing

The heat of solution ΔH_s (T, r) is the difference in enthalpy for a reaction where 1 mole of solute (either a gas or a solid) is dissolved in r moles of liquid solvent at a constant temperature T.

As the value of r increases significantly, the change in enthalpy (ΔH_s) tends to reach a maximum value, called the heat of solution, when the dilution is infinite.

2.22.1. *Reactive process balances (Felder et al. 2020)*

Heat of reaction, ΔH_r (T, P): this refers to the change in enthalpy that occurs when stoichiometric amounts of reactants react completely in a single reaction to generate products at the same temperature and pressure.

If V_A is the stoichiometric coefficient of a reactant or reaction product A (+ve for the product, -ve for the reactant) and $n_{A,r}$ moles of A are consumed or generated at $T = T_o$ and $P = P_o$:

$$\Delta H = \frac{\Delta H_r\ (T_o, P_o)}{|V_A|} . n_{A,r}$$

$$\varepsilon = \frac{n_{A,out} - n_{A,in}}{|V_A|} = \frac{n_{A,r}}{|V_A|}$$

– If ΔH_r (T, P) is negative, the reaction is exothermic at T and P.

– If ΔH_r (T, P) is positive, the reaction is endothermic.

– At low and moderate pressures. ΔH_r (T, P) is nearly independent of pressure.

– The value of the heat of reaction depends on how the stoichiometric equation is written:

$CH_4 (g) + 2O_2 (g) \rightarrow CO_2 (g) + 2H_2O$ (l), ΔH_r (25°C) = - 890.3 kJ/mol,

$2CH_4 (g) + 4O_2 (g) \rightarrow 2CO_2 (g) + 4H_2O$ (l), ΔH_r (25°C) = - 1780.6 kJ/mol.

– The value of the heat of reaction depends on the states of aggregation (gas, liquid or solid) of the reactants and products:

$CH_4(g) + 2O_2(g) \rightarrow CO_2(g) + 2H_2O$ (l), ΔH_r (25°C) = - 890.3 kJ/mol,

$CH_4(g) + 2O_2(g) \rightarrow CO_2(g) + 2H_2O$ (g), ΔH_r (25°C) = - 802.3 kJ/mol.

– The standard heat of reaction, ΔH_r, is the heat of reaction when both reactants and products are at 25°C and 1 atm.

– If a reaction takes place in a closed reactor at constant volume, the heat released or absorbed is determined by the change in internal energy between reactants and products. The internal energy of a reaction, ΔU_r(T), is given by:

$$\Delta U_r(T) = U_{products} - U_{reactants}$$

If stoichiometric quantities of reactants react completely at T, V_i is the stoichiometric coefficient of the ith gaseous reactant or product, an ideal gas is valid, and specific volumes of liquid and solid reactants and products are negligible compared with those of the gases, then:

$$\Delta Ur(T) = \Delta Hr(T) - RT\left(\sum_{\substack{product\\gases}} |V_i| - \sum_{\substack{reactant\\gases}} |V_i|\right)$$

2.23. Heat of reaction measurement

The measurement can be made with a colorimeter. A colorimeter is a sealed apparatus immersed in a liquid in a well-insulated container. The change in liquid temperature can be quantified and used to determine the energy released or absorbed during the reaction. The value of ΔH_r can then be calculated based on this energy and the known heat capacities of the reactants and products.

Hess's law: if the stoichiometric equation for reaction 1 can be derived from algebraic operations (e.g. multiplication by constants, addition and subtraction) on stoichiometric equations 2, 3, etc., then the heat of reaction ΔH_{r1} can be obtained by applying the same operations to the heat of reaction ΔH_{r2}, ΔH_{r3}, ΔH_{r4}, etc.

2.23.1. *Formation reactions and heats of formation*

Formation reactions, on the one hand, are chemical reactions that result in the formation of a compound or molecule. Heats of formation, on the other hand, are the amount of heat energy released or absorbed during the formation of a compound or molecule.

The formation reaction of a compound refers to the reaction in which the compound is formed from its elemental constituents in their natural state. For example, oxygen is represented by the symbol O_2, not O. The standard heat of production of a compound, ΔH_f, is the enthalpy change that occurs when one mole of the compound is formed at a reference temperature and pressure (25°C, 1 atm):

$$\Delta Hr = \sum_i V_i \Delta H_{ji} = \left(\sum_{products} |V_i| \Delta H_{ji} - \sum_{reactants} |V_i| \Delta H_{ji}\right).$$

The standard heats of formation of all elemental species should be set to zero.

Heat of combustion

The ΔH_c, or standard heat of combustion, refers to the amount of heat released when a material reacts with oxygen to produce specific products. For example, the reaction involves carbon dioxide (CO_2) in its gaseous state and water (H_2O) in its liquid state, with both the reactants and the products present at a temperature of 25°C and a pressure of 1 atmosphere.

The standard heat of reactions involving only combustible material and combustion products can be determined using tabulated standard heat of combustion values:

$$\Delta Hr = \sum_i V_i \Delta H_c = \left(\sum_{products} |V_i| \Delta H_c - \sum_{reactants} |V_i| \Delta H_c\right).$$

If any of the reactants or products consist of combustion byproducts such as CO_2, $H_2O(l)$, SO_2, etc., their enthalpy change, ΔH_c, should be set to zero.

2.23.2. *Energy balances for reactive processes*

Standard operating procedure:

(a) Draw and annotate a flow chart.

(b) Material balances and phase equilibrium relationships, such as Raoult's law, should be used to determine the quantities of stream components and flow rates as much as possible.

(c) Select relevant reference states for calculating specific enthalpy or internal energy changes (ΔH, ΔU, or ΔH). Then, input the obtained value into the corresponding energy balance equation and perform the necessary calculations.

There are two commonly used approaches for selecting reference states in enthalpy calculations and for calculating particular enthalpies and ΔH.

2.23.3. *Method for determining the heat of reaction (Felder et al. 2020)*

This approach is typically used when there is only one reaction for which the change in enthalpy (ΔH_r) is known.

(a) Perform the material balance calculations for the reactor as much as possible.

(b) Select appropriate reference states to calculate specific enthalpy. Typically, the most suitable options are the reactant and product species at a temperature of

25°C and a pressure of 1 atm, where the heat of reaction is known. For non-reacting species, any suitable temperature can be chosen, such as the temperature at the inlet or outlet of the reactor or the reference condition specified in an available enthalpy table.

(c) For a single reaction in a continuous process, calculate the extent of the reaction $\mathcal{E}$, which is independent of the species:

$$\mathcal{E} = \frac{n_{A,out} - n_{A,in}}{|V_A|} = \frac{n_{A,r}}{|V_A|}$$

(d) Prepare the inlet–outlet enthalpy table.

(e) Calculate the enthalpy of each unknown stream component, H_i, as ΔH for the species going from its reference state to the process state, and insert the enthalpies in the table.

(f) Calculate ΔH for the reactor:

$$\Delta H = \mathcal{E}\Delta H_r + \textstyle\sum_{out} n_i H_i - \sum_{in} n_i H_i \text{ (single reaction)}$$

$$\Delta H = \textstyle\sum_{reaction} \mathcal{E}_i \Delta H_{ri} + \sum_{out} n_i H_i - \sum_{in} n_i H_i \text{ (multiple reactions)}$$

(g) Substitute the calculated value of ΔH in the energy balances:

$Q - W_s = \Delta H + \Delta E_k + \Delta E_p$ for an open system and complete the required calculations.

Heat of formation method (Felder et al. 2020)

This approach is often more favorable for many reactions and single reactions where the enthalpy change (ΔH_r) is not easily accessible.

(a) Perform the material balance calculations for the reactor as much as possible.

(b) Select appropriate reference states to calculate enthalpy. This differentiates the previous procedure from the current one. For example, the options should consist of the elemental species that make up the reactants and products in their respective states at a temperature of 25°C and a pressure of 1 atmosphere. Chemical species such as C(S) and H_2(g), as well as other non-reacting species, can be present at any desired temperature.

(c) Create an inlet–outlet enthalpy table by including the known molar amounts or flow rates for all components in the input and output streams.

(d) Determine the value of each unspecified specific enthalpy:

$\Delta H_i = \Delta H_{fi} + \int_{25}^{T} Cp(T)\, dT$ where T is temperature of the stream.

(e) Determine the change in enthalpy (ΔH) for the reactor:

$$\Delta H = \sum_{out} n_i H_i - \sum_{in} n_i H_i$$

The use of the heat of reaction is unnecessary if the elements are selected as references. The heat of reaction is calculated by subtracting the heat of formation of the reactants (included in H_{in}) from the heat of formation of the products in H_{out}.

(f) Replace the calculated value of ΔH in the energy balance equation and perform the necessary calculations.

2.23.4. *Thermochemistry of solutions*

The standard heat of formation of a solution refers to the enthalpy change that occurs when the solutes and solvents combine to form the solution at a temperature of 25°C. If a solution has a mole ratio of n moles of solvent per mole of solute, then:

$$\Delta H_{f,\ solution} = \Delta H_{f,\ solute} + \Delta H_s\ (n)$$

where ΔH_s is the heat of the solution.

2.23.5. *Characteristics and properties of fuels*

– **Solid fuels**: these fuels, such as coal, consist of a mixture of carbon (C), water (H_2O), non-combustible ash, HCs, sulfur and coke. Coke is the solid residue that remains when coal or petroleum is burned, causing volatile elements to evaporate and HCs to decompose. In addition to coal and petroleum, small amounts of wood and solid waste (trash) can also contribute to solid fuels.

– **Liquid fuels**: they primarily consist of HCs derived from the distillation of crude oil, coal tars and shale oil.

– **Gaseous fuels**: the primary component of natural gas is methane (CH_4), which makes up 80–95% of the gas. The remaining portion consists of propane (C_3H_8), ethane (C_2H_6) and small amounts of other gases. Additionally, light HCs are derived from the processing of petroleum or coal, as well as from acetylene and hydrogen.

2.23.6. *Heating value*

The heating value of a combustible substance is equal to the negative value of its standard heat of combustion. The higher heating value, also known as the total or gross heating value, is represented by -ΔH_c and includes $H_2O(l)$ as a combustion product. On the contrary, the lower heating value, or the net heating value, is based on $H_2O(V)$ as a product.

Given that ΔH_c is consistently negative, it follows that the heating value is positive:

$$HHV = LHV + n\,\Delta H_v\,(H_2O,\ 25°C)$$

$$\Delta H_v(H_2O,\ 25°C) = 44.013\ kJ/mol$$

$$= 18{,}934\ BTU/lb\text{-}mole$$

$$HV = \sum x_i\,(HV)_i\,;$$

$(HV)_I$ = heating value of the ith combustible substance.

2.23.7. *Adiabatic flame temperature*

The adiabatic flame temperature (T_{ad}) is the maximum temperature achieved during fuel combustion under ideal conditions.

It occurs when:

– all the energy released during combustion is used to heat the products;

– no heat is lost to the surroundings (adiabatic process);

– the reaction reaches completion.

This temperature represents the theoretical upper limit for combustion product temperatures in a given system.

The temperature is known as the adiabatic flame temperature, denoted as T_{ad}:

$$\sum n_i\,H_i(T_{ad}) = -n_f \Delta H_c + \sum_{in} n_i\,H_i(T_{feed})$$

2.23.8. *Ignition*

It is the rapid acceleration of an oxidation reaction when the temperature of the reactant mixture exceeds a critical point, known as the ignition temperature.

Key concepts include the following:

– Ignition lag: time between reaching the ignition temperature and the actual ignition.

– Autoignition temperature: minimum temperature for spontaneous fuel ignition (e.g. H_2: 400°C, CH_4: 540°C).

– *Flammability limits:*

(a) Lower flammability limit: minimum fuel concentration for ignition.

(b) Upper flammability limit: maximum fuel concentration for ignition.

– *Explosion range:* composition range between the lower and upper flammability limits where ignition is possible.

Outside the flammability limits, the mixture cannot ignite or explode, even with an ignition source.

– *Flash point:* the flash point of a liquid is the lowest temperature at which it emits sufficient vapor to form a flammable mixture with air near its surface. This property is crucial for assessing fire hazards caused by volatile liquids.

For example:

– Gasoline flash point: –43°C.

– Ethanol flash point: 13°C (Himmelblau and Riggs 2012; Felder et al. 2020).

2.24. Energy balance problem (Felder et al. 2020)

Let us consider a continuous process involving a chemical reactor and a heat exchanger, as shown in Figure 2.2.

A gas-phase reaction A $\rightarrow$ B + C occurs in an adiabatic plug flow reactor. The product stream is then cooled in a heat exchanger before further processing.

Using the following information, perform an energy balance analysis of the system:

1. Feed stream: 100 kmol/h of pure A at 25°C and 1 atm.
2. The reactor operates at 2 atm.
3. The reaction is 80% complete.
4. The heat of reaction (ΔH_{rxn}) = 50 kJ/mol (endothermic).
5. The feed is preheated to 300°C before entering the reactor.
6. The heat exchanger cools the product stream to 100°C.
7. Cp values (assumed constant):
   - $Cp_A = 30$ J/(mol·K);
   - $Cp_B = 25$ J/(mol·K);
   - $Cp_C = 35$ J/(mol·K).

Calculate the following:

(a) The temperature of the reactor effluent.

(b) The heat duty of the heat exchanger.

(c) The overall energy balance of the system.

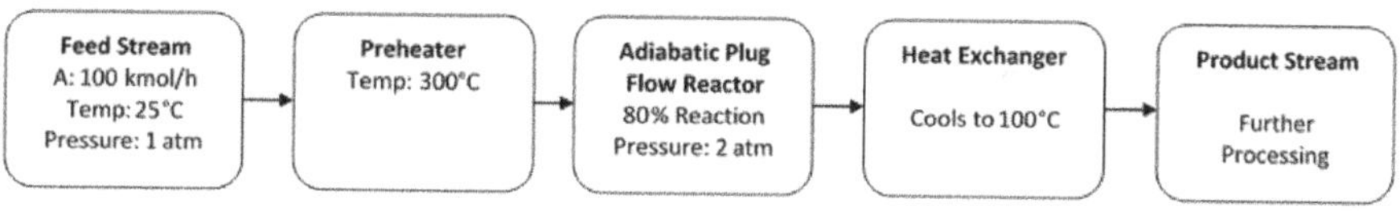

Figure 2.2. *Energy balance for adiabatic reactor*

Solution

1. Define the system boundaries.

Our system includes the preheater, the reactor and the heat exchanger.

2. Preheater calculations.

Energy required to heat the feed from 25°C to 300°C:

$$Q\_preheat = n_A \times Cp_A \times (T_{final} - T_{initial}).$$

$$Q\_preheat = 100 \text{ kmol/h} \times 30 \text{ J/(mol·K)} \times (300 - 25) \text{ K}.$$

$$Q\_preheat = 825{,}000 \text{ kJ/h}.$$

3. Reactor calculations.

Extent of reaction:

$$\xi = 0.80 \times 100 \text{ kmol/h} = 80 \text{ kmol/h}.$$

Molar flow rates after the reaction:

$$n_{A\_out} = 20 \text{ kmol/h}.$$

$$n_{B\_out} = n_{C\_out} = 80 \text{ kmol/h}.$$

Energy consumed by the reaction:

$$Q_{rxn} = \xi \times \Delta H_{rxn}.$$

$$Q_{rxn} = 80 \text{ kmol/h} \times 50 \text{ kJ/mol} = 4{,}000{,}000 \text{ kJ/h}.$$

4. Reactor effluent temperature.

Energy balance in the reactor: $Q_{in} + Q_{rxn} = Q_{out}$

$$Q_{in} = n_{A\_in} \times Cp_A \times (300 - 25).$$

$$Q_{out} = (n_{A\_out} \times Cp_A + n_{B\_out} \times Cp_B + n_{C\_out} \times Cp_C) \times (T_{out} - 25).$$

Solving for T_{out}:

$$T_{out} = 25 + (Q_{in} - Q_{rxn})/(n_{A\_out} \times Cp_A + n_{B\_out} \times Cp_B + n_{C\_out} \times Cp_C).$$

$$T_{out} \approx 452°C.$$

5. Heat exchanger calculations.

Heat removed in the heat exchanger:

$$Q_{HX} = (n_{A\_out} \times Cp_A + n_{B\_out} \times Cp_B + n_{C\_out} \times Cp_C) \times (452 - 100).$$

$$Q_{HX} \approx 2{,}156{,}000 \text{ kJ/h (heat removed)}.$$

6. Overall energy balance.

Energy in:

– Feed enthalpy: 100 kmol/h × 30 J/(mol·K) × (25 - 25) K = 0 kJ/h.

– Preheater: 825,000 kJ/h.

Energy out:

– Product stream enthalpy: (20 × 30 + 80 × 25 + 80 × 35) kJ/(kmol·K) × (100 – 25) K = 562,500 kJ/h.

– Heat removed in the heat exchanger: 2,156,000 kJ/h.

Energy accumulated = Energy in – Energy out. Energy accumulated = (0 + 825,000) – (562,500 + 2,156,000) = –1,893,500 kJ/h.

The negative value indicates that additional energy (1,893,500 kJ/h) must be supplied to the system to maintain the energy balance, which is consistent with the endothermic nature of the reaction.

2.25. References

Felder, R.M., Rousseau, R.W., Bullard, L.G. (2020). *Elementary Principles of Chemical Processes*. John Wiley & Sons, New York.

Himmelblau, D.M. and Riggs, J.B. (2012). *Basic Principles and Calculations in Chemical Engineering*. FT Press, London.

Pritchard, P.J. and Mitchell, J.W. (2016). *Fox and McDonald's Introduction to Fluid Mechanics*. John Wiley & Sons, New York.

3

Laws of Thermodynamics

Thermodynamics is a scientific discipline that encompasses the fundamental rules and principles that control the transmission and conversion of energy, as well as the alterations in attributes of things resulting from these processes. The laws are derived from empirical observations of natural phenomena. The fundamental principles, which are derived from the laws, are established by logical deductions that follow the laws. Thermodynamics is a scientific field that studies the links between the properties of substances and how they change during physical processes. Thermodynamics is a crucial subject that serves as a fundamental basis for all engineering disciplines. This is because technical processes and advancements in engineering include the transfer and conversion of energy. In this chapter, we will explore the principles of thermodynamics and its associated consequences. The chapter revolves around physical principles, mathematical equations and practical examples to show their applications. This will not only enhance students' comprehension of physical principles but also help them to overcome common misconceptions when it comes to understanding laws and their practical implementations.

3.1. Introduction

Thermodynamics constitutes the scientific and engineering discipline concerned with the principles governing heat, energy and work. A majority of processes within this domain involve the conversion of heat into work and vice versa. At its core, thermodynamics encompasses the transformation of one form of energy to another.

In the realm of mechanics, the generation of heat is observed when mechanical work is executed by a mechanical machine, exemplifying the conversion of mechanical work into mechanical energy. Conversely, in this context, steam is applied to facilitate mechanical work. Analogously, within the domain of electric

Chapter written by Girish GUPTA and Neha DHIMAN.

current, the migration or flow of electrons within a conductor from one atom to another results in the production of electrical energy.

From a chemical perspective, the manifestation of chemical energy occurs through its storage or release during a chemical reaction. Exothermic reactions generate chemical energy in the form of heat, whereas endothermic reactions store energy in the form of heat.

Hence, thermodynamics can be precisely defined as the examination of the interconversion of work or heat or the transformation of energy into its diverse form within a system. In the context of chemical processes entailing the conversion of energy from one form to another, this field is specifically termed chemical thermodynamics.

Thermodynamics is the branch of science and engineering that deals with heat, energy and work. Most of the processes involve the conversion of heat into work or vice versa. The simplest form of heat is energy. In other words, we could say that thermodynamics deals with the conversion of one form of energy into another form of energy.

For instance, in a mechanical field, heat is produced when work is done by the mechanical machine, which shows the conversion of mechanical work into mechanical energy. On the other hand, heat in the form of steam is introduced to the steam engine to perform mechanical work. Similarly, in terms of electric current, the movement or flow of electrons in a conductor from one atom to another produces an electrical form of energy.

From a chemical point of view, chemical energy is stored or released during a chemical reaction. Chemical energy is produced in the form of heat for exothermic reactions, and energy is stored in the form of heat for endothermic reactions.

Therefore, thermodynamics can be defined as the study of the inter-conversion of work and heat or the inter-conversion of energy into its various forms in a system. Dealing with the chemical processes involving the transformation of energy from one form to another is referred to as chemical thermodynamics.

3.2. Thermodynamical systems

3.2.1. *System*

A thermodynamics system refers to a "macroscopic" segment of the universe comprised of a substantial number of particles. Certain thermodynamic properties, such as temperature, pressure and volume, are not inherently attributed to individual particles. Hence, the investigation of these properties necessitates an examination of the system as a whole. Additionally, a thermodynamics system can be characterized as a specific region of the universe subject to study, enclosed within a defined boundary and isolated from its surroundings.

3.2.2. *Surrounding*

Surroundings encompass all entities within the universe outside the designated system. The exchange of energy occurs between the system and its surroundings, maintaining the total energy of the universe which is constant. This principle reflects the conservation of energy, i.e. energy is neither created nor destroyed

Universe = System + Surrounding

3.2.3. *System boundary*

The thermodynamic system and its surroundings are demarcated by a flexible or impermeable partition, commonly referred to as a boundary. Alternatively, it can be conceptualized as the enclosed limit around the system through which energy, in the form of heat, either enters or exits the system. A tangible example of a real boundary system is a closed glass container containing water, whereas a system with a cylinder fitted with a piston containing gas represents a flexible boundary system.

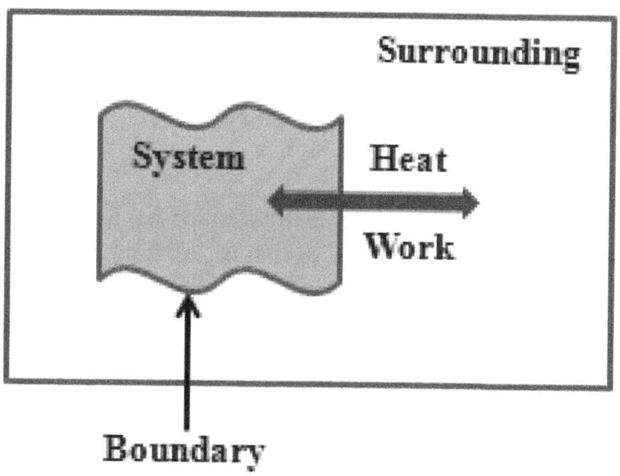

Figure 3.1. *Thermodynamical system*

3.3. Types of system

The classification of a thermodynamics system is based on the exchange of mass and heat energy between the system and its surroundings. This categorization encompasses three primary types: open system, closed system and isolated system.

– Open system: an open system is characterized by an open boundary that facilitates the transfer of both heat and mass with the surroundings as depicted in Figure 3.2. An illustrative example of an opened system is a glass of hot water, whereas the exchange of heat in the form of energy and mass in the form of vapors occurs between the systems and its surroundings.

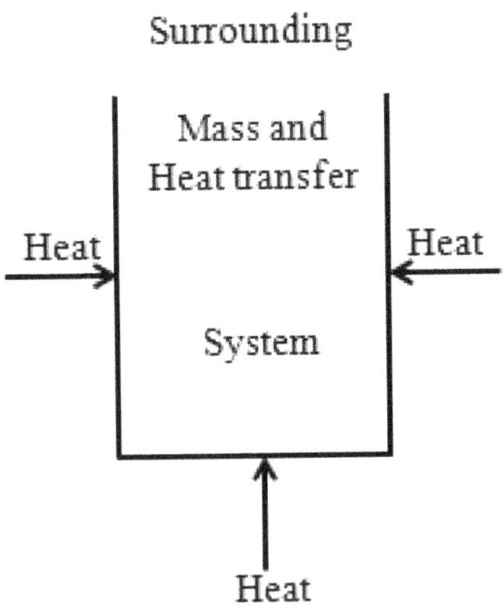

Figure 3.2. *Open system*

– Closed system: a closed system is characterized by a tightly sealed boundary without insulation, allowing solely the exchange of heat in the form of energy. Figure 3.3 illustrates such a system where no mass transfer occurs between the system and its surroundings. An exemplar of a closed system is a tightly sealed steel flask containing hot water, wherein heat is exclusively transferred through the sealed container's wall into the surroundings, while there is no mass transfer across the boundary.

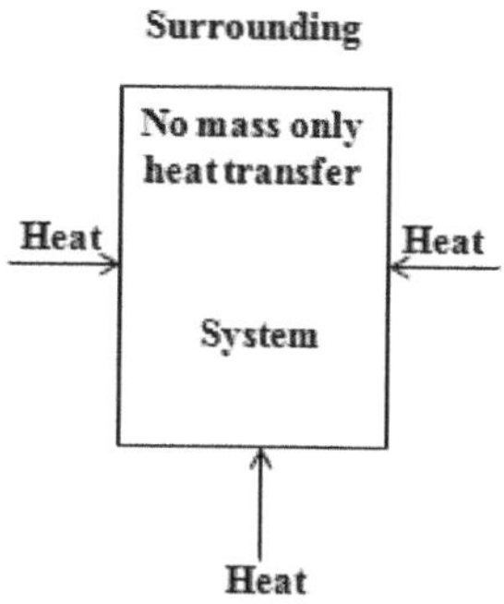

Figure 3.3. *Closed system*

– Isolated system: an isolated system is characterized by a thoroughly sealed and well-insulated boundary, preventing both heat and mass transfer to the surroundings, as illustrated in Figure 3.4. An isolated system example is a closed and well-insulated kettle containing hot water, wherein neither matter nor energy is exchanged through the boundary with the surroundings.

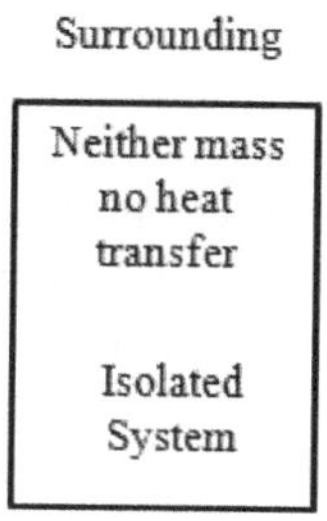

Figure 3.4. *Isolated system*

3.4. Macroscopic system and properties

A macroscopic system is defined as a system composed of a substantial number of particles or bulk entities, such as atoms, ions and molecules.

The characteristics of a system that can be quantified on a large scale of particles such as those associated with the bulk system are referred to as macroscopic properties. In contrast, properties associated with individual atoms are considered as microscopic properties, while properties of the study of molecules in the aggregate are categorized as macroscopic properties.

Several macroscopic properties include temperature, pressure, volume, composition, density, viscosity, surface tension, etc. For instance, to measure the volume of gas in a cylinder, the consideration of the total number of gas molecules within the cylinder is essential. Macroscopic properties can be categorized into two types: extensive properties and intensive properties.

– Extensive properties: these properties are characteristics of a system that are entirely contingent on the quantity and size of the matter present, and their values vary with changes in the size and quantity of matter. Examples of extensive properties include mass, volume, weight, length, internal energy and enthalpy. Consideration of two cylinders, one with a capacity of 10 liters and the other with 20 liters, illustrates this concept. The cylinder with a 20-liter capacity contains a greater quantity of matter compared to the 10-liter capacity cylinder, highlighting how the volume of matter is contingent upon the size and quantity of matter.

– Intensive properties: the properties of a system that have no dependence on the size and amount of matter, for example, temperature, pressure, viscosity, density, melting point boiling point, etc., are all intensive properties. Suppose we have taken 10 mL of water at 20°C in a beaker and double its volume by adding water at a similar temperature, then the resultant temperature will be the same.

Additionally, there is a fact that for a specified unit amount of matter, extensive properties may be converted into intensive properties, for example, mass and volume of matter are extensive properties but density, which is the ratio of mass per unit volume, is considered to be an intensive property because of the specified unit volume. Similarly, concentration is an intensive property as the concentration of solution is not dependent on the quantity of matter (it will be the same throughout the solution). Concentration is defined as the ratio of the number of moles per unit volume, where the number of moles and volume are extensive properties of matter.

In short, we can say that the ratio of two extensive properties will give rise to the intensive property of matter.

3.5. State of system and state variables

The state of the system can be defined when a macroscopic system has a definite set of values of its macroscopic properties, and the system is said to be in a definite state. Change in any one of the macroscopic properties will also give rise to a change in the state of the system. For example, if the temperature of the gas molecules is increased in a cylinder fitted with a piston, then expansion in gas molecules leads to an increase in the volume of the system. Similarly, if pressure on a piston is increased, the volume of gas will decrease and the state of the system will be changed (Smith et al. 2017).

A particular set of values of macroscopic properties required to determine the state of the system are known as state variables. Among all of the macroscopic properties of a system, temperature, pressure, volume and concentration are the main state variables. Except for these, all other variables are independent. If these four variables have a defined set of values, then the other variables automatically attain a definite set of values, For instance, according to the ideal gas equation for 1 mole of gas, PV = RT. Suppose V changes in the above ideal gas equation, the value of P and T depends on P. Therefore, V is the state variable and P and T are the state functions.

3.6. Zeroth law of thermodynamics

The zeroth law of thermodynamics is based on the assumption of the law of thermal equilibrium. It states that when two systems are separately in thermal equilibrium with a third system, then the two systems are also in thermal equilibrium when placed in contact with each other (Smith et al. 2017).

For example, let us assume the two systems A and B, which are separated from each other by an insulated or adiabatic wall that does not allow any transfer of heat between them. The third system C is separated from both systems A and B by a conducting wall which allows the flow of heat (Exell 2018; Jawale et al. 2023).

According to the zeroth law, when A and B systems are in thermal equilibrium with each other, there must be a thermodynamic variable having the same value for both systems, i.e. temperature is the same for both systems which leads to attaining a thermal equilibrium between them. If systems A and B are in equilibrium with system C separately, i.e. $T_A = T_C$ and $T_B = T_C$, this gives rise to $T_A = T_B$ which states that systems A and B are in equilibrium with each other. All three systems A, B and C have the same temperature T.

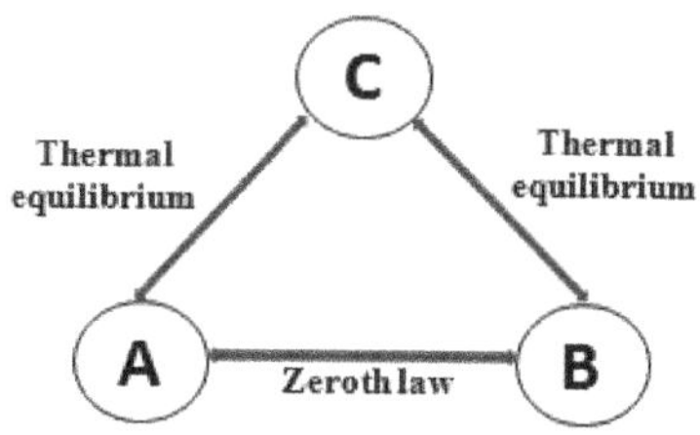

Figure 3.5. *Zeroth law of thermodynamics*

3.6.1. *Applicability of zeroth law*

Temperature is one of the important measurable quantities that helps to predict whether two systems are in equilibrium or not. Heat always flows from the hot body towards the cold body until equilibrium is attained between them. Therefore, the zeroth law states that no heat flows between two bodies after the attainment of thermal equilibrium.

The widely used application of the zeroth law compares the temperatures of two different bodies. For the measurement of the temperature of any body, a reference body must be required. Measurement of temperature using a thermometer is a common example that is working based on the zeroth law of thermodynamics. If the thermometer is allowed to be in thermal equilibrium with body I and body II, respectively, then the temperature readings of the two bodies give the comparison of hotness between them.

The other example also illustrates the zeroth law of thermodynamics, i.e. if we place two beakers on a table, one contains hot water and the other is filled with cold water. After a few hours, both the beakers come into thermal equilibrium at room temperature. Therefore, we can say that after the attainment of equilibrium of two beakers of water at room temperature, they both will also be in thermal equilibrium with each other.

3.7. Heat

Heat is a form of energy that is exchanged between two bodies when both are at different temperatures (Smith et al. 2017). When one body A is at a higher temperature and the other at a lower temperature, then the flow of heat can be determined as heat is lost from body A to B, whereas when body A is at a lower temperature than body B, then heat is gained by body A from B and the amount of heat which is lost or gained is expressed by the term "Q".

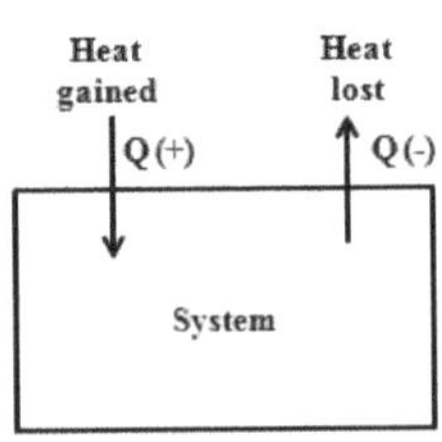

Figure 3.6. *Sign convention of heat*

If heat is lost, it is denoted with a negative sign. If heat is gained, it is denoted with a positive sign.

The unit of heat is calorie (cal) which is defined as "heat desired to raise the temperature of 1gm of water by 1°C or 1K". The SI unit of heat or energy is joule and 1 joule = 0.2390 calories.

3.8. Work

Work done is also associated with the exchange of heat. Let us consider an example of a cylinder fitted with a piston containing gas. If gas expands and the piston moves outwards, then work is said to be done by the gas and taken as negative. If gas is compressed with external pressure, the piston moves inward and work is said to be done on the gas and is taken as positive (Dahm and Visco 2014; Smith et al. 2017)

$$W_{expansion} = negative = -P\Delta V \quad [3.1]$$

$$W_{compression} = positive = +P\Delta V \quad [3.2]$$

3.9. State functions and path functions

State functions can be defined as those thermodynamical properties of a system whose value does not depend on the pathway taken to travel through initial to the final state (Zohuri and McDaniel 2019). State function is dependent only upon the initial and final states of the system. The simplest way to understand the concept of a state function is an integral method that only deals with the function along with its lower and upper limits.

Internal energy (U), pressure, temperature, volume, mass, enthalpy, density, etc., are the state functions

$$\text{Integral of internal energy, } \int_1^2 dU = U_2 - U_1 = \Delta U \quad [3.3]$$

Path functions can be defined as those properties whose value depends upon the pathway taken to achieve the final state by traveling through the initial state of the system, for example, work and heat are path functions that only depend on the path followed.

3.10. First law of thermodynamics

This is based on the law of conservation of energy which states that the total energy of a closed system always remains the same. It can neither be created nor destroyed during any process, and it can only change from one form to another (Smith et al. 2017).

The mathematical definition of the first law of thermodynamics states that when a particular amount of heat is introduced to a given system, then the change in internal energy will be equal to the sum of heat given and the external work done on the system (Exell 2018; Jawale et al. 2023), i.e.

$$\Delta U = Q + W \quad [3.4]$$

If work is done by the system, then,

$$\Delta U = Q - W \quad [3.5]$$

3.11. Internal energy is a state function

Let us suppose a system that is changing its state concerning pressure and volume. Let the initial and final states of the system be A and B and let us assume that U_A and U_B are the internal energies of the system concerning states A and B, respectively. Now, the change in internal energy can be given as

$$\Delta U = U_B - U_A \quad [3.6]$$

Any path can be taken to travel through the initial state A to final state B. Therefore, let us suppose that the system changes its state from A to B by using path I and back from B to A using path II and the change in internal energy associated with two paths are ΔU and ΔU', respectively. If ΔU and ΔU' are not equal, then suppose ΔU > ΔU'. This implies if the same cycle is repeated again and again, the energy is continuously generated and increased in each cycle.

However this is contrary to the first law of thermodynamics according to which energy is neither created nor destroyed. Therefore, ΔU must be equal to ΔU'; hence, the change in internal energy is a state function that only depends upon the initial and final states of the system and is independent of the pathway followed (Smith et al. 2017).

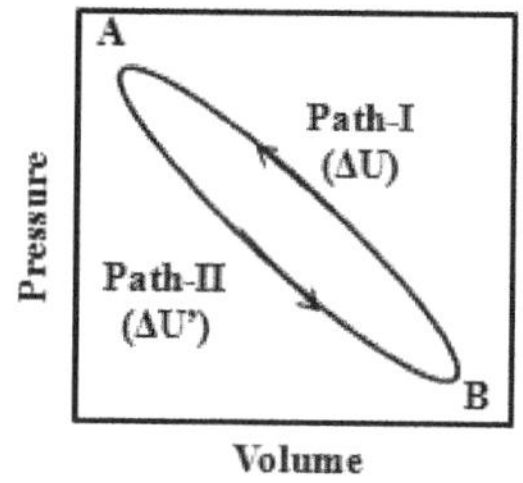

Figure 3.7. *Internal energy change for direct and reverse paths*

3.12. Work and heat are path functions

Consider that a system changes its state concerning pressure and volume by using three different paths, i.e. paths I, II and III (Figure 3.8).

Let us assume that the work done by the system is given as

$$W = -PdV \tag{3.7}$$

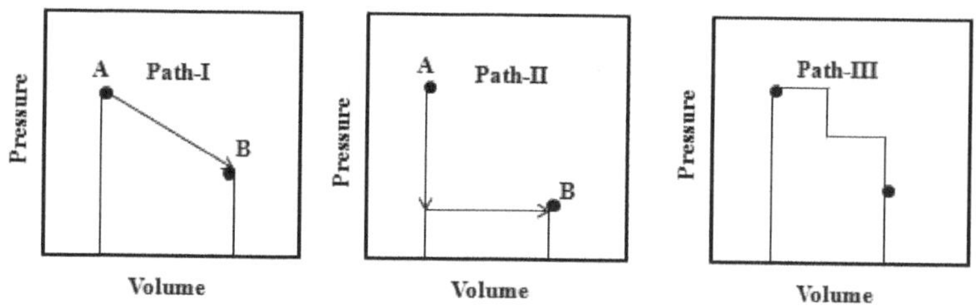

Figure 3.8. *Work done for different paths followed*

Where P is the external pressure, dV is the small change in the volume of the system. Therefore, the total work done can be determined for the change of volume from V_A to V_B using the integral of equation [3.7].

Whereas the graphical method shows that the magnitude of work done for the three paths followed can be determined by calculating the area under the curve. For each P–V diagram, the area under the curve is different, which shows that work done is dependent on the path followed and independent of the initial and final states.

Hence, the work done is not an exact differential

$$\int_A^B dW \neq W_B - W_A \quad [3.8]$$

According to the first law of thermodynamics, heat is given as

$$Q = \Delta U - W \quad [3.9]$$

ΔU is a state function that is dependent on the path taken, whereas W is the path function. Therefore, (ΔU – W) also depends on the path followed. Hence, heat is also not an exact differential

$$\int_A^B dQ \neq Q_B - Q_A \quad [3.10]$$

3.13. Second law of thermodynamics

The second law of thermodynamics posits that processes unfold in a specific direction, emphasizing that energy possesses both quality and quantity. In contrast, the first law imposes no constraints on the direction of a process, and adherence to the first law does not inherently ensure the occurrence of the process. Consequently, an additional overarching principle, the second law, becomes essential to ascertain the feasibility of a given process (Smith et al. 2017; Jawale et al. 2023).

A process is deemed feasible when it concurrently adheres to both the first and second laws of thermodynamics.

The second law further underscores the qualitative aspects of energy. Engineers prioritize the preservation of energy quality as illustrated in the given example where the energy stored in a hot container (higher temperature) exhibits high quality (capacity to perform work) compared to the energy present in the surroundings (lower temperature). It states that heat cannot flow itself from a body at a lower temperature to a body at a higher temperature (Koretsky 2013; Smith et al. 2017).

Additionally, the second law is instrumental in establishing theoretical limits for the performance of widely utilized engineering systems, including heat engines and refrigerators.

3.14. Thermal energy reservoir

Thermal energy reservoirs are theoretical entities characterized by a substantial thermal energy capacity (mass x specific heat), enabling them to furnish or absorb finite quantities of heat without experiencing any temperature change. Examples of thermal reservoirs include lakes, rivers, the atmosphere and oceans.

A two-phase system can be effectively conceptualized as a reservoir, as it can absorb and release a substantial amount of heat while maintaining a constant temperature.

A reservoir that supplies energy in the form of heat is termed a source, while one that absorbs energy in the form of heat is referred to as a sink.

3.15. Heat engines

Heat engines are devices designed to transform heat into work. While various types of heat engines exist, they share common characteristics:

– They all receive heat from high-temperature sources such as an oil furnace or nuclear reactor.

– They convert a portion of this received heat into work.

– They expel the remaining waste heat to a low-temperature sink.

– They operate in a cylindrical manner.

Thermal efficiency is quantified as the ratio of the heat input converted into the network output expressed by the formula:

Efficiency = net work output/heat input

The thermal efficiencies of work-producing devices exhibit low values. Conventional spark ignition automobile engines typically achieve a thermal efficiency of approximately 20%, diesel approximately 30% and power plants in the order of 40%. The prospect of conserving the rejecting heat (Qout) in a power cycle is not feasible, as the cycle cannot be completed without the cooling process in the condenser. In every heat engine, there is an evitable energy loss as some heat must be transferred to a low-temperature reservoir to finalize the cycle, even in an idealized cycle.

3.16. The second law: Kelvin–Plank statement

Any cyclic device receiving heat solely from a single reservoir and producing a net amount of work is inherently impossible. Stated differently, no heat engine can achieve a thermal efficiency of 100%.

3.17. The second law: Clausius' statement

This states that heat cannot flow itself from a body at a lower temperature to a body at a higher temperature without any input of work. It is not possible to construct a device or cyclic heat engine.

3.18. Refrigerator and heat pumps

In natural processes, heat naturally moves from regions of high temperature to those of lower temperature. However, the spontaneous occurrence of the reverse process is not possible. To facilitate the transfer of heat from a lower-temperature region to a higher-temperature one, a specialized device known as the refrigerator is employed. Refrigerators operate cyclically, utilizing working fluids termed refrigerants.

Heat pumps, on the other hand, transfer heat from a low-temperature medium to a higher-temperature one. Refrigerators and heat pumps share fundamental design principles, differing only in their intended objectives. A refrigerator is designed to maintain a refrigerated space at low temperatures, while a heat pump extracts heat from a lower-temperature source and supplies it to a warmer medium.

3.19. References

Dahm, K.D. and Visco Jr, D.P. (2014). *Fundamentals Chemical Engineering Thermodynamics*, SI edition. Cengage Learning, London.

Exell, R.H.B. (2018). A personal view of macroscopic equilibrium thermodynamics. *J. Sustain. Energy Environ.*, 9, 21–24.

Jawale, S., Deokar, S., Avhad, H., Mulani, S., Sir, P.B. (2023). Review on laws of thermodynamics. *Int. Res. J. Mod. Eng. Technol. Sci.*, 5(11), 2281–2285.

Koretsky, M.D. (2013). *Engineering and Chemical Thermodynamics*, 2nd edition. John Wiley & Sons, New York.

Smith, J.M., Van Ness, H.C., Abbott, M.M., Swihart, M.T. (2017). *Introduction to Chemical Engineering Thermodynamics*, 8th edition. McGraw Hill, New York.

Zohuri, B. and McDaniel, P. (2019). First law of thermodynamics. In *Thermodynamics in Nuclear Power Plant Systems*, 2nd edition. Springer, Cham.

4

Thermodynamic Properties and Equations of States

A set of balance equations based on the conservation of mass, energy and entropy are available. These equations are helpful in calculating energy changes in the system, the amount of heat or work required in a particular process, and in finding the properties of a species when a change in the state of the substance takes place. These equations, which determine the thermodynamic state of a system, are typically expressed in terms of internal energy, enthalpy and entropy rather than pressure, temperature or volume, which are easily measurable.

In the present case, no gas or liquid act as an ideal gas or solution across the entire pressure and temperature range. Also, data on thermodynamic properties are not readily available. Consequently, the calculation of multiple thermodynamic properties of real gases or liquids is required in thermodynamic computations.

According to the phase rule, if the values of a certain number of intensive properties of the system are specified, the values of all other thermodynamic intensive properties of the system are also fixed. On the other hand, the phase rule does not provide any knowledge of how these values for these other attributes may be determined or estimated.

In most cases, volumetric data is frequently used to evaluate thermodynamic parameters, such as internal energy and enthalpy, which are used to determine the heat and work needs of any industrial process. All thermodynamic parameters of interest for pure or mixtures of substances can be computed from the thermal and volumetric measurement data. While volumetric measurements provide information on how thermodynamic qualities vary with pressure or density at a constant

---

Chapter written by Dipaloy DATTA.

temperature, thermal measurements (heat capacities) provide information on how some thermodynamic properties vary with temperature for a specific phase, whether solid, liquid or gas.

At first, it is required to develop the fundamental property relations from the first and second laws of thermodynamics, which pave the way for the mathematical structure of thermodynamics. These equations will help us to calculate the values of enthalpy and entropy from the volumetric and heat capacity data.

4.1. Thermodynamic property relations of pure substances

The following represents the first law of thermodynamics for a closed system and a reversible process involving n moles of a substance:

$$d(nU) = dQ_{rev} + dW_{rev}, \text{ where } dW_{rev} = -Pd(nV) \text{ and } dQ_{rev} = Td(nS) \qquad [4.1]$$

$$d(nU) = Td(nS) - Pd(nV) \qquad [4.2]$$

where U = molar internal energy, S = molar entropy, V = molar volume, T = absolute temperature, P = pressure, Q = heat and W = work.

Equation [4.2] is derived from the first and second laws of thermodynamics and contains only properties of the system (U, T, S, P and V). These properties are state variables and do not depend on the type of process that was followed to reach a particular state of a system. Therefore, any process in a closed system of constant mass that produces a differential change from one equilibrium state to another may be described in equation [4.2].

As per the definition, other thermodynamic relations are as follows:

$$\text{Enthalpy: } H \equiv U + PV \qquad [4.3]$$

$$\text{Helmholtz energy: } A \equiv U - TS \qquad [4.4]$$

$$\text{Gibbs energy: } G \equiv H - TS \qquad [4.5]$$

The following thermodynamic relations are obtained with the help of equations [4.2]–[4.5].

$$d(nH) = Td(nS) + (nV)dP \qquad [4.6]$$

$$d(nA) = -Pd(nV) - (nS)dT \qquad [4.7]$$

$$d(nG) = (nV)dP - (nS)dT \qquad [4.8]$$

Equations [4.2] and [4.6]–[4.8] are the fundamental property relations for a homogeneous fluid of constant composition.

4.1.1. *Criteria of exactness*

According to the Gibbs phase rule, the thermodynamic state can be fully defined by any two intensive thermodynamic parameters. Therefore, if a variable, z, is expressed in terms of two other variables, x and y, i.e. $z = z(x, y)$, and z is a continuous differential function, then the total differential of z is defined as:

$$dz = \left(\frac{\partial z}{\partial x}\right)_y dx + \left(\frac{\partial z}{\partial y}\right)_x dy \qquad [4.9]$$

Say, $M = \left(\frac{\partial z}{\partial x}\right)_y$ and $N = \left(\frac{\partial z}{\partial y}\right)_x$

Then, we can write,

$$dz = M\,dx + N\,dy \qquad [4.10]$$

Further differentiating, we obtain

$$\left(\frac{\partial M}{\partial y}\right)_x = \frac{\partial^2 z}{\partial y \partial x} \text{ and } \left(\frac{\partial N}{\partial x}\right)_y = \frac{\partial^2 z}{\partial x \partial y} \qquad [4.11]$$

Since the second-order derivatives are the same, then

$$\left(\frac{\partial M}{\partial y}\right)_x = \left(\frac{\partial N}{\partial x}\right)_y \qquad [4.12]$$

Equation [4.12] must be satisfied for equation [4.10] to be an exact differential expression. This is the requirement of the criterion of exactness.

4.1.2. *Maxwell equations*

The thermodynamic properties such as U, H, A and G [4.2] and [4.6]–[4.8] are functions of two intensive properties. The criterion of exactness, equation [4.12], is used to get the following relationships, known as Maxwell's equations:

$$\left(\frac{\partial T}{\partial V}\right)_S = -\left(\frac{\partial P}{\partial S}\right)_V \quad [4.13]$$

$$\left(\frac{\partial T}{\partial P}\right)_S = \left(\frac{\partial V}{\partial S}\right)_P \quad [4.14]$$

$$\left(\frac{\partial P}{\partial T}\right)_V = \left(\frac{\partial S}{\partial V}\right)_T \quad [4.15]$$

$$\left(\frac{\partial V}{\partial T}\right)_P = -\left(\frac{\partial S}{\partial P}\right)_T \quad [4.16]$$

All of the thermodynamic properties cannot be measured, and thus, they must be presented in terms of measurable variables. It can be seen that equations [4.13]–[4.16] are represented in variables such as T, P, V and S only. When S appears in any equation, that equation can be converted into P, V or T, which are measurable quantities.

Equations [4.2] and [4.6]–[4.8] can be used to develop other thermodynamic relations relating to thermodynamic properties (equations [4.17]–[4.20])

$$T = \left(\frac{\partial U}{\partial S}\right)_V ; P = -\left(\frac{\partial U}{\partial V}\right)_T \quad [4.17]$$

$$T = \left(\frac{\partial H}{\partial S}\right)_P ; V = \left(\frac{\partial H}{\partial P}\right)_S \quad [4.18]$$

$$P = -\left(\frac{\partial A}{\partial V}\right)_T ; \; S = -\left(\frac{\partial A}{\partial T}\right)_V \quad [4.19]$$

$$V = \left(\frac{\partial G}{\partial P}\right)_T ; S = -\left(\frac{\partial G}{\partial T}\right)_P \quad [4.20]$$

4.1.3. *Enthalpy and entropy as functions of temperature and pressure*

H and S are frequently required in calculating work and heat and are of greater thermodynamic interest. In any thermodynamic system, T and P can be directly measured. Therefore, they need to be expressed in terms of T and P

Say, $H = H(T,P)$

Hence, $$dH = \left(\frac{\partial H}{\partial T}\right)_P dT + \left(\frac{\partial H}{\partial P}\right)_T dP \quad [4.21]$$

The first derivative term on the right-hand side is the heat capacity at constant pressure:

$$\left(\frac{\partial H}{\partial T}\right)_P = C_P \quad [4.22]$$

For n = 1, equation [4.7] becomes $dH = TdS + VdP$. Dividing the equation by dP at constant temperature will give

$$\left(\frac{\partial H}{\partial P}\right)_T = T\left(\frac{\partial S}{\partial P}\right)_T + V \quad [4.23]$$

Using fourth Maxwell's equation [4.16], the above equation becomes:

$$\left(\frac{\partial H}{\partial P}\right)_T = V - T\left(\frac{\partial V}{\partial T}\right)_P \quad [4.24]$$

Now, the partial derivatives obtained from equations [4.22] and [4.24] are substituted in equation [4.21], resulting in the following:

$$dH = C_P dT + \left[V - T\left(\frac{\partial V}{\partial T}\right)_P \right] dP \qquad [4.25]$$

Again, say, $S = S(T,P)$

$$dS = \left(\frac{\partial S}{\partial T}\right)_P dT + \left(\frac{\partial S}{\partial P}\right)_T dP \qquad [4.26]$$

For n = 1, equation [4.7] becomes $dH = TdS + VdP$. Dividing the equation by dT at constant pressure will give

$$\left(\frac{\partial H}{\partial T}\right)_P = T\left(\frac{\partial S}{\partial T}\right)_P \text{ or } \left(\frac{\partial S}{\partial T}\right)_P = \frac{C_P}{T} \qquad [4.27]$$

The pressure derivative of the entropy results directly from equation [4.16]:

$$\left(\frac{\partial S}{\partial P}\right)_T = -\left(\frac{\partial V}{\partial T}\right)_P \qquad [4.28]$$

$$dS = C_P \frac{dT}{T} - \left(\frac{\partial V}{\partial T}\right)_P dP \qquad [4.29]$$

These are general equations relating the properties of homogeneous fluids of constant composition to temperature and pressure.

In a similar fashion, the variation of the U, H and S can also be represented in terms of T, P or V.

Internal energy as function of pressure:

$$\left(\frac{\partial U}{\partial P}\right)_T = -T\left(\frac{\partial V}{\partial T}\right)_P - P\left(\frac{\partial V}{\partial P}\right)_T \qquad [4.30]$$

Internal energy as functions of temperature and volume:

$$dU = C_V dT + \left[T\left(\frac{\partial P}{\partial T}\right)_V - P \right] dV \qquad [4.31]$$

Entropy as functions of temperature and volume:

$$dS = C_V \frac{dT}{T} + \left(\frac{\partial P}{\partial T}\right)_V dV \qquad [4.32]$$

The above equations show the relationship of T and V with U and S for homogeneous fluids of constant composition.

4.1.4. *Volume expansivity and isothermal compressibility*

An equation of state may be expressed as $f(T,P,V)=0$. For example, let V be considered as a function of T and P, $V=V(T,P)$, then,

$$dV = \left(\frac{\partial V}{\partial T}\right)_P dT + \left(\frac{\partial V}{\partial P}\right)_T dP \qquad [4.33]$$

The partial derivatives on the right-hand side of equation [4.33] have some physical interpretation and meanings and give two new thermodynamic properties of substances. These are defined as follows:

– Volume expansivity: $\beta \equiv \frac{1}{V}\left(\frac{\partial V}{\partial T}\right)_P$ [4.34]

– Isothermal compressibility: $\kappa \equiv -\frac{1}{V}\left(\frac{\partial V}{\partial P}\right)_T$ [4.35]

Now, equation [9.33] can be expressed as:

$$\frac{dV}{V} = \beta dT - k dP \qquad [4.36]$$

The isotherms in a P–V diagram on the left side are generally very steep and closely spaced in the liquid phase. Therefore, the values of both the derivatives

$[(\partial V/\partial T)_P$ and $(\partial V/\partial P)_T]$ have insignificant values, resulting in smaller values of both β and κ. This is the characteristic behavior of incompressible fluids in the fluid mechanics where $\beta \sim 0$ and $\kappa \sim 0$. As there is no real fluid that is actually incompressible, the equation of state cannot be applied to such fluids. For such fluids, the volume does not change with a change in temperature and pressure. With the exception of liquid water between 0°C and 4°C, β is always positive for liquids, and κ is also positive.

β and κ are weak functions of T and P near the critical point. For the small changes in T and P, considering both β and κ are constants, equation [4.36] finds the following form of equation [4.37]:

$$In\frac{V_2}{V_1}\beta(T_2-T_1)-k(P_2-P_1) \quad [4.37]$$

Change in internal energy, enthalpy and entropy in terms of β and κ

$$dU = C_V dT + \left(\frac{\beta}{k}T - P\right)dV \quad [4.38]$$

$$\left(\frac{\partial U}{\partial P}\right)_T = (kP - \beta T)V \quad [4.39]$$

$$dH = C_P dT + (1-\beta T)V\,dP \quad [4.40]$$

$$dS = \frac{C_V}{T}dT + \frac{\beta}{k}dV \quad [4.41]$$

$$dS = C_P\frac{dT}{T} - \beta V\,dP \quad [4.42]$$

4.1.5. *Acentric factor*

Pitzer introduced the acentric factor (ω) in 1955 and analyzed the vapor-pressure curves of several pure components. The acentric factor is a property of any pure substance such as molecular weight, critical temperature, critical pressure, critical volume, etc. The Clausius–Clapeyron equation can be used to deduce the mathematical relationship between vapor pressure and absolute temperature. The following equation represents this relationship:

$$\ln P^{sat} = A - \frac{B}{T} \qquad [4.43]$$

It has been observed experimentally that the compressibility factors (*Z*) for different fluids exhibit similar trends, and their values are approximately the same at a reduced temperature, $T_r = \frac{T}{(T_C)}$, and reduced pressure, $P_r = \frac{P}{(P_C)}$. Hence, all of the fluids will show equal deviation from the ideal gas behavior. This is referred to as the two-parameter theorem of corresponding states. This statement is quite valid for simple fluids (SF) (say argon, krypton and xenon), but some systematic variations are noted for more complex fluids.

To introduce the characteristics of the molecular structure of fluids and to define the characteristics of complex fluids, a third parameter known as the acentric factor was introduced, which formed the basis for the three-parameter theorem of corresponding states. The statement is: "all fluids with the same value of ω will show approximately the equal value of Z at the same T_r and P_r. Further, they will exhibit the same extent of deviation from the ideal gas behavior".

Equation [4.43] can now be plotted as log P_r^{sat} versus $1/T_r$ for any pure chemical species with a slope as follows:

$$\frac{d \log P_r^{sat}}{d(1/T_r)} = S \qquad [4.44]$$

Pitzer observed that all the experimental data on vapor pressure fall on the same line for the SF. This line also passes through a point where log $P_r^{sat} = -1$ at $T_r = 0.7$. Now, as per the three-parameter theorem of corresponding states, the location of lines for other complex fluids can be fixed, and the difference

[$\log P_r^{sat}$ (SF) – $\log P_r^{sat}$] with respect to the SF evaluated at $T_r = 0.7$ will give the value of the acentric factor as:

$$\omega \equiv -1.0 - \log\left(P_r^{sat}\right)_{T_r = 0.7} \qquad [4.45]$$

Noble gases (SF) are taken as a reference for this estimation, and the value of their acentric factor is zero. It measures the non-sphericity (acentricity) of the molecules, and nonspherical molecules have values above zero.

If the Causius–Clayperon equation is valid, then for the estimation of the acentric factor, *Edmister's formula* can be used as given:

$$\omega = \frac{3}{7}\left(\frac{\theta}{1-\theta}\right)\log P_c - 1 \qquad [4.46]$$

where $\theta = T_n/T_c$, T_n = normal boiling point, P_c = critical temperature.

For any fluid, the compressibility factor, Z, can be written as:

$$Z(P_r, T_r) = Z^0(P_r, T_r) + \omega Z'(P_r, T_r) \qquad [4.47]$$

where Z_0 is the compressibility factor of spherical molecules and Z' is the correction for non-sphericity.

4.1.6. *Specific heat at constant volume and constant pressure*

Specific heat or specific heat capacity is the amount of heat required to raise the temperature of a unit mass/mol of a substance by 1°C and is represented as:

$$C = \frac{1}{m}\frac{dQ}{dT} \qquad [4.48]$$

When the heat transfer to a sample is done while keeping the volume constant, it is termed the specific heat at constant volume (C_v)

$$dU = \delta Q - PdV \qquad [4.49]$$

$$dU = \delta Q \quad [4.50]$$

$$U = U(\mathrm{T, V}) \quad [4.51]$$

$$dU = \left(\frac{\partial U}{\partial T}\right)_V dT + \left(\frac{\partial U}{\partial V}\right)_T dV \quad [4.52]$$

$$dU = \left(\frac{\partial U}{\partial T}\right)_V dT \quad [4.53]$$

$$\left(\frac{\partial Q}{\partial T}\right)_V = \left(\frac{\partial U}{\partial T}\right)_V \quad [4.54]$$

$$C_V \equiv \left(\frac{\partial U}{\partial T}\right)_V \quad [4.55]$$

When the heat transfer to a sample is done while keeping the pressure constant, it is defined as the specific heat at constant pressure (C_P)

$$dU = \delta Q - PdV \quad [4.56]$$

$$\delta Q = dU + PdV \quad [4.57]$$

$$\delta Q = dU + d(PV) \quad [4.58]$$

$$\delta Q = d(U + PV) \quad [4.59]$$

$$\delta Q = dH \quad [4.60]$$

$$H = H(\mathrm{T, P}) \quad [4.61]$$

$$dH = \left(\frac{\partial H}{\partial T}\right)_P dT + \left(\frac{\partial H}{\partial P}\right)_T dP \quad [4.62]$$

$$dH = \left(\frac{\partial H}{\partial T}\right)_P dT \qquad [4.63]$$

$$\left(\frac{\partial Q}{\partial T}\right)_P = \left(\frac{\partial H}{\partial T}\right)_P \qquad [4.64]$$

$$C_P \equiv \left(\frac{\partial H}{\partial T}\right)_P \qquad [4.65]$$

The following equations give the difference and ratio of two specific heats, i.e. C_p and C_v:

$$C_P - C_V = T\left(\frac{\partial P}{\partial T}\right)_V \left(\frac{\partial V}{\partial T}\right)_P \qquad [4.66]$$

$$\gamma = \frac{C_P}{C_V} = \frac{\left(\frac{\partial V}{\partial P}\right)_T}{\left(\frac{\partial V}{\partial P}\right)_S} = \frac{\kappa_T}{\kappa_S} = \frac{\text{Isothermal compressibility}}{\text{Isentropic compressibility}} \qquad [4.67]$$

The change in the internal energy and enthalpy can be obtained using the following equations:

$$Q = \Delta U = \int_{T_1}^{T_2} C_V dT \quad \text{for a constant volume process} \qquad [4.68]$$

$$Q = \Delta H = \int_{T_1}^{T_2} C_P dT \quad \text{for a constant pressure process} \qquad [4.69]$$

The dependency of heat capacities on temperature must be found to estimate the change in ΔU and ΔH from equations [4.68] and [4.69]. Generally, this is given by an empirical equation of the form:

$$\frac{C_P}{R} = A + BT + CT^2 + DT^{-2} \quad [4.70]$$

Here, A, B, C and D are the constants, and describe the characteristics of a particular substance. C_p / R is dimensionless, and the unit of C_p will be fixed by choosing the unit of R.

The estimation of the integral $\int_{T_1}^{T_2} C_P dT$ is achieved by substituting the expression of C_p. Integration is performed between temperature limits of T_1 and T_2, and the following expression is obtained, which is conveniently used to determine the integral:

$$\int_{T_1}^{T_2} \frac{C_P}{R} dT = AT_1(\tau - 1) + \frac{B}{2} T_1^2(\tau^2 - 1) + \frac{C}{3} T_1^3(\tau^3 - 1) + \frac{D}{T_1}\left(\frac{\tau - 1}{\tau}\right) \quad [4.71]$$

where $\tau \equiv \frac{T_2}{T_1}$

For the given values of the temperatures, the values of heat (Q) or change in enthalpy (ΔH) can be determined.

The real gases at low pressures or as $P \rightarrow 0$ may be treated as ideal gases, and equation [4.70] can be used to estimate the values of heat capacities at constant pressure. The ideal gas heat capacities are denoted by C_P^{ig} and C_V^{ig}, and are very distinct for gases of different types. With the increase in temperature, their values increase smoothly to a limiting value but do not change with a change in pressure.

For ideal gas mixtures of constant composition, individual gas molecules will not influence other molecules present in the mixture. Therefore, each gas in the mixture will exist independently, and the properties will not be affected by the presence of different molecules. The mole-fraction-weighted average of the heat capacities of the respective species can be considered to obtain the mixture heat capacity of ideal gases. For 1 mol of an ideal gas mixture comprising unlike gases, the molar heat capacity of the mixture may be obtained using the following equation:

$$C_{P_{mixture}}^{ig} = \sum_{i=1}^{i=n} y_i C_{P_i}^{ig} \quad [4.72]$$

where $C_{P_i}^{ig}$ are the molar heat capacities of pure gas in the ideal gas state and y_i denotes the mole fraction of a gas.

4.1.7. *Clapeyron equation*

The lines on the PT diagram show the different boundaries between phases for a pure substance. The change from one phase to another phase takes place at constant temperature and pressure where these boundary lines are crossed. As with internal energy, enthalpy and entropy, it can be seen that the molar or specific volume of saturated vapor is entirely different from its saturated liquid value at the same temperature and pressure. The molar or specific Gibbs energy is an exception, as it remains constant for a pure species through phase transitions (such as melting, evaporation or sublimation).

Let us consider that phase 1 of a pure species is in equilibrium with phase 2 of a pure species at a temperature T and at a pressure P. The pressure corresponding to the temperature is the saturation pressure of the equilibrium system. At constant T and P, equation [4.8] reduces to $d(nG) = 0$. Now, for this equilibrium system, we can write that

$$dG^1 = dG^2 \quad [4.73]$$

where G^1 and G^2 are the molar or specific Gibbs energies of the individual phases. Substituting the expressions for dG^1 and dG^2 as given by equation [4.11] and upon subsequent rearrangement gives:

$$\frac{dP}{dT} = \frac{S^2 - S^1}{V^2 - V^1} = \frac{\Delta S^{12}}{\Delta V^{12}} \quad [4.74]$$

When a unit quantity of a pure chemical species is transferred from phase 1 to phase 2 at the equilibria T and P, a change in entropy (ΔS^{12}) and volume (ΔV^{12}) occurs. Furthermore, at equilibrium, ΔS^{12} and ΔH^{12} (latent heat or enthalpy of phase transition) can be related as follows:

$$\Delta H^{12} = T\Delta S^{12} \quad [4.75]$$

Now, $\Delta S^{12} = \Delta H^{12} / T$ can be substituted in equation [4.74] to get the following equation:

$$\frac{dP^{sat}}{dT} = \frac{\Delta H^{12}}{T\Delta V^{12}} \qquad [4.76]$$

This equation is known as the Clapeyron equation for a phase transition.

Again, for a particular case of the liquid–vapor equilibrium at low vaporization pressures, equation [4.76] can be represented in a more simplified form by considering the following assumptions:

i) The molar volume of the vapor phase is much higher than the molar volume of the liquid phase, i.e. $V^v >> V^l$.

ii) The vapor phase is an ideal gas, i.e. $\Delta V^{lv} \approx V^v = \frac{RT}{P^{sat}}$.

With these assumptions, equation [4.76] takes this form:

$$\Delta H^{lv} = -R\frac{d \ ln\, P^{sat}}{d(1/T)} \qquad [4.77]$$

The above equation is the Clausius–Clapeyron equation that is used to estimate the latent heat of vaporization from the slope of $1nP^{sat}$ versus $1/T$ curve. It is observed that when the experimental data were plotted between $\ln P^{sat}$ versus $1/T$, they produced almost straight lines for many substances, which indicates that the value of latent heat of vaporization (ΔH^{lv}) is nearly constant over a temperature range and independent of T. In the actual case, ΔH^{lv} is not constant and decreases monotonically with the increase in the temperature from the triple point to the critical point, where its value is zero. The applicability and validity of the Clausius–Clapeyron equation lie when the experimental data are obtained at low vaporization pressures.

4.1.8. *Relation between vapor pressure and temperature*

The Clausius–Clapeyron equation can now be integrated to get the following forms which are applicable over a limited range of temperatures:

$$\ln\frac{P_2^{sat}}{P_1^{sat}} = -\frac{\Delta H^{lv}}{R}\left(\frac{1}{T_2}-\frac{1}{T_1}\right) \qquad [4.78]$$

$$\text{or } \ln P^{sat} = A - \frac{B}{T} \quad [4.79]$$

where A and B are constants for a given species.

These equations give an approximate estimate of the vapor-pressure data over the entire temperature range from the triple point to the critical point and when there is a requirement for vapor-pressure data to be found out by interpolation in the small range of temperatures.

More commonly, the Antoine equation can be used to estimate vapor-pressure data more satisfactory and has the following form:

$$\ln P^{sat} = A - \frac{B}{T + C} \quad [4.80]$$

Again, the values of the constants A, B and C are readily available for a large number of species, and they are valid and need to be used for the temperature range mentioned.

More equations are available which are applicable over a wide range of temperatures, such as:

$$\text{Wagner equation: } \ln P^{sat} = \frac{A\tau + B\tau^{1.5} + C\tau^{3} + D\tau^{6}}{1-\tau} \text{ and } \tau \equiv 1 - T_r \quad [4.81]$$

$$\text{Riedel equation: } \ln P^{sat} = A + \frac{B}{T} + C\ln T + DT^{6} \quad [4.82]$$

$$\text{Harlecher–Braun equation: } \ln P^{sat} = A + \frac{B}{T} + C\ln T + \frac{DP^{sat}}{T^{2}} \quad [4.83]$$

4.1.9. *Residual property*

The residual property is defined as the difference between the molar property (such as V, U, H, S or G) and the ideal gas property (such as V^{ig}, U^{ig}, H^{ig}, S^{ig} or G^{ig}) at the same T and P. The property represents the deviation of the molar properties from their values at the ideal gas state and is denoted as:

$$M^R \equiv M - M^{ig} \qquad [4.84]$$

where M is the molar thermodynamic property and M^{ig} is the molar property at the ideal gas state at the same T and P.

For example, the residual volume is defined as:

$$V^R \equiv V - V^{ig} = V - \frac{RT}{P} = \frac{ZRT}{P} - \frac{RT}{P} = \frac{RT}{P}(Z-1) \qquad [4.85]$$

Generally, the value of molar Gibbs free energy cannot be determined directly from the experimental measurements. The relation for G has an important significance and can be used to generate relationships for other thermodynamic properties. Therefore, from the definition, the residual molar Gibbs energy is written as follows:

$$G^R \equiv G - G^{ig} \qquad [4.86]$$

Now, G can be represented as a dimensionless form as G/RT, and after differentiation, will get the form as:

$$d\left(\frac{G}{RT}\right) = \frac{dG}{RT} - \frac{G}{RT^2}dT \qquad [4.87]$$

Now, substituting $G = H - TS$ and $dG = VdP - SdT$ in equation [4.87] and after rearrangement, the following form of the equation is obtained:

$$d\left(\frac{G}{RT}\right) = \frac{V}{RT}dP - \frac{H}{RT^2}dT \qquad [4.88]$$

For an ideal gas, the equation takes the form

$$d\left(\frac{G^{ig}}{RT}\right) = \frac{V^{ig}}{RT}dP - \frac{H^{ig}}{RT^2}dT \qquad [4.89]$$

Taking the difference between equations [4.88] and [4.89], the expression for G^R is obtained

$$d\left(\frac{G^R}{RT}\right)=\frac{V^R}{RT}dP-\frac{H^R}{RT^2}dT \quad [4.90]$$

It is the fundamental property relation in terms of residual properties of fluids at constant composition, which is used to find the following equations for residual volume and residual enthalpy:

$$\frac{V^R}{RT}=\left[\frac{\partial\left(G^R/RT\right)}{\partial P}\right]_T \quad [4.91]$$

$$\frac{H^R}{RT}=-T\left[\frac{\partial\left(G^R/RT\right)}{\partial T}\right]_P \quad [4.92]$$

Similarly, the residual equations for internal energy and entropy follow

$$\frac{U^R}{RT}=\frac{H^R}{RT}-\frac{PV^R}{RT} \quad [4.93]$$

$$\frac{S^R}{R}=\frac{H^R}{RT}-\frac{G^R}{RT} \quad [4.94]$$

Equation [4.90] at a constant temperature can be written as:

$$d\left(\frac{G^R}{RT}\right)=\frac{V^R}{RT}dP \quad [4.95]$$

Integrating the above equation from zero pressure (ideal gas state) to any pressure P gives:

$$\frac{G^R}{RT}=\int_0^P\frac{V^R}{RT}dP=\int_0^P(Z-1)\frac{dP}{P} \quad [4.96]$$

Here, $\frac{G^R}{RT} = 0$ at $P = 0$, an ideal gas state.

Expression [4.92] gives the value of residual enthalpy and entropy at constant temperature as:

$$\frac{H^R}{RT} = -T\int_0^P \left(\frac{\partial Z}{\partial T}\right)_P \frac{dP}{P} \qquad [4.97]$$

$$\frac{S^R}{R} = -T\int_0^P \left(\frac{\partial Z}{\partial T}\right)_P \frac{dP}{P} - \int_0^P (Z-1)\frac{dP}{P} \qquad [4.98]$$

The value of the compressibility factor, $Z = PV / RT$, and partial differentials in equations [4.97] and [4.98] can be estimated using the experimental pressure–volume–temperature data.

Therefore, H and S can be calculated by adding the corresponding ideal gas and residual properties as:

$$H = H^{ig} + H^R \quad S = S^{ig} + S^R \qquad [4.99]$$

The expressions for H^{ig} and S^{ig} are obtained from the following equations from an ideal gas state (at T_0 and P_0, reference state) to the ideal gas state at T and P:

$$H^{ig} = H_0^{ig} + \int_{T_0}^T C_P^{ig}\, dT \quad S^{ig} = S_0^{ig} + \int_{T_0}^T C_P^{ig} \frac{dT}{T} - R\ln\frac{P}{P_0} \qquad [4.100]$$

The changes in the enthalpy and entropy between states 1 and 2 are determined using equation [4.101]

$$\Delta H_{12} = \Delta H_{12}^{ig} + H_2^R - H_1^R \quad \Delta S_{12} = \Delta S_{12}^{ig} + S_2^R - S_1^R \qquad [4.101]$$

4.1.10. *Fugacity and its coefficient*

For one mole of an ideal gas, equation [4.8] at a constant temperature can be expressed as:

$$dG_i^{ig} = V_i^{ig} dP = RTd \ln P \quad [4.102]$$

For a real fluid, an analogous expression can be written for the change in the molar Gibbs free energy of a pure species i as:

$$dG_i = RTd \ln f_i \quad [4.103]$$

where f_i is the pure species fugacity.

If equations [4.102] and [4.103] are compared, then it is obvious that $f_i^{ig} = P$. Now, taking the difference of these equations and integrating from 0 (an ideal gas state) to P, we can get:

$$G_i - G_i^{ig} = RT \ln \frac{f_i}{P} \text{ or } G_i^R = RT \ln \frac{f_i}{P} = RT \ln \varphi_i \quad [4.104]$$

where $\varphi = \frac{f_i}{P}$, a dimensionless quantity is known as the fugacity coefficient.

Using equations [4.96] and [4.104], the following expression can be used for the estimation of the value of fugacity coefficient:

$$\ln \varphi_i = \int_0^P (Z_i - 1) \frac{dP}{P} \text{ at constant temperature} \quad [4.105]$$

The value of f_i for a pure species i in the compressed liquid state can be calculated using the following expression:

$$f_i = f_i^{sat} \exp \frac{V_i^l \left(P - P_i^{sat}\right)}{RT} \quad [4.106]$$

The exponential term is the Poynting factor, which shows the effect of pressure on the fugacity.

4.1.11. *Excess property*

The deviation of the thermodynamic properties of a real gas from the ideal gas is given by the residual property. Similarly, the excess property (M^E) is defined as the deviation of an actual liquid solution property from its liquid solution property as the ideal solution

$$M^E \equiv M - M^{id} \quad [4.107]$$

4.2. Equations of states

Equations of states are relations between pressure, volume and temperature. Generally, pressure and temperature can be measured quite accurately, and equations of states can be used to estimate the molar volume

$$f(P,V,T)=0 \quad [4.108]$$

The simplest form of this type is known as the ideal gas equation of states and is given by

$$PV = RT \quad [4.109]$$

For a real gas, we have the following form of equation of state:

$$PV = ZRT \quad [4.110]$$

4.2.1. *Virial equations of states*

These are the equations of states for gases where the compressibility factor, Z, is represented in the form of a power series as the function of molar volume or pressure

$$Z = \frac{PV}{RT} = 1 + \frac{B}{V} + \frac{C}{V^2} + \frac{D}{V^3} + \ldots \quad [4.111]$$

$$Z = \frac{PV}{RT} = 1 + B'P + C'P^2 + D'P^3 + \ldots \quad [4.112]$$

where B, B', C, C', D, D', etc., are the virial coefficients. These coefficients are solely functions of temperature for the pure components, and they are independent of density or pressure. They also represent the interaction between two bodies, three bodies, etc.

For convenience, it is also expressed in the form of density ($\rho = 1/V$) as:

$$Z = \frac{PV}{RT} = 1 + B\rho + C\rho^2 + D\rho^3 + ... \quad [4.113]$$

An extended form of the virial equation is given by the Benedict/Webb/Rubin equation:

$$Z = 1 + \left(B_0 - \frac{A_0}{RT} - \frac{C_0}{RT^3}\right)\frac{1}{V} + \left(b - \frac{a}{RT}\right)\frac{1}{V^2} + \frac{a\alpha}{RTV^5} + \frac{\beta}{RT^3V}\left(1 + \frac{\gamma}{V^2}\right)\exp\left(\frac{-\gamma}{V^2}\right) \quad [4.114]$$

where $A_0, B_0, C_0, a, b, \alpha, \beta$ and γ are all constants for a given fluid. This complex equation is used in the petroleum and natural gas industries for light hydrocarbons and a few other commonly encountered gases.

Bender equation for Z is given by:

$$Z = 1 + \frac{B}{V} + \frac{C}{V^2} + \frac{D}{V^3} + \frac{E}{V^4} + \frac{F}{V^5} + \left(G + \frac{H}{V^2}\right)\frac{1}{V^2}\exp\left(\frac{-\alpha}{V^2}\right) \quad [4.115]$$

where B, C, D, E, F, G, H and α are constants for a specific fluid.

4.2.2. *Cubic equations of states*

Virial equations can only be used for gases but not for liquids. The cubic equations of states are applicable to both liquids and gases. In this, the very first cubic equation of state was proposed by van der Waals in 1873 to estimate both liquid and vapor molar volumes at a given T and P

$$P = \frac{RT}{V - b} - \frac{a}{V^2} \quad [4.116]$$

This equation does not estimate accurate values, but it creates a basis for the more precise recent equation of state.

All modern cubic equations of state have the special form of the following type:

$$P = \frac{RT}{V-b} - \frac{\theta(V-\eta)}{(V-b)\left(V^2+kV+\lambda\right)} \tag{4.117}$$

Here, b, θ, k, λ and η are the parameters of the equation, and they generally depend on the temperature and composition (for mixtures).

Name	Year	Parameters
van der Waals	1873	$\theta = \frac{27R^2T_C^2}{64P_C}$; $b = \frac{RT_C}{8P_C}$; $\eta = b$; $k = 0$; $\lambda = 0$
Clausius	1880	$\theta = \frac{27R^2T_C^3}{64P_C}$; $b = V_C - \frac{RT_C}{4P_C}$; $\eta = b$; $k = \frac{3RT_C}{4P_C} - V_C$; $\lambda = \left(\frac{3RT_C}{8P_C} - V_C\right)^2$
Berthelot	1899	$\theta = \frac{27R^2T_C^2}{64}$; $b = \frac{RT_C}{8}$; $\eta = b$; $k = 0$; $\lambda = 0$
Redlich–Kwong	1949	$\theta(T) = a/\sqrt{T}$; $a = \frac{0.42748R^2T_C^{2.5}}{P_C}$; $b = \frac{0.08664RT_C}{P_C}$; $\eta = b$; $k = b$; $\lambda = 0$
Soave–Redlich–Kwong	1972	$\theta = a(T)$; $a(T) = \frac{0.4274(RT_C)^2}{P_C}\alpha$; $\sqrt{\alpha} = 1 + m\left(1 - \sqrt{\frac{T}{T_C}}\right)$; $m = 0.48 + 1.57\omega - 0.176\omega^2$; $b = \frac{0.08664RT_C}{P_C}$; $\eta = b$; $k = b$; $\lambda = 0$
Peng–Robinson	1976	$\theta = a(T)$; $a(T) = \frac{0.42724(RT_C)^2}{P_C}\alpha$; $\sqrt{\alpha} = 1 + m\left(1 - \sqrt{\frac{T}{T_C}}\right)$; $m = 0.37464 + 1.5226\omega - 0.26992\omega^2$; $b = \frac{0.0778RT_C}{P_C}$; $\eta = b$ $k = 2b$; $\lambda = -b^2$

Table 4.1. *Equations of states and their parameters*

4.3. References

Ahuja, P. (2008). *Chemical Engineering Thermodynamics*. PHI Learning, Delhi [Online]. Available at: https://www.phindia.com/Books/BookDetail/9788120336377/chemical-engineering-thermodynamics-ahuja.

Prausnitz, J.M., Lichtenthaler, R.N., de Azevedo, E.C. (1999). *Molecular Thermodynamics of Fluid-Phase Equilibria*. Prentice Hall, Upper Saddle River.

Rao, Y.V.C. (2004). *An Introduction to Thermodynamics*. Universities Press, Hyderabad.

Sandler, S.I. (2017). *Chemical, Biochemical, and Engineering Thermodynamics,* 5th edition. John Wiley & Sons, New York [Online]. Available at: https://www.wiley.com/en-dk/Chemical%2C+Biochemical%2C+and+Engineering+Thermodynamics%2C+5th+Edition-p-9781119321286.

Smith, J.M., Van Ness, H.C., Abbott, M.M., Swihart, M.T. (2022). *Introduction to Chemical Engineering Thermodynamics*, 9th edition. McGraw Hill, New York [Online]. Available at: https://www.mheducation.com/highered/product/introduction-to-chemical-engineering-thermodynamics-smith.html.

5

Thermodynamics and Phase Equilibrium

Thermodynamics is a fundamental element in the realm of chemical engineering, impacting nearly every aspect of the field. It is crucial for the advancement of chemical processes, the enhancement of energy efficiency and the prediction of phase equilibria. Engineers depend on thermodynamic principles to steer them in process design. This chapter presents a brief exploration of thermodynamic concepts, along with vapor–liquid, liquid–liquid and solid–liquid-phase equilibrium.

5.1. Introduction

Thermodynamics is an indispensable component of chemical engineering, permeating virtually every aspect of the field. It is integral to the design of chemical processes, the optimization of energy efficiency and the prediction of phase equilibria.

In the realm of process design, thermodynamic principles provide engineers with invaluable guidance in determining the feasibility of reactions, optimizing operating conditions and devising efficient separation processes such as distillation and extraction. By employing energy balances rooted in thermodynamic principles, engineers can thoroughly analyze and optimize energy consumption, thereby fostering the development of more sustainable and cost-effective processes. Furthermore, thermodynamics plays a pivotal role in assessing safety considerations by evaluating potential hazards, including runaway reactions.

Chapter written by Vyomesh M. PARSANA, Abhishek GUPTA, Dharmesh SUR and Kedar JOSHI.
For a color version of all figures in this chapter, see www.iste.co.uk/kumararya/chemical engineering1.zip.

The understanding of phase equilibria is essential in chemical engineering as it governs the behavior of substances in different processes, influencing design, operation and optimization. It is critical for designing separation processes such as distillation, absorption and extraction, which are integral to industries ranging from petroleum refining to pharmaceutical manufacturing. By predicting and controlling phase behavior, engineers can optimize process conditions, improve product purity and maximize yield, leading to more efficient and cost-effective operations. Additionally, the knowledge of phase equilibria is crucial for ensuring the safety and reliability of processes, as deviations from expected phase behavior can result in operational challenges and safety hazards. Ultimately, phase equilibria serve as a cornerstone of chemical engineering, providing the basis for the design and optimization of processes across various industries.

This chapter provides an overview of phase equilibrium and the associated concepts of thermodynamics. The initial section of the chapter focuses on determining the intensive state of a system using the phase rule. The fundamental laws of thermodynamics clarify the basic principles concerning heat, work, energy and entropy, as well as the direction of their spontaneous transfer. Furthermore, the chapter explores different properties including total, specific, molar, partial, excess and residual.

Additionally, the chapter discusses the vapor–liquid equilibrium (VLE) in detail, including equilibrium-defining equations, basic terminologies, bubble/dew-point calculations, azeotropes, models for data reduction, prediction and measurement of VLE data, and thermodynamic consistency tests. Finally, the basics of liquid–liquid and solid–liquid equilibrium have been discussed briefly.

5.2. Fundamentals of thermodynamics

5.2.1. *Phase rule and degree of freedom*

A qualitative treatment of systems in equilibrium is provided by the phase rule. It was developed by J. W. Gibbs in 1876. The degree of freedom of a multiphase system at equilibrium refers to the fixed number of independent variables required to establish its intensive state

$$F + \pi = 2 + N \qquad [5.1]$$

where F represents the degree of freedom, π stands for the number of phases and N indicates the number of chemical substances present.

The relationship between the degree of freedom F, the number of phases π and the number of components in a system N is expressed by this equation, which is valid for systems in equilibrium. It is important to note that the phase rule cannot be applied to reacting systems.

Upon fixing the temperature, pressure and phase composition of a system, the system's intensive state at equilibrium is determined. Although these variables are not all independent, they form the basis of the phase rule. The phase rule outlines the necessary number of variables from this set that must be provided to calculate all remaining intensive variables and, consequently, the system's intensive state.

5.2.2. *Fundamental laws and related equations*

5.2.2.1. *Zeroth law of thermodynamics*

Zeroth law says "when there is thermal equilibrium between body A and body C and between body B and body C independently, then there is also thermal equilibrium between body A and body B".

In simpler terms, this law states that two entities are in thermal equilibrium if they are also in thermal equilibrium with a third entity.

This law is essential to understanding temperature and proving that it is a measurable quantity. In essence, it establishes the foundation for the transitive characteristic of thermal balance. The zeroth law, despite being developed subsequent to the first and second laws, plays a crucial role in comprehending temperature scaling.

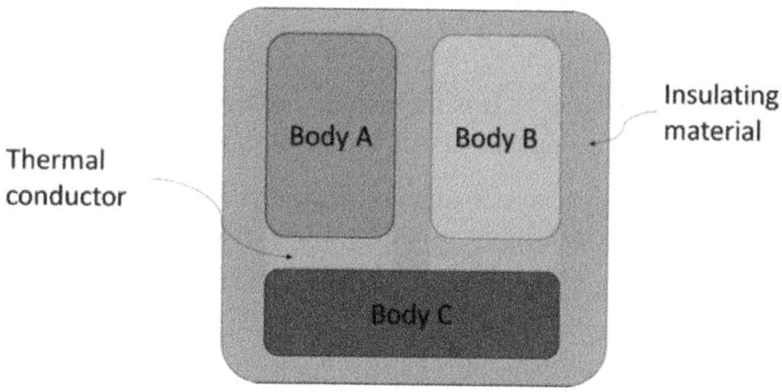

Figure 5.1. *Zeroth law of thermodynamics*

5.2.2.2. *First law of thermodynamics*

The first law of thermodynamics, also known as the law of energy conservation, states that "energy cannot be created or destroyed within an isolated system".

It is impossible to create energy; however, it can be converted from one form to another. To put it another way, the universe's total energy is constant. If the system gains energy, it can only mean that the surrounding has lost it, and vice versa

$$\Delta E_{system} + \Delta E_{surrounding} = \Delta E_{universe} = 0 \qquad [5.2]$$

Close systems do not permit the transfer of mass across its boundaries. Therefore, energy transfer is only possible in the form of work and heat

$$\Delta E_{surrounding} = \pm Q \pm W \qquad [5.3]$$

For both heat (Q) and work (W), according to modern sign conventions, transfer from the surrounding to the system is considered to be positive.

The mathematical representation of the first law of thermodynamics is commonly stated as:

$$\Delta U = Q + W \qquad [5.4]$$

In the equation, ΔU stands for the variation in the system's internal energy, Q denotes the heat that is either added to the system or removed from the system, and W represents the work carried out on the system or by the system.

5.2.2.3. *Second law of thermodynamics*

The second law of thermodynamics can be articulated in multiple ways, and each way offers a unique viewpoint on its applications and ramifications. Among the crucial claims are the following:

1) *Kelvin–Planck statement*[1]: "It is impossible for any device that operates on a cycle to receive heat from a single reservoir and produce a net amount of work".

This statement, which is a cornerstone of thermodynamics and supports the idea of heat engine efficiency, states that waste heat will always be generated to some extent during any practical heat engine operation. It typically defines the limitations of heat engines.

2) *Clausius statement*: "It is impossible to construct a device that operates in a cycle and produces no effect other than the transfer of heat from a lower temperature body to a higher temperature body".

1 The Kelvin–Planck statement is one of the fundamental principles of thermodynamics, which is named after Lord Kelvin and Max Planck.

In simpler terms, heat cannot move on its own from a colder object to a hotter one without outside interference.

This law has prominent application in the study of various thermodynamic cycles.

3) *Entropy statement*: "The statement asserts that in every spontaneous process, the entropy of the universe (or an isolated system) rises" Entropy serves as a gauge of the disorder or randomness within a system, and the second law indicates that natural processes have a tendency to transition toward states of increased entropy.

4) *Carnot principle*: the assertion is made that "a Carnot engine operating between two identical heat reservoirs cannot be outperformed in terms of efficiency by any other engine". This statement sets a standard for the highest efficiency achievable by a heat engine working between two temperature reservoirs.

5.2.2.4. *Third law of thermodynamics*

"In accordance with the Third Law of Thermodynamics, a system's entropy reaches a minimum or zero value when its temperature reaches absolute zero (0 Kelvin)". Stated differently, reaching absolute zero temperature cannot be accomplished with any finite number of processes.

This law has several implications:

Absolute entropy: "At absolute zero temperature, the entropy of an ideal crystalline substance is precisely zero", serving as a benchmark for entropy calculations.

According to the third law, which establishes a relationship between temperature and entropy, a system's disorder or randomness reduces with decreasing temperature until it achieves a minimum value at absolute zero.

Understanding the behavior of systems at very low temperatures, as those found in cryogenics and the study of condensed matter physics, is made easier by this law.

5.2.3. *Chemical potential and equilibrium*

In thermodynamics, chemical potential is a notion that denotes a substance's potential energy to go through a phase shift or a chemical reaction. It is defined as the differential change in the mixture's Gibbs function in a given phase for each differential change in component "i" in the same phase, while temperature, pressure and the quantities of all other components remain constant

$$\mu_i = \left(\frac{\partial G}{\partial n_i}\right)_{P,T,n_j} \quad [5.5]$$

Chemical potential is important for phase equilibrium and related properties.

Let us consider a binary system with two phases α and β. If a closed system satisfies the following equation, then it is said to be in equilibrium:

$$\sum_i \mu_i^\alpha dn_i^\alpha + \sum_i \mu_i^\beta dn_i^\beta = 0 \quad [5.6]$$

where n_i signifies the variation in the number of moles of a component "i" in a phase.

A simpler way to represent the condition of phase equilibrium can be derived from the above equation by substituting $dn_i^\alpha = -dn_i^\beta$

$$\sum_i (\mu_i^\alpha - \mu_i^\beta) dn_i^\alpha = 0 \quad [5.7]$$

$$\therefore \mu_i^\alpha = \mu_i^\beta = \cdots = \mu_i^\pi \quad [5.8]$$

Phase equilibrium is achieved if each chemical component has the same chemical potential in each of the phases.

Equilibrium is significantly affected by the chemical potential. When a liquid and its vapor reach equilibrium, it is essential for the chemical potential of the substance in the liquid phase to match the chemical potential in the vapor phase. This balance prevents any exchange of particles between the phases, thereby maintaining equilibrium.

This concept also applies to other systems, including chemical reactions that are in an equilibrium. The chemical potentials of the reactants and products must match for a reaction to be in equilibrium. The chemical equilibrium constant characterizes this condition.

In general, chemical potential aids in predicting the direction and magnitude of chemical reactions as well as phase transitions by offering a quantitative insight of how systems operate at equilibrium.

5.2.4. *Total and specific properties*

A "specific property" refers to a characteristic of a substance or system that is expressed on a unit–mass or unit–mole basis. Specific property (M) is not dependent on mass. It is intensive in nature (for example, specific heat capacity and specific volume).

A "total property" refers to a property of a system that represents the overall quantity or value of that property for the entire system, rather than on a per-unit basis. When the number of moles of a certain quantity is multiplied by a specific property, it yields the total property ($\underline{M}$).

The relation between the total and specific properties is straightforward as per the following:

$$\underline{M} = nM \qquad [5.9]$$

Differential form of the above equation is

$$d\underline{M} = ndM + Mdn \qquad [5.10]$$

Now, let us consider a property M that changes according to two independent variables X and Y, as well as the molar compositions,

$$\underline{M} = \underline{M}(X, Y, n_1, n_2, \dots) \qquad [5.11]$$

Differential form of the above equation is

$$d\underline{M} = d(nM) = \left(\frac{\partial(nM)}{\partial X}\right)_{Y,n} dX + \left(\frac{\partial(nM)}{\partial Y}\right)_{X,n} dY + +\Sigma\left(\frac{\partial(nM)}{\partial n_i}\right)_{X,Y,n_j} dn_i \qquad [5.12]$$

In practical applications, numerical values of chemical potentials are required due to their vital role in the identification of equilibrium states. Special symbols are used to represent the derivatives of the energy functions U, H, A and G with regard to the number of moles and are called chemical potentials

$$\mu_i = \left(\frac{\partial \underline{M}}{\partial n_i}\right)_{X,Y,n_j} \qquad [5.13]$$

The fundamental thermodynamic property relations are as follows:

$$d\underline{U} = Td\underline{S} - Pd\underline{V} + \Sigma\mu_i dn_i \qquad [5.14]$$

$$d\underline{H} = Td\underline{S} + \underline{V}dP + \Sigma\mu_i dn_i \qquad [5.15]$$

$$d\underline{A} = -\underline{S}dT - Pd\underline{V} + \Sigma\mu_i dn_i \qquad [5.16]$$

$$d\underline{G} = -\underline{S}dT + \underline{V}dP + \Sigma\mu_i dn_i \qquad [5.17]$$

The chemical potential can be expressed using any of the four fundamental groups of properties,

$$\mu_i = \left(\frac{\partial \underline{U}}{\partial n_i}\right)_{\underline{S},\underline{V},n_j} = \left(\frac{\partial \underline{H}}{\partial n_i}\right)_{\underline{S},P,n_j} = \left(\frac{\partial \underline{A}}{\partial n_i}\right)_{T,\underline{V},n_j} = \left(\frac{\partial \underline{G}}{\partial n_i}\right)_{T,P,n_j} \qquad [5.18]$$

The partial molal Gibbs energy, which is the derivative of Gibbs energy with respect to the number of moles, is the definition of the chemical potential that is most commonly recognized

$$\mu_i = \bar{G}_i = \left(\frac{\partial \underline{G}}{\partial n_i}\right)_{T,P,n_j} \qquad [5.19]$$

5.2.5. *Partial molar properties*

In the realm of thermodynamics, partial molar properties serve as crucial indicators of how a component behaves within a mixture. These properties gauge the alteration in the total property $(\underline{M})$ upon the addition of a minute quantity of component "i" to a finite solution at constant temperature and pressure

$$\bar{M}_i = \left(\frac{\partial \underline{M}}{\partial n_i}\right)_{T,P,n_j} \qquad [5.20]$$

The chemical potential is simply another term for the partial molar Gibbs energy $(\mu_i = \underline{G}_i)$.

5.2.6. *Excess and residual properties*

Excess property (M^E) is the deference between the actual value and value it would give assuming mixture as ideal solution for any specific property

$$M^E = M - M^{id} \qquad [5.21]$$

There is another similar terminology called residual property (M^R); it is the deference between the actual value and the value it would give assuming the mixture as ideal gas for any specific property

$$M^R = M - M^{ig} \qquad [5.22]$$

Table 5.1 shows a summary of all discussed properties and the relations among them.

Property	Relations
Specific property	M
Total property	$\underline{M} = nM$
Partial molar property	$\bar{M}_i = \left(\frac{\partial(nM)}{\partial n_i}\right)_{T,P,n_j} = \left(\frac{\partial \underline{M}}{\partial n_i}\right)_{T,P,n_j}$
Excess property	$M^E = M - M^{id}$
Residual property	$M^R = M - M^{ig}$

Table 5.1. *Thermodynamic properties: symbols and relations among them*

5.3. Vapor–liquid equilibrium

5.3.1. *Equilibrium in general; and vapor–liquid equilibrium*

The state of equilibrium in a system is achieved when there are no variations in its macroscopic properties over time.

The chemical potential of every component in each phase will be identical during the state of equilibrium. Due to equal chemical potential, there is no force that might cause a change in properties.

When vapor and liquid phases are in intimate contact with each other and if temperature, pressure and composition are unchanged over time, then we can say that VLE (vapor–liquid equilibrium) is established.

5.3.2. *Types of VLE data/diagrams*

The following variables are typically used to represent VLE data:

– temperature (T);

– pressure (P);

– liquid-phase composition (x_i);

– vapor-phase composition (y_i).

Comprehensive knowledge of VLE data and diagrams is crucial in understanding the behavior of mixtures of substances in equilibrium between their vapor and liquid phases. The following are some useful types of VLE data and diagrams:

– *T–x–y diagram:* when pressure is assumed to be constant (with $T - x - y$ as variables), then this type of VLE data is called the isobaric VLE data.

– *P–x–y diagram:* when temperature is assumed to be constant (with $P - x - \mathrm{y}$ as variables), then this type of VLE data is called the isothermal VLE data.

– *x–y diagram:* the composition of the liquid and vapor phases at equilibrium for a specific temperature and pressure is shown in the $x - \mathrm{y}$ diagram. Each point on the graph corresponds to a particular composition at equilibrium. Typically, the x-axis shows the mole fraction of one component in the liquid phase (x), and the y-axis shows the mole fraction of the same component in the vapor phase (y).

– *P–T diagram:* a pressure–temperature diagram, commonly referred to as a P–T diagram, is a visual depiction illustrating the phase characteristics of a substance in relation to pressure and temperature.

Figure 5.2 represents isothermal ($P - x - y$) VLE data, and Figure 5.3 represents isobaric ($T - x - y$) VLE data. The solid lines in plots are indicating saturated liquid lines or bubble-point lines where liquid starts boiling. The dotted lines in plots are called saturated vapor lines or dew-point lines where vapor starts condensing. For both plots with an increase in isobaric pressure and isothermal temperature, the area enclosed by bubble and dew lines decreases. The phase envelope is the region delineated by the curved lines on both plots, where two phases coexist. Figure 5.4 represents the x – y diagram for different systems. The point where the $x - y$ diagram intersects the 45° line is the azeotropic point. At azeotrope, the liquid and vapor phases have the same composition and hence no separation is achieved. The azeotropes are discussed in detail in section 5.3.5. Figure 5.5 represents a P–T diagram. In the diagram, the solid lines denote the saturated liquid lines or bubble-point lines, signifying the initiation of boiling for the liquid. Meanwhile, the dotted lines in plots are referred to as the saturated vapor lines or dew-point lines, indicating the beginning of condensation for the vapor.

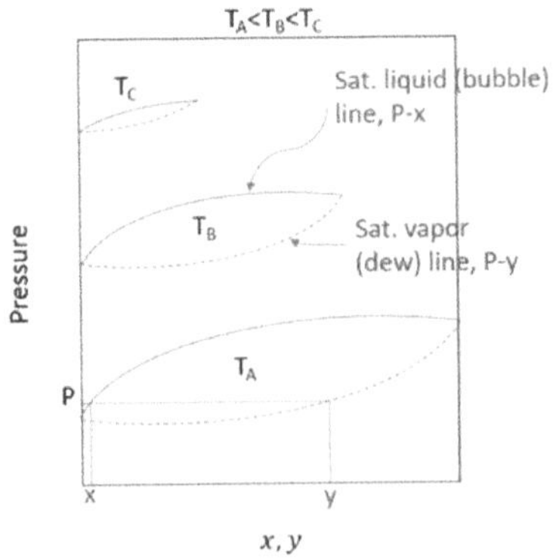

Figure 5.2. *Isothermal ($P - x - y$) VLE plot*

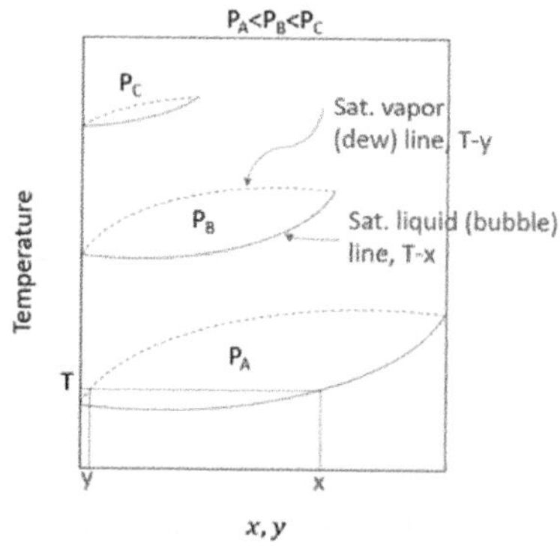

Figure 5.3. *Isobaric* $(T - x - y)$ *VLE plot*

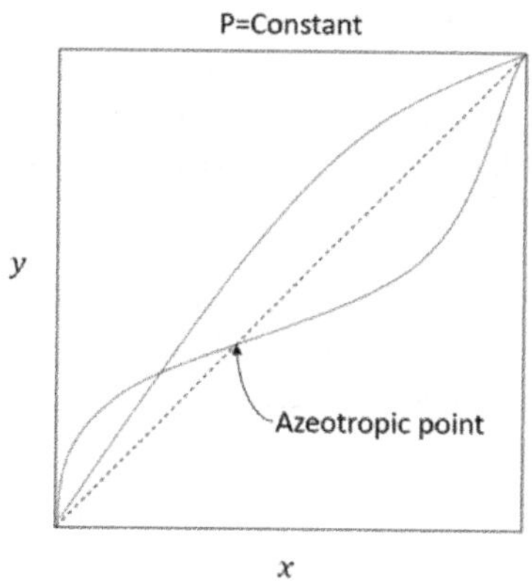

Figure 5.4. $x - y$ *VLE plot*

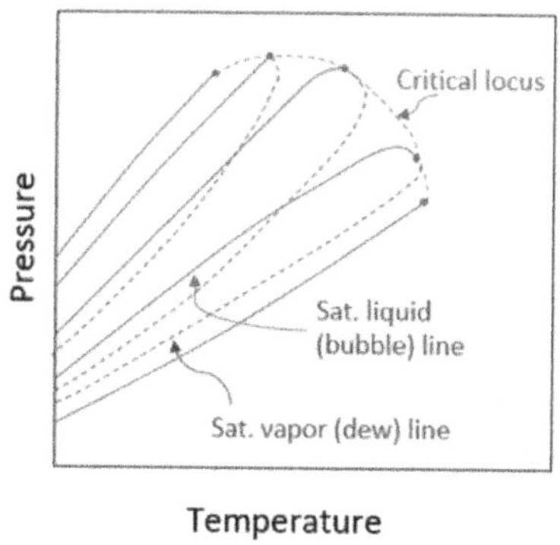

Figure 5.5. $P - T$ *VLE plot*

5.3.3. *Basic laws and properties*

5.3.3.1. *Raoult's law*

Raoult's law offers a straightforward equation to represent VLE data, assuming ideal behavior for both the vapor and liquid phases. The equation is expressed as follows:

$$y_i P = x_i P_i^{\text{sat}} \qquad (i = 1, 2, \dots, N) \qquad [5.23]$$

where y_i is the vapor-phase composition of component "i", x_i is the liquid-phase composition of component "i", P is the total pressure and P_i^{sat} is the saturation pressure of component "i" (often called the vapor pressure of component "i"). A notable point about Raoult's law is that the term $y_i P$ is the partial pressure.

This law applies when there are resemblances in the interactions among molecules of the same component (solute–solvent and solvent–solvent interactions) and the interactions among molecules of different components in the solution. Stated differently, Raoult's law holds true for perfect solutions in which there is no departure from ideality.

Hence, this law can only be applied under conditions of low to moderate pressure and to components that are chemically similar, such as mixtures of isomers or adjacent members of homologous series.

Raoult's law is commonly used when analyzing colligative properties, such as the decrease in vapor pressure, elevation in boiling point and depression in freezing point, which manifest when nonvolatile solutes are introduced into a solvent.

Moreover, non-ideal solutions exhibit deviations from Raoult's law, encompassing positive deviations (yielding vapor pressures higher than anticipated) and negative deviations (yielding vapor pressures lower than anticipated), owing to the intermolecular interactions between distinct components.

5.3.3.2. *Modified Raoult's law*

Modified Raoult's law serves as an extension to Raoult's law by accounting for the deviations from ideality observed in solutions. While Raoult's law assumes ideal behavior, characterized by comparable interactions between solute–solvent and solvent–solvent molecules, modified Raoult's law accommodates and addresses these deviations from the ideal behavior.

Here, only the vapor phase is assumed to ideal, and liquid phases to non-ideality is described by the activity coefficient (γ_i)

$$y_i P = x_i \gamma_i P_i^{\text{sat}} \qquad (i = 1, 2, \ldots, N) \tag{5.24}$$

where γ_i is the activity coefficient of component "i" in the liquid phase.

This law can be used in systems with dissimilar components since it takes into account the non-ideality of the liquid phase.

5.3.3.3. *K-value*

In the context of VLE, the K-value, which is also known as the equilibrium constant or distribution coefficient, denotes the proportion between the concentration of a particular component in the vapor phase and its concentration in the liquid phase at equilibrium for a given temperature and pressure.

The K-value for component "i" can be mathematically expressed as:

$$K_i = \frac{y_i}{x_i} \tag{5.25}$$

where K_i symbolizes the K-value for component "i", y_i represents the mole fraction of component "i" in the vapor phase and x_i indicates the mole fraction of component "i" in the liquid phase.

The K-value is a fundamental parameter that provides valuable information about how components distribute between the vapor and liquid phases at equilibrium

$$\begin{array}{l} \textit{Components distribution between} \\ \textit{the vapor and liquid phases} \\ \textit{based on } K-\textit{value} \end{array} = \begin{cases} K_i > 1, \textit{component 'i' favors the vapor phase,} \\ K_i = 1, \textit{component 'i' is equally distributed} \\ \qquad \textit{between the two phases, and} \\ K_i < 1, \textit{component 'i' favors the liquid phase} \end{cases}$$

When Raoult's law is valid,

$$K_i = \frac{P_i^{sat}}{P} \ (or\ y_i P = x_i P_i^{sat}) \tag{5.26}$$

When modified Raoult's law is applicable,

$$K_i = \frac{\gamma_i P_i^{sat}}{P} \ (or\ y_i P = x_i \gamma_i P_i^{sat}) \tag{5.27}$$

The significance of K-values cannot be overstated in phase equilibrium calculations, especially in separation processes such as distillation. Achieving optimal product yields and purities hinges on a deep understanding of the component distribution between liquid and vapor phases.

5.3.4. *Gamma-phi formulation*

The gamma-phi formulation is a thermodynamic methodology employed for the purpose of modeling and examining phase equilibrium in systems with multiple components, specifically in VLE.

The modified Raoult's law incorporates the activity coefficient to address non-idealities in the liquid phase, yet it is constrained by assuming ideality in the vapor phase. This limitation is circumvented by introducing the vapor-phase fugacity coefficient.

For a component "i" in the mixture, the mathematical expression for gamma-phi formulation is

$$y_i \Phi_i P = x_i \gamma_i P_i^{\text{sat}} \qquad (i = 1, 2, \ldots, N) \tag{5.28}$$

where Φ_i is the vapor-phase fugacity coefficient of component "i".

Mathematically,

$$\Phi_i \equiv \frac{\hat{\phi}_i^v}{\phi_i^{\text{sat}}} \exp\left[-\frac{V_i^l (P - P_i^{\text{sat}})}{RT}\right] \tag{5.29}$$

The exclusion of the Poynting factor, which is denoted by an exponential function, at low to moderate pressures, results in a minimal error due to its deviation from unity by a small fraction of a percent.

Consequently, this equation is frequently simplified

$$\Phi_i \equiv \frac{\hat{\phi}_i^v}{\phi_i^{\text{sat}}} \tag{5.30}$$

In brief, the gamma-phi formulation serves as a theoretical framework to comprehend and predict phase equilibria in complex mixtures. It achieves this by establishing a relationship between the fugacity coefficient and the activity coefficient, which in turn reflects the non-ideal behavior exhibited by the system.

5.3.5. *Azeotropic mixtures*

Azeotropic mixtures exhibit a captivating phenomenon in VLE, wherein the composition of the vapor phase mirrors that of the liquid phase at a specific temperature and pressure.

This distinctive behavior results in a constant boiling point for the mixture, rendering the separation of its constituents through simple distillation a challenging task.

The classification of azeotropic mixtures includes two categories: positive azeotropes and negative azeotropes.

Positive azeotropes, also referred to as maximum pressure azeotropes, are characterized by a minimum boiling azeotrope, where the boiling point of the mixture is lower than that of its individual components (Figure 5.6).

A common illustration of positive azeotropes is the ethanol–water mixture.

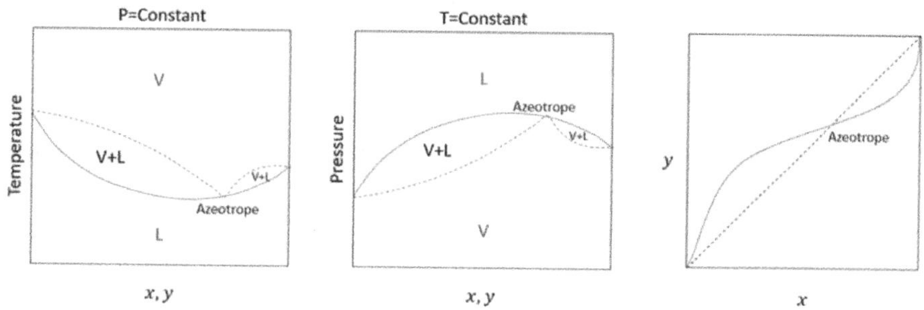

Figure 5.6. *Minimum boiling azeotrope*

Negative azeotropes, also referred to as minimum pressure azeotropes or maximum boiling azeotropes, manifest when the boiling point of a mixture surpasses the boiling point of its individual components (Figure 5.7).

An illustration of a negative azeotrope is the mixture of water and hydrochloric acid.

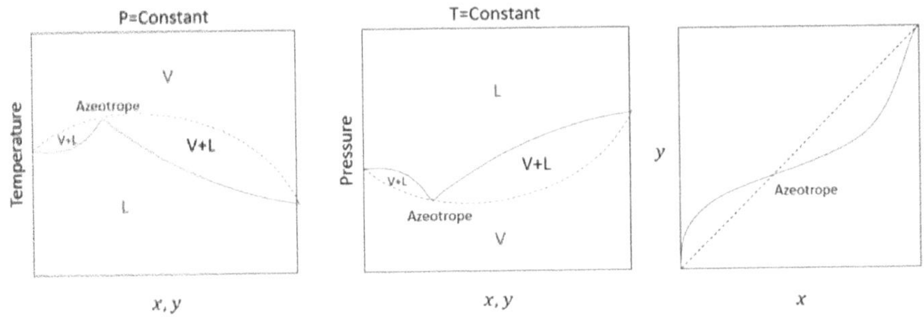

Figure 5.7. *Maximum boiling azeotrope*

Azeotropic mixtures present difficulties for distillation procedures since the constituents cannot be effectively separated using conventional methods.

To get over this problem and successfully separate the mixture, methods including pressure swing distillation, azeotropic distillation or adding an additional component (azeotrope breaker) method can be employed.

5.3.6. *Classification of VLE calculations*

A variety of VLE calculations can be used to describe the vapor–liquid partitioning, depending on the information available.

Seven different calculations can be performed, namely: *bubble-point pressure* (BPP), *dew-point pressure* (DPP), *bubble-point temperature* (BPT), *dew-point temperature* (DPT), *composition calculation* (XY), *isothermal flash* (ITF) and *adiabatic flash* (ABF).

The details of these calculations are listed below:

– BPP: calculate $\{y_i\}$ and P; given $\{x_i\}$ and T;

– DPP: calculate $\{x_i\}$ and P; given $\{y_i\}$ and T;

– BPT: calculate $\{y_i\}$ and T; given $\{x_i\}$ and P;

– DPT: calculate $\{x_i\}$ and T; given $\{y_i\}$ and P;

– XY: calculate $\{x_i\}$ and $\{y_i\}$; given P and T;

– ITF: calculate $\{x_i\}$, $\{y_i\}$, V/F, and Q; given P, T^{in}, and T^{out};

– ABF: calculate $\{x_i\}$, $\{y_i\}$, V/F, and T^{out}; given P, T^{in}, and $Q = 0$.

One of the most frequently performed calculations is bubble-point pressure (BPP) calculations which can be carried out with the following steps (for modified Raoul's law):

1) read $\{x_i\}$ and T;

2) calculate $P_i^{sat}$and γ_i;

3) calculate $P = x_i\gamma_i P_i^{sat}$;

4) calculate $y_i = \frac{x_i\gamma_i P_i^{sat}}{P}$.

Similarly, dew-point pressure (DPP) calculations can be carried out with the following steps:

1) read $\{y_i\}$ and T;

2) calculate P_i^{sat};

3) guess $\{x_{guess,i}\}$ and calculate γ_i;

4) calculate $P = \frac{1}{\Sigma_i \frac{y_i}{\gamma_i P_i^{sat}}}$;

5) calculate $x_{cal,i} = \frac{P y_i}{\gamma_i P_i^{sat}}$;

6) check if $\{x_{guess,i}\} = \{x_{cal,i}\}$, if not then change it and repeat steps 3–6 until it becomes equal or up to the desired accuracy.

Both BPT and DPT involve iterative calculations, but they can be efficiently performed with the assistance of computer tools.

5.3.7. *Flash calculations*

An integral feature of VLE is the flash calculations, in which a liquid experiences partial evaporation at equilibrium into vapor and liquid phases upon lowering the pressure. This phenomenon, referred to as "flashing", takes place only when the initial pressure is equal to or higher than the bubble-point pressure.

The overall composition of a system, which consists of one mole of nonreacting chemical species, is described by the set of mole fractions $\{z_i\}$. The liquid phase, with mole fractions $\{x_i\}$, is referred to as L, while the vapor phase, with mole fractions $\{y_i\}$, is denoted as $\mathcal{V}$.

The material balance equations can be expressed as:

$$\mathcal{L} + \mathcal{V} = 1 \tag{5.31}$$

$$z_i = x_i\mathcal{L} + y_i\mathcal{V} \qquad (i = 1, 2, \dots, N) \tag{5.32}$$

Combining these equations to eliminate L gives:

$$z_i = x_i(1 - \mathcal{V}) + y_i\mathcal{V} \tag{5.33}$$

The K-value, also referred to as K_i, is an essential factor in flash calculations. It is a practical parameter obtained by dividing y_i by x_i ($K_i = \frac{y_i}{x_i}$).

By substituting $x_i = \frac{y_i}{K_i}$ into the equation and solving for y_i, we can obtain the desired result

$$y_i = \frac{z_i K_i}{1+\mathcal{V}(K_i - 1)} \qquad (i = 1, 2, \dots, N) \tag{5.34}$$

As $\sum_i y_i = 1$, summing this expression over all species gives us a single equation in which, for known K-values, the only unknown is $\mathcal{V}$

$$\sum_i \frac{z_i K_i}{1+\mathcal{V}(K_i - 1)} = 1 \tag{5.35}$$

To resolve the above problem, a commonly employed method is to identify a suitable value for $\mathcal{V}$, ranging from 0 to 1, which fulfils the given equation.

5.3.8. *Activity coefficients models to correlate VLE data*

Data reduction is utmost critical to correlate VLE data. Various models are available to correlate VLE data. These models are based on excess Gibbs energy (G^E)

$$G^E = RT \sum_i x_i ln\, \gamma_i \tag{5.36}$$

Basically, these models will take $\{x_i\}$ and T as input and there are binary interaction parameters (BIP) that define the model for a particular system and gives $\{\gamma_i\}$ as output.

A list of these models, along with their equations is shown in Table 5.2.

Name	G^E	Binary parameters	$\ln\gamma_1$ and $\ln\gamma_2$
Van Laar	$G^E = \dfrac{Ax_1x_2}{x_1(A/B)+x_2}$	A, B	$RT\ln\gamma_1 = A\left(1+\dfrac{A\,x_1}{B\,x_2}\right)^{-2}$
			$RT\ln\gamma_2 = B\left(1+\dfrac{B\,x_2}{A\,x_1}\right)^{-2}$
Wilson	$\dfrac{G^E}{RT} = -x_1\ln(x_1+\Lambda_{12}x_2) - x_2\ln(x_2+\Lambda_{21}x_1)$	$\Lambda_{12}, \Lambda_{21}$	$\ln\gamma_1 = -\ln(x_1+\Lambda_{12}x_2) - x_2\left(\frac{\Lambda_{12}}{x_1+\Lambda_{12}x_2} - \frac{\Lambda_{21}}{\Lambda_{21}x_1+x_2}\right)$
			$\ln\gamma_2 = -\ln(x_2+\Lambda_{21}x_1) - x_1\left(\dfrac{\Lambda_{12}}{x_1+\Lambda_{12}x_2} - \dfrac{\Lambda_{21}}{\Lambda_{21}x_1+x_2}\right)$
NRTL[2]	$\dfrac{G^E}{RT} = x_1x_2\left(\dfrac{\tau_{21}G_{21}}{x_1+x_2G_{21}} + \dfrac{\tau_{12}G_{12}}{x_2+x_1G_{12}}\right)$ Where, $\tau_{12} = \frac{\Delta g_{12}}{RT}$, $\tau_{21} = \frac{\Delta g_{21}}{RT}$, $\ln G_{12} = -\alpha_{12}\tau_{12}$, $\ln G_{21} = -\alpha_{21}\tau_{21}$	$\Delta g_{12}, \Delta g_{21}, \alpha_{12}$	$\ln\gamma_1 = x_2^2\left[\tau_{21}\left(\dfrac{G_{21}}{x_1+x_2G_{21}}\right)^2 + \dfrac{\tau_{12}G_{12}}{(x_2+x_1G_{12})^2}\right]$
			$\ln\gamma_2 = x_1^2\left[\tau_{12}\left(\dfrac{G_{12}}{x_2+x_1G_{12}}\right)^2 + \dfrac{\tau_{21}G_{21}}{(x_1+x_2G_{21})^2}\right]$

2 Non-random two-liquid.

Name	G^E	Binary parameters	$\ln\gamma_1$ and $\ln\gamma_2$
UNIQUAC[3]	$G^E = G^E_{\text{combinatorial}} + G^E_{\text{residual}}$ $\frac{G^E_{\text{combinatorial}}}{RT} = x_1\ln\frac{\Phi_1}{x_1} + x_2\ln\frac{\Phi_2}{x_2} + \frac{z}{2}\left(q_1x_1\ln\frac{\theta_1}{\Phi_1} + q_2x_2\ln\frac{\theta_2}{\Phi_2}\right)$ $\frac{G^E_{\text{residual}}}{RT} = -q_1x_1\ln[\theta_1 + \theta_2\tau_{21}] - q_2x_2\ln[\theta_2 + \theta_1\tau_{12}]$ $\Phi_1 = \frac{x_1r_1}{x_1r_1 + x_2r_2}, \Phi_2 = \frac{x_2r_2}{x_1r_1 + x_2r_2}$ $\theta_1 = \frac{x_1q_1}{x_1q_1 + x_2q_2}, \theta_2 = \frac{x_2q_2}{x_1q_1 + x_2q_2}$ $\ln\tau_{21} = -\frac{\Delta u_{21}}{RT}, \ln\tau_{12} = -\frac{\Delta u_{12}}{RT}$ r and q are pure component area and volume parameters, and $z = 10$	$\Delta u_{12}, \Delta u_{21}$	$\ln\gamma_i = \ln\frac{\Phi_i}{x_i} + \frac{z}{2}q_i\ln\frac{\theta_i}{\Phi_i} + \Phi_j\left(l_i - \frac{r_i}{r_j}l_j\right) - q_i\ln(\theta_i + \theta_j\tau_{ji}) + \theta_jq_i\left(\frac{\tau_{ji}}{\theta_i+\theta_j\tau_{ji}} - \frac{\tau_{ij}}{\theta_j+\theta_i\tau_{ij}}\right)$ where $i = 1, j = 2$ or $i = 2, j = 1$ $l_i = \frac{z}{2}(r_i - q_i) - (r_i - 1)$ $l_j = \frac{z}{2}(r_j - q_j) - (r_j - 1)$

Table 5.2. *Excess Gibbs energy-based models for VLE data reduction*[4]

3 UNIversal QUAsi Chemical.

4 Poling, B.E. (2001). *The Properties of Gases and Liquids*, 5th edition. McGraw-Hill, New York.

5.3.9. *Prediction and measurement of vapor–liquid equilibrium*

5.3.9.1. *Estimation of VLE data by group contribution methods*

The feasibility of experimentation is sometimes limited by the expensive nature of the required setup or by the physical properties of binary mixtures and safety concerns. However, even in the absence of these obstacles, predictions can offer valuable insights for experiment design. Various group contribution-based methods can be employed to predict VLE data. Some group contribution-based predictive methods are listed below:

– ASOG (Analytical Solution of Groups) (Wilson and Deal 1962);

– UNIFAC (UNIQUAC Functional-group Activity Coefficients) (Fredenslund et al. 1975);

– Modified UNIFAC (Weidlich and Gmehling 1987);

– COSMO-RS (COnductor such as Screening MOdel for Real Solvents) (Klamt 1995);

– UNIFAC-IL (UNIFAC-Ionic Liquid) (Lei et al. 2009).

Equation of state-based predictive methods are:

– PPR78 (Predictive Peng–Robinson 1978; Jaubert and Mutelet 2004).

– PSRK (Predictive Soave–Redlich–Kwong) (Holderbaum and Gmehling et al. 1991).

ASOG is one of the earliest proposed methods for the prediction of VLE data, where activity coefficients for a given composition are determined at a temperature and then they are used to correlate VLE data. The UNIFAC method works similarly, and additionally, it can be used for a wide range of compounds. A modification in the UNIFAC method was proposed in 1975 which had better temperature-dependent correlations. UNIFAC-IL can be applied while dealing with ionic liquids.

PPR78 and PSRK are the equation of state-based methods that also incorporate gas-phase fugacity. As UNIFAC is widely used, the following is the in-depth description of the method.

5.3.9.2. *UNIFAC group contribution method*

The UNIQUAC functional-group activity coefficient (UNIFAC) method, originally developed by Fredenslund et al. (1975), is based on behavior/interaction between groups. For instance, VLE data for methyl tert-butyl ether (MTBE) and acetic acid can be predicted using this method. Initially, the functional groups are identified in both the compounds ($-O-CH_3$, $>C<$, $-CH_3$ and $-COOH$ as shown in Figure 5.8). Then, BIPs of the involved groups are used to calculate activity coefficient and eventually VLE data.

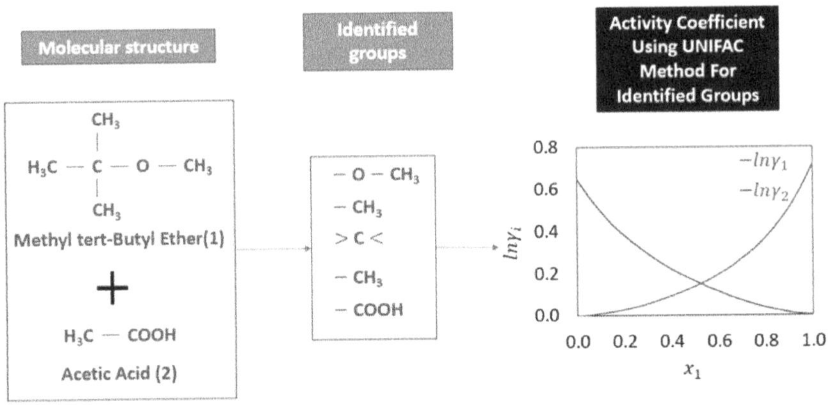

Figure 5.8. *Methodology of group contribution methods*

$$\ln \gamma_i = \ln \gamma_{i(\text{Combinatorial})} + \ln \gamma_{i(\text{Residual})} \tag{5.37}$$

The UNIFAC method calculates activity coefficient based on temperature and composition. Combinatorial and residual are the main parts of UNIFAC calculations. The combinatorial part accounts for the shape and size of molecules, while the residual part accounts for groups and interaction between groups involved in the system

$$\ln \gamma_{i(\text{Combinatorial})} = \ln\frac{\phi_i}{x_i} + \frac{z}{2} q_i \ln\frac{\theta_i}{\phi_i} + l_i - \frac{\phi_i}{x_i}\sum_j x_j l_j \tag{5.38}$$

where

$$l_i = \frac{z}{2}(r_i - q_i) - (r_i - 1),\quad \theta_i = \frac{q_i x_i}{\sum_j q_j x_j},\quad r_i = \sum_k v_k^{(i)} R_k,\quad q_i = \sum_k v_k^{(i)} Q_k \tag{5.39}$$

$$\ln \gamma_{i(\text{Residual})} = \sum_k v_k^{(i)}\left(\ln \Gamma_k - \ln \Gamma_k^{(i)}\right) \tag{5.40}$$

where

$$\ln \Gamma_k = Q_k\left(1 - \ln\left(\sum_m \theta_m \Psi_{mk}\right) - \sum_m \left(\frac{\theta_m \Psi_{km}}{\sum_n \theta_n \Psi_{nm}}\right)\right)$$

$$\theta_m = \frac{Q_m X_m}{\sum_n Q_n X_n}$$

$$X_m = \frac{\sum_j v_m^{(i)} x_j}{\sum_j \sum_n v_n^{(i)} x_j}$$

$$\Psi_{mn} = \exp\left(-\frac{U_{mn} - U_{nm}}{RT}\right) = \exp\left(-\frac{a_{mn}}{T}\right) \qquad [5.41]$$

Once the $T - x - y$ relation is established, then the following steps can be used to calculate isobaric VLE data:

1) read isobaric P;

2) read $\{x_i\}$;

3) assume initial T, T_{guess};

4) calculate $\{\gamma_i\} = fun(T_{guess}, \{x_i\})$;

5) calculate $P_{cal} = x_i \gamma_i P_i^{sat}$;

6) check if $P_{cal} = P$, if not then change T_{guess} and repeat steps 3–6;

7) calculate $y_i = \frac{x_i \gamma_i P_i^{sat}}{P}$.

5.3.9.3. *Experimental measurement of VLE data*

Chemical engineers and physical chemists have devised numerous experimental and computational methods to thermodynamically measure and analyze VLE data.

The accurate detection of VLE in practical systems poses a significant challenge for experimentalists.

Isobaric and isothermal are the two main classifications employed in the measurement of VLE data. The design of the separation train of components can be enhanced more effectively with isobaric VLE data.

Low-pressure vapor–liquid equilibrium (LPVLE) and high-pressure vapor–liquid equilibrium (HPVLE) are the two main types of isobaric VLE measuring devices.

As early as 1909, records of the development of VLE tools and procedures exist. For both HPVLE and LPVLE measurements, several studies have provided comprehensive experimental protocols and tools, such as those by Abbott (1986), Malanowski (1982a, 1982b), Hala (1967) and Raal and Mühlbauer (1994a, 1994b).

The measurement of LPVLE has been the sole subject of many papers. LPVLE was divided into four main categories by Hala (1967): static, dew and bubble points, circulation and distillation. Malanowski (1982a, 1982b) classified circulation techniques further according to the thermodynamic conditions of the phases that were to be recirculated. He divided the dynamic circulating VLE stills into the subsequent groups:

1) methods for vapor recirculation;
2) methods for condensate recirculation:
   a) liquid condensate recirculation,
   b) revaporized condensate recirculation;
3) recirculation of liquid phase and vapor condensate.

A dynamic VLE still, commonly employed in chemical engineering and process industries, is known as a dynamic VLE still or VLE apparatus. Its purpose is to investigate and analyze real-time VLE (P–T–x–y) data.

However, the determination of temperature, pressure and the compositions of both liquid and vapor results is an "overdetermination" of VLE, as any of these variables can be calculated using the Gibbs–Duhem equation.

So, an ebulliometer that measures only P–T–x data can be a more economical choice as we will only measure the bubble point and dew point, which can be predicted using data reduction and modeling techniques.

5.3.10. *The Gibbs–Duhem equation and thermodynamic consistency*

The Gibbs–Duhem equation establishes a requirement on activity coefficients that might not be upheld by a set of experimental VLE data

$$\int_{x_1=0}^{x_1=1} \ln\frac{\gamma_1}{\gamma_2}\,dx_1 = \int_{P(x_1=0)}^{P(x_1=1)} \frac{v^E}{RT}\,dP - \int_{T(x_1=0)}^{T(x_1=1)} \frac{h^E}{RT^2}\,dT \qquad [5.42]$$

If a set of VLE data follows the abovementioned equation, then the VLE data can be considered thermodynamically consistent.

There are various consistency tests for VLE data available (Wisniak et al. 2017); a few of them are listed below:

- area test;
- the Herington test;
- the Redlich–Kister test;
- point-to-point test (slope test);
- the Van Ness point-to-point test;
- the L–W test of Wisniak.

Selection of the test to be adopted for consistency check will depend on the type of data at hand ($T - x$ or $T - x - y$). The Herington test was derived from the VLE data of only a limited number systems.

Hence, it is less reliable. However, other area tests can be used. Here, the Redlich–Kister area test is discussed, which can be used for both $T - x$ and $T - x - y$ data.

5.3.10.1. *The Redlich–Kister test*

Like most other tests, this test proposed by Redlich and Kister (1948) also relies on the Gibbs–Duhem equation. The fundamental concept is that if the total area of the plot $\ln\frac{\gamma_1}{\gamma_2}$ vs x_1 equals zero under constant pressure and temperature, then the VLE data passes the test

$$\int_{x_1=0}^{x_1=1} \ln\frac{\gamma_1}{\gamma_2}\,dx_1 = 0 \qquad [5.43]$$

However, a more suitable approach is to compute the positive (A^+) and negative (A^-) areas separately and assess the following criteria for consistency verification:

$$D = 100\frac{\left|\int_0^1 \ln\frac{\gamma_1}{\gamma_2}dx_1\right|}{\int_0^1\left|\ln\frac{\gamma_1}{\gamma_2}\right|dx_1} = 100\frac{\left||A^+|-|A^-|\right|}{|A^+|+|A^-|} \leq 10 \qquad [5.44]$$

The VLE data passes the thermodynamic consistency test if the value of D is less than 10.

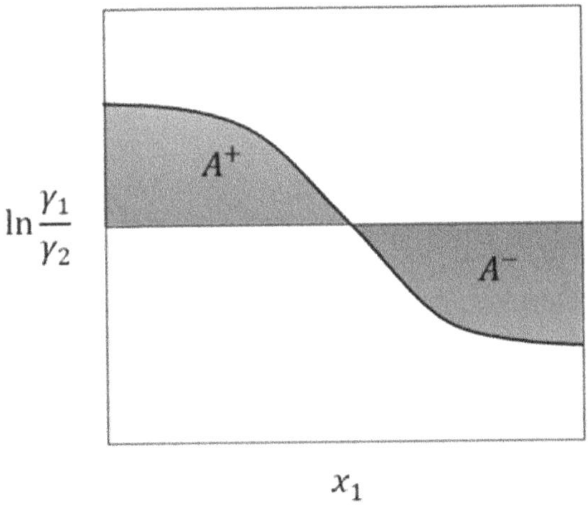

Figure 5.9. *Graphical representation of the Redlich–Kister test*

5.4. Solid–liquid equilibrium

Solid–liquid equilibrium is the term used to describe the condition in which a substance exists in both its solid and liquid states at the same time, under specific temperature and pressure conditions. This equilibrium is achieved when the rates of melting and freezing are balanced. At this equilibrium point, the chemical potential of the substance in its solid phase is equal to that in its liquid phase.

The calculations for solid–liquid equilibrium (SLE) follow the same initial steps as liquid–liquid equilibrium (LLE) and VLE calculations, involving the equalization of fugacities.

At equilibrium, the fugacity of each component is identical in both the liquid and solid phases

$$\hat{f}_i^L = \hat{f}_i^S \qquad (i = 1, 2, \ldots, N) \tag{5.45}$$

where T and P are assumed to be constant. Like LLE, each $\hat{f}_i$ is removed and replaced with an activity coefficient. Hence,

$$x_i \gamma_i f_i^L = z_i \gamma_i f_i^S \tag{5.46}$$

$$x_i \gamma_i = z_i \gamma_i \delta_i \tag{5.47}$$

where $\delta_i = \frac{f_i^S}{f_i^L = exp\left(\frac{\Delta H_i^{SL}}{R}\left(\frac{1}{T_{m,i}} - \frac{1}{T}\right)\right)}$, $\Delta H_{m,i}$ is the enthalpy of fusion/melting of component "i", and $T_{m,i}$ is the melting temperature of component "i".

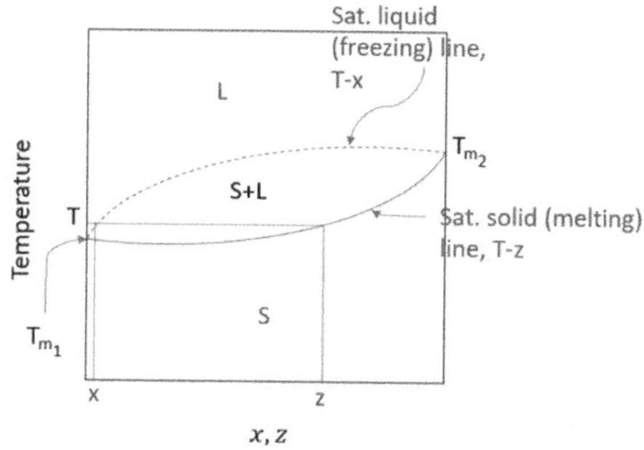

Figure 5.10. *SLE solution equilibrium diagram: ideal vapor and liquid system*

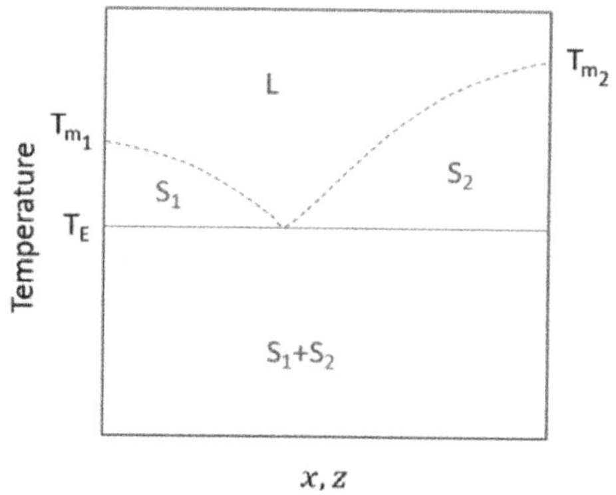

Figure 5.11. *SLE solution equilibrium diagram: ideal liquid system with immiscible solids*

Two types of equilibria between liquids and solids hold significant importance:

– solution equilibria, which involve liquids and solids of distinct chemical species;

– melt equilibria, which involve molten and solid states of the same chemical species.

Activity coefficients quantitatively express behavior, with liquid-phase coefficients derived from VLE or LLE measurements applicable to liquid phases in liquid–solid equilibria after temperature adjustments.

The accuracy of describing solubilities in practice is generally reliable, but when it comes to melt equilibria, it becomes more complex due to various factors such as:

– the presence of multiple crystalline forms in solids depending on temperature and pressure;

– intermolecular compounds may be present;

– solid phases that can be either pure or partially miscible;

– the temperatures associated with melting equilibria may exceed the safe range for extrapolation based on VLE or LLE measurements.

The intricacy of melt equilibria can be considerably greater than that of multiliquid–vapor equilibria, potentially stemming from various factors.

5.5. Liquid–liquid equilibrium

5.5.1. *Introduction*

Liquid–liquid equilibrium (LLE) is the state in which two immiscible liquid phases coexist under specific temperature and pressure conditions. Equilibrium is reached when the chemical potentials of the constituents in each phase are equal.

At equilibrium, the fugacity of each component is identical in both the liquid phases (α and β)

$$\hat{f}_i^\alpha = \hat{f}_i^\beta \qquad (i = 1, 2, \ldots, N) \qquad [5.48]$$

With the incorporation of activity coefficients, the equation transforms into:

$$x_i^\alpha \gamma_i^\alpha f_i^\alpha = x_i^\beta \gamma_i^\beta f_i^\beta \qquad [5.49]$$

In the case where each pure species exists as a liquid at the temperature of the system, $f_i^\alpha = f_i^\beta = f_i$

$$x_i^\alpha \gamma_i^\alpha = x_i^\beta \gamma_i^\beta \qquad [5.50]$$

Activity coefficients (γ_i^α and γ_i^β) are obtained from the same excess Gibbs energy functions as discussed in section 5.3.8.

5.5.2. *Binary LLE system*

A binary system will consist of two components and two liquid phases. According to the phase rule (equation [5.1]),

$\pi = 2$ phases for LLE,

$$F = 2 - 2 + N = N \qquad [5.51]$$

Consider the specific scenario of binary LLE, either at a constant pressure or at reduced temperatures low enough where the impact of pressure on the activity coefficients can be disregarded

$$\begin{matrix} x_1^\alpha \gamma_1^\alpha = x_1^\beta \gamma_1^\beta \\ and \\ x_2^\alpha \gamma_2^\alpha = x_2^\beta \gamma_2^\beta \; or \; (1 - x_1^\alpha)\gamma_2^\alpha = \left(1 - x_1^\beta\right)\gamma_2^\beta \end{matrix} \begin{cases} \gamma_1^\alpha, \gamma_2^\alpha = f(x_1^\alpha, T) \\ and \\ \gamma_1^\beta, \gamma_2^\beta = f\left(x_1^\beta, T\right) \end{cases} \qquad [5.52]$$

with one independent mole fraction per phase.

Fixing one of the three variables (x_1^α, x_1^β, T) allows the solution of the above equations.

When pressure is constant or when the impact of pressure is insignificant, binary LLE can be simply represented on a solubility diagram. This diagram is a graphical representation of temperature (T) versus the mole fraction (x_1) of one component.

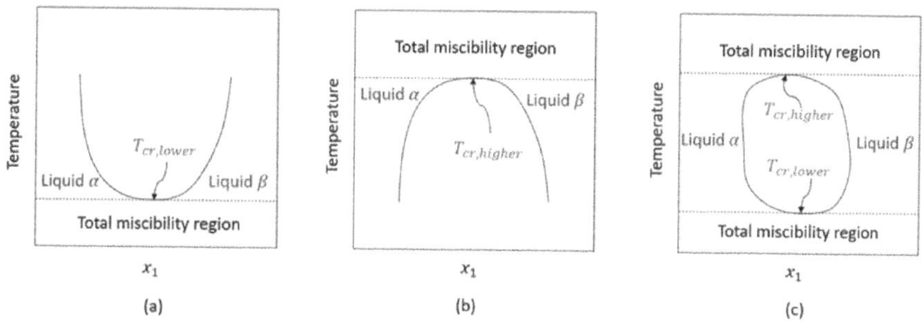

Figure 5.12. *Liquid–liquid equilibrium: binary T–x plots*

Generally, there are three types of *LLE* systems as shown in Figure 5.12. As indicated in Figure 5.12(a), the first type includes systems which have lower critical temperature, below which the binary system is completely miscible and above which there are two liquid phases present. A common example of this type of behavior is the water–trimethylamine system.

The second type of system is the opposite of the first, where the above higher critical temperature binary system is completely miscible and below it the system will form two liquid phases (Figure 5.12(b)). The phenol–water system will exhibit this behavior.

The third type of system will have both higher and lower critical temperatures, meaning that it will show two liquid phases between this temperature range (Figure 5.12(c)). A common example of this type of behavior is the water–nicotine system.

5.6. References

Abbott, M.M. (1986). Low-pressure phase equilibria: Measurement of VLE. *Fluid Phase Equilibria*, 29, 193–207.

Deal, C.H. and Derr, E.L. (1968). Group contributions in mixtures. *Industrial & Engineering Chemistry*, 60(4), 28–38.

Fredenslund, A., Jones, R.L., Prausnitz, J.M. (1975). Group-contribution estimation of activity coefficients in nonideal liquid mixtures. *AIChE Journal*, 21(6), 1086–1099.

Hala, P. (1967). *Fried, and Vilim, "Vapor-Liquid Equilibrium"*. Pergamon Press, Oxford.

Holderbaum, T. and Gmehling, J.P.S.R.K. (1991). PSRK: A group contribution equation of state based on UNIFAC. *Fluid Phase Equilibria*, 70(2–3), 251–265.

Jaubert, J.N. and Mutelet, F. (2004). VLE predictions with the Peng–Robinson equation of state and temperature dependent k_{ij} calculated through a group contribution method. *Fluid Phase Equilibria*, 224(2), 285–304.

Klamt, A. (1995). Conductor-like screening model for real solvents: A new approach to the quantitative calculation of solvation phenomena. *The Journal of Physical Chemistry*, 99(7), 2224–2235.

Lei, Z., Zhang, J., Li, Q., Chen, B. (2009). UNIFAC model for ionic liquids. *Industrial & Engineering Chemistry Research*, 48(5), 2697–2704.

Malanowski, S. (1982a). Experimental methods for vapor-liquid equilibria. Part I. Circulation methods. *Fluid Phase Equilibria*, 8(2), 197–219.

Malanowski, S. (1982b). Experimental methods for vapor-liquid equilibria. Part II. Dew-and bubble-point method. *Fluid Phase Equilibria*, 9(3), 311–317.

Raal, J.D. and Mühlbauer, A.L. (1994a). The measurement of high pressure vapor-liquid-equilibria: Part I: Dynamic methods. *Developments in Chemical Engineering and Mineral Processing*, 2(2–3), 69–87.

Raal, J.D. and Mühlbauer, A.L. (1994b). The measurement of high pressure vapor-liquid-equilibria: Part II: Static methods. *Developments in Chemical Engineering and Mineral Processing*, 2(2–3), 88–104.

Redlich, O. and Kister, A.T. (1948). Algebraic representation of thermodynamic properties and the classification of solutions. *Industrial & Engineering Chemistry*, 40(2), 345–348.

Smith, J.M., Van Ness, H.C., Abbott, M.M. (2005). *Introduction to Chemical Engineering Thermodynamics*, 7th edition. McGraw-Hill, New York.

Walas, S.M. (2013). *Phase Equilibria in Chemical Engineering*. Elsevier Science, Kent.

Weidlich, U. and Gmehling, J. (1987). A modified UNIFAC model. 1. Prediction of VLE, hE, and gamma infinite. *Industrial & Engineering Chemistry Research*, 26(7), 1372–1381.

Wilson, G.M. and Deal, C.H. (1962). Activity coefficients and molecular structure. Activity coefficients in changing environments-solutions of groups. *Industrial & Engineering Chemistry Fundamentals*, 1(1), 20–23.

Wisniak, J., Ortega, J., Fernández, L. (2017). A fresh look at the thermodynamic consistency of vapor–liquid equilibria data. *The Journal of Chemical Thermodynamics*, 105, 385–395.

5.7. Further reading

Elliott, J.R. and Lira, C.T. (2012). *Introductory Chemical Engineering Thermodynamics*, 2nd edition. Prentice Hall, Upper Saddle River.

Poling, B.E., Prausnitz, J.M., O'Connell, J.P. (2001). *The Properties of Gases and Liquids*, 5th edition. McGraw-Hill, New York.

Prausnitz, J.M., Lichtenthaler, R.N., de Azevedo, E.G. (1999). *Molecular Thermodynamics of Fluid-Phase Equilibria*, 3rd edition. Prentice Hall, Upper Saddle River.

PART 2

Fluid Mechanics and Transport Phenomena

6

Basics of Flow Through Equipment

6.1. Fundamental laws of fluid mechanics

Flow in equipment is crucial for chemical industry processes in order to achieve the maximum profit with the lowest environmental and financial cost. The aim of this chapter is to provide a tutorial for flow in equipment. The first part deals with the derivation of mass, energy and momentum conservation laws for flow by using fundamental principles of both continuum mechanics and thermodynamics. The modification of these conservation laws leading to practical applications in process equipment calculations was also presented. Directions for complex technical equipment calculations, including pressure drop flow in pipes and relative equipment, heat exchangers, packed or fluidized beds, etc., are also given. Finally, tutorial CFD calculations are presented and recent advancements in the area are reviewed.

6.1.1. *The general balance laws of continuum physics*

This chapter closely follows recent developments as described in the book by Jou et al. (2010). Assuming the continuum hypothesis is true, we posit that for each extensive quantity B, the following holds: we consider a portion of the matter of mass M occupying a volume V limited by a boundary Σ. Assuming that the continuum hypothesis holds, we posit that for any extensive quantity, B, the following condition is met:

Chapter written by Devyani THAPLIYAL, Kshitij TEWARI, Chitresh Kumar BHARGAVA, Avinash CHANDRA, Pramita SEN, Rahul KUMAR, Amit K. THAKUR, George D. VERROS and Raj Kumar ARYA.

$$\lim_{\Delta M \to 0} \frac{\Delta B}{\Delta M} = b, \quad \lim_{\Delta M \to 0} \frac{\Delta B}{\Delta V} = \rho b \quad \text{or} \quad B = \int_V \rho b \mathrm{d}V,$$

where ρ represents the overall mass density, while b represents the precise value of B in relation to each unit of mass. The balance equation for every amount b is expressed in its most general form as:

$$\int_{V(t)} \frac{\partial}{\partial t}(\rho b)\mathrm{d}V = -\int_{\Sigma(t)} \boldsymbol{J}^b \cdot \boldsymbol{n}\mathrm{d}\Sigma + \int_{V(t)} \sigma^b \mathrm{d}V \tag{6.1}$$

Here, ∂/∂t denotes the local or Eulerian time derivative, the symbol "." represents the scalar product, Jb is the quantity B flowing per unit area and unit time through the boundary Σ, and n is the outward-pointing unit normal to Σ. The term σb represents the rate of production (> 0) or destruction (< 0) per unit volume of B. The flux Jb is a vector (or tensor) if b is a scalar (or vector), and σb has the same tensorial rank as b. Assuming that equation [6.1] holds for any volume V and that the integrands are continuous functions of position, the following local form of the balance equation can be derived:

$$\frac{\partial}{\partial t}(\rho b) = -\nabla \cdot \boldsymbol{J}^b + \sigma^b, \tag{6.2}$$

after use is made of Gauss' theorem,

$$\int \boldsymbol{J}^b \cdot \boldsymbol{n}\mathrm{d}\Sigma = \int \nabla \cdot \boldsymbol{J}^b \mathrm{d}V, \tag{6.3}$$

where $\boldsymbol{\nabla}$ denotes the nabla operator. When the value of σ^b is equal to zero, the quantity b is considered to be conserved.

6.1.2. *The one-component system*

The first law of thermodynamics, which represents the equilibrium of the overall energy, is formulated for an electrically neutral system as follows:

$$\frac{\mathrm{d}E}{\mathrm{d}t} = \frac{\mathrm{d}W}{\mathrm{d}t} + \frac{\mathrm{d}Q}{\mathrm{d}t} \tag{6.4}$$

The equation E = U + k represents the total energy, which is a combination of the internal energy U and the kinetic energy k. In addition, W represents the work done by both body forces per unit mass F and contact forces T, while Q represents the heat exchanged with the external environment across the boundary. To be more precise, the variables U, k, dQ/dt and dW/dt are provided:

$$U = \int_{V(t)} \rho u \mathrm{d}V,\ K = \frac{1}{2}\int_{V(t)} \rho \boldsymbol{v}\cdot\boldsymbol{v}\mathrm{d}V, \tag{6.5a}$$

$$\frac{dQ}{\mathrm{d}t} = -\int_{\Sigma(t)} \boldsymbol{q}\cdot\boldsymbol{n}\mathrm{d}\Sigma,\ \frac{\mathrm{d}W}{\mathrm{d}t} = -\int_{\Sigma(t)} \boldsymbol{T}\cdot\boldsymbol{v}\mathrm{d}\Sigma + \int_{V(t)} \rho\boldsymbol{F}\cdot\boldsymbol{v}\mathrm{d}V, \tag{6.5b}$$

$$\frac{\mathrm{d}}{\mathrm{d}t}\int_{V(t)} b\mathrm{d}V = \int_{V(t)} \frac{\partial b}{\partial t}\mathrm{d}V + \int_{\Sigma(t)} b\boldsymbol{v}\cdot\boldsymbol{n}\mathrm{d}\Sigma.$$

where T is related to the pressure tensor P by means of Cauchy's relation T = –P n. The notation used in equations [6.5a] and [6.5b] is traditional, where u represents the specific internal energy, v represents the velocity field and q represents the heat flux vector. Let us use the Reynolds transport theorem, which can be expressed for any arbitrary amount b as

After substitutions of expressions [6.5] in the first law [6.4], we are led to the local form of the energy balance

$$\rho(\dot{u}+\boldsymbol{v}\cdot\dot{\boldsymbol{v}})+\left(u+\frac{1}{2}\boldsymbol{v}\cdot\boldsymbol{v}\right)(\dot{\rho}+\rho\nabla\cdot\boldsymbol{v}) = -\nabla\cdot\boldsymbol{q}-\mathbf{P}^T:\nabla\boldsymbol{v}-\boldsymbol{v}\cdot(\nabla\cdot\mathbf{P})+\rho\boldsymbol{F}\cdot\boldsymbol{v}, \tag{6.6}$$

where superscript T means transposition, an upper dot stands for the material or Lagrangian time derivative (i.e. $\mathrm{d}/\mathrm{d}t = \partial/\partial t + \boldsymbol{v}\cdot\nabla$) and a colon stands for the scalar product vector

$$\mathbf{A}:\mathbf{B} = \sum_{i,j} A_{ij}B_{ji}.$$

According to the Galileo principle, equation [6.6] must be invariant with respect to the transformation $v \rightarrow v + v_0$, where v_0 is a constant and uniform velocity. After substitution in equation [6.6] of v by $v + v_0$ and the subtraction of equation [6.6], we obtain

$$\frac{1}{2}\boldsymbol{v}_0 \cdot \boldsymbol{v}_0(\dot{\rho} + \rho\nabla \cdot \boldsymbol{v}) + \boldsymbol{v}_0 \cdot [(\dot{\rho} + \rho\nabla \cdot \boldsymbol{v})\boldsymbol{v} + \rho\dot{\boldsymbol{v}} + \nabla \cdot \mathbf{P} - \rho\boldsymbol{F}] = 0. \qquad [6.7]$$

This relation could be invariant with respect to v_0 on the condition that the following equations are satisfied:

$$\dot{\rho} = -\rho\nabla \cdot \boldsymbol{v}, \qquad [6.8]$$

$$\rho\dot{\boldsymbol{v}} = -\nabla \cdot \mathbf{P} + \rho\boldsymbol{F}. \qquad [6.9]$$

With these results in mind, equation [6.6] reduces to

$$\rho\dot{u} = -\nabla \cdot \boldsymbol{q} - \mathbf{P}^T : \nabla\boldsymbol{v}. \qquad [6.10]$$

It is also convenient to express the balance of mass (equation [6.8]) in terms of the specific volume υ ($= 1/\rho$). This leads to

$$\rho\dot{\upsilon} = \nabla \cdot \mathbf{v} \qquad [6.11]$$

Relations [6.8]–[6.10] refer to the principles governing the conservation of mass, momentum and internal energy. These principles are expressed in the Lagrange representation. The conservation of mass is also referred to as the continuity equation.

Quantity, b	Flux, $\boldsymbol{J}^b$	Source, σ^b
Mass	$\rho\boldsymbol{v}$	0
Momentum	$\mathbf{P} + \rho\boldsymbol{v}\boldsymbol{v}$	$\rho\boldsymbol{F}$
Internal energy	$\boldsymbol{q} + \rho u\boldsymbol{v}$	$-\mathbf{P}^T : \nabla\boldsymbol{v}$

Table 6.1. *Fluxes and supply terms (sources)*

In terms of the local time derivative (Euler representation), equations [6.8]–[6.10] take the form

$$\frac{\partial \rho}{\partial t} = -\nabla \cdot (\rho \boldsymbol{v}), \qquad [6.12]$$

$$\frac{\partial (\rho \boldsymbol{v})}{\partial t} = -\nabla \cdot (\mathbf{P} + \rho \boldsymbol{v}\boldsymbol{v}) + \rho \boldsymbol{F}, \qquad [6.13]$$

$$\frac{\partial (\rho u)}{\partial t} = -\nabla \cdot (\boldsymbol{q} + \rho u \boldsymbol{v}) - \mathbf{P}^T : \nabla \boldsymbol{v} \qquad [6.14]$$

Here, vv denotes a dyadic product. Equations [6.12]–[6.14] are particularly useful as they enable the identification of various fluxes and source terms associated with mass, momentum and energy, as outlined in Table 6.1. It is evident that mass is conserved. The terms P and ρ vv represent the conductive and convective transport of momentum, respectively. In the absence of external (or body) forces, momentum is conserved.

Similarly, q describes the transport of internal energy due to conduction, while ρ uv accounts for the contribution from convection, and the term PT: ∇ v represents the internal energy supply. By adding equation [6.13], after scalar multiplication by v, to equation [6.14], we obtain the balance equation for total energy:

$$\frac{\partial}{\partial t}\left[\rho\left(u + \tfrac{1}{2}\boldsymbol{v}\cdot\boldsymbol{v}\right)\right] = -\nabla \cdot \left[\boldsymbol{q} + \rho\left(u + \tfrac{1}{2}\boldsymbol{v}\cdot\boldsymbol{v}\right)\boldsymbol{v} + \mathbf{P}\cdot\boldsymbol{v}\right] + \rho \boldsymbol{F}\cdot\boldsymbol{v}, \qquad [6.15]$$

from which it follows that in the absence of external force, the total energy is conserved.

Please note that this chapter is restricted to a simple isotropic one-component system. For a complete analysis, including charged systems, elastic contributions, mixtures of components or chemical reactions, please see Groot and Mazur (1984), Kuiken (1994) and Jou et al. (2010).

6.1.3. *Nonequilibrium thermodynamics and flow*

The discipline of irreversible thermodynamics offers a comprehensive foundation for the macroscopic depiction of processes. It can be regarded as a subdivision of statistical physics that has practical uses in the study of transport phenomena, electromagnetic theory and other related fields. The primary concern in the study of irreversible processes in thermodynamics is that it can be initially approached as a continuum theory, where the state parameters of the theory are treated as field variables, meaning they are continuous functions of spatial coordinates and time (Verros 2007).

Irreversible thermodynamics emerged in the previous century as a potent theoretical framework for comprehending the underlying principles governing various physical phenomena and industrial processes, such as transport phenomena and material manufacturing (Verros 2007).

Irreversible thermodynamics is based on three fundamental axioms that extend beyond the principles of equilibrium thermodynamics (Kondepudi and Prigogine 2003; Demirel 2007; Rastogi 2007):

– The equilibrium thermodynamic relations can be applied to systems out of equilibrium, as long as the gradients remain small (quasi-equilibrium postulate).

– All fluxes J within the system can be expressed as linear relationships involving the thermodynamic forces X, known as the linearity postulate (**J = LX**).

– In the absence of magnetic fields, and assuming linearly independent fluxes or thermodynamic forces, the coefficient matrix L in the flux-force relations is symmetric. This is referred to as the Onsager reciprocal relations (ORR).

To begin the application of nonequilibrium thermodynamics axioms, it is necessary to determine the terms related to entropy production. Subsequently, the linearity assumption is used to deduce fundamental laws from these words. The objective of this part is to deduce fundamental principles of rheology, including Newton's law of viscosity. Viscosity, which refers to the ability of a liquid to flow, is a characteristic of liquids that falls within the field of rheology. The term rheology originates from the Greek word "rheos", which signifies the concept of flow.

A tangential resistance, known as shear stress, is created due to molecular processes when two layers of liquid flow at different rates. The shear stress is attributed to the internal friction of the liquid, or alternatively, the liquid transmits shear forces due to its internal friction. The shear forces exerted on a liquid in motion cause continual deformation. The stress magnitude is determined by the rate of shear deformation and the viscosity μ, which represents the sluggishness of the liquid.

In the preceding part, we have identified the words for internal energy production in flow as $P^T : \nabla v$ (see Table 1.1). In the absence of intrinsic rotational motions and external couples, the conservation of the angular momentum implies the symmetry of the pressure tensor (de Groot and Mazur 1984): $P = P^T$.

In order to calculate entropy production, we have to resort to equilibrium thermodynamics. By applying the quasi-equilibrium postulate, the Gibbs equation could be applied:

$$Tds = du + pd\upsilon \quad [6.16]$$

In terms of time derivatives, the above equation can be written as:

$$T\dot{s} = \dot{u} + p\dot{\upsilon}$$

Moreover, the above equation could be multiplied by density ρ as:

$$\rho T\dot{s} = \rho\dot{u} + \rho p\dot{\upsilon} \quad [6.17]$$

Without the loss of generality, the pressure tensor can be decomposed into an equilibrium part p and a viscous part

$$\mathbf{P}^v : \mathbf{P} = p\mathbf{I} + \mathbf{P}^v \quad [6.18]$$

According to Groot and Mazur (1984) in assuming that the equilibrium part of the total sensor is a scalar, we restrict the discussion to nonelastic fluids. For an elastic medium, the equilibrium "pressure" tensor is the elastic stress tensors.

Furthermore, the viscous pressure tensor $\mathbf{P}^v$ can be split into a scalar bulk viscous pressure p^v and a traceless deviator $\overset{0}{\mathbf{P}}{}^v$ according to

$$\mathbf{P}^v = p^v\mathbf{I} + \overset{0}{\mathbf{P}}{}^v ; p^v = \frac{1}{3} Tr\mathbf{P}^v \quad [6.19]$$

An alternative approach is to decompose the velocity gradient, which is a tensor, into its symmetric and antisymmetric components

$$\nabla \boldsymbol{v} = \mathbf{V} + \mathbf{W}, \quad [6.20]$$

with

$$\mathbf{V} = \tfrac{1}{2}\left[\nabla v + (\nabla v)^T\right], \mathbf{W} = \tfrac{1}{2}\left[\nabla v - (\nabla v)^T\right],$$

or in Cartesian coordinates

$$V_{ij} = \frac{1}{2}\left(\frac{\partial v_j}{\partial x_i} + \frac{\partial v_i}{\partial x_j}\right), W_{ij} = \frac{1}{2}\left(\frac{\partial v_j}{\partial x_i} - \frac{\partial v_i}{\partial x_j}\right).$$

The rate of deformation tensor V may be further decomposed as

$$\mathbf{V} = \tfrac{1}{3}(\nabla \cdot \mathbf{v})\mathbf{I} + \overset{0}{\mathbf{V}} \qquad [6.21]$$

where I is the identity tensor and $\overset{0}{\mathbf{V}}$ is the deviatoric traceless tensor.

By further using equations [6.18]–[6.21] and applying standard tensor theory (Groot and Mazur 1984) such as $\nabla \cdot \mathrm{v} = \nabla \mathrm{v} : \mathrm{I}$, $\overset{0}{\mathrm{P}^{v}} : \mathrm{I} = 0$, and the fact that the scalar product of a symmetric and antisymmetric tensor vanishes, we could directly show that

$$-\frac{\mathbf{P} \ :\nabla \mathbf{v}}{T} = -\frac{1}{T} p^{v} \nabla \cdot \mathbf{v} - \frac{1}{T} \overset{0}{\mathbf{P}^{v}} : \overset{\mathbf{0}}{\mathbf{V}} \qquad [6.22]$$

where use has also been made of the mass conservation law [6.11]: $\rho \dot{\upsilon} = \nabla \cdot \mathbf{v} = \mathbf{V} : \mathbf{I}$ and the Gibss–Duhem equation (equations [6.16] or [6.17]).

A detailed derivation can be found in Groot and Mazur (1984) and Jou et al. (2010).

Please note that the total contribution of viscous phenomena to the entropy production has thus been split up into two parts. The second part is related to the rate of change of specific volume.

Let us apply the linearity postulate to the above equation by identifying the fluxes and the forces as shown in Table 1.2 to derive the fundamental laws of Newton and Stokes.

The first part relates to the shear stress phenomena arising from the internal friction of the liquid, indicating that the liquid transmits shear forces due to this internal friction. For instance, a moving liquid undergoes continuous deformation as a result of the shear forces acting on it.

	Newton law of viscosity	**Stokes law**
Flux (J)	$\overset{0}{\mathrm{P}}{}^{v}$	pv
Force (X)	$-T^{-1}\overset{0}{\mathrm{V}}$	$-T^{-1}\nabla \cdot \mathbf{v}$
L	Shear viscosity: $\mu = L_v T^{-1}$	Bulk viscosity: $\zeta = L_b T^{-1}$
Laws	$\overset{0}{\mathbf{P}}{}^{v} = -2\mu \overset{0}{\mathbf{V}}$	$p^v = -\zeta \nabla \cdot \mathbf{v}$

Table 6.2. *Newton's and Stokes' laws (Jou et al. 2010)*

By combining the above laws, we could derive the following equation:

$$P^v = -\zeta\left(\nabla \cdot \mathrm{v}\right)\mathrm{I} - 2\mu \overset{0}{V} \qquad [6.23]$$

From the above equation, the classical Newton–Stokes constitutive equation is derived by reintroducing V:

$$P^v = -\zeta\left(\nabla \cdot \mathrm{v}\right)\mathrm{I} - 2\mu \mathrm{V} - \frac{2}{3}\left(\nabla \cdot \mathrm{v}\right)\mathrm{I} \qquad [6.24]$$

By introducing equations [6.24] into [6.9] and by further using equations [6.18] and [6.19], the momentum balance (equation of motion) takes the following form (Groot and Mazur 1984)

$$\rho \frac{d\mathbf{v}}{dt} = -\nabla p + \nabla \cdot \left[\mu\left(\nabla \mathrm{v} + (\nabla \mathrm{v})^T\right) - \frac{2}{3}(\nabla \cdot \mathrm{v})\mathbf{I}\right] + \nabla \cdot \left[\zeta(\nabla \cdot \mathrm{v})\mathbf{I}\right] + \rho \mathbf{F} \qquad [6.25]$$

Typically, the study of fluid mechanics focuses on analyzing equation [6.9] and the continuity equation (equation [6.8]) along with a suitable thermodynamic state equation. Hence, the aforementioned equation is linked with the continuity equation

$$\frac{d\rho}{dt} = -\rho\nabla \cdot \mathrm{v} \qquad [6.26]$$

The symbol d/dt stands for the Langrangian derivative

$$\mathrm{d}/\mathrm{d}t = \partial/\partial t + v \cdot \nabla$$

We now discuss briefly the three most common simplifications of the equation of motion:

– For constant density ρ, the terms in the momentum equation that contain the divergence of the velocity $\nabla \cdot \mathrm{v}$ disappear because the continuity equation is written as $\nabla \cdot \mathrm{v} = 0$. Please note that $\nabla^2 \mathrm{v} = \nabla \cdot \left(\nabla \mathrm{v} + (\nabla \mathrm{v})^T \right)$. Therefore, for constants μ and ρ, the equation of motion leads to the very famous Navier–Stokes equation, first developed from molecular arguments by Navier and from continuum arguments by Stokes (Bird et al. 2002): $\rho \frac{d\mathrm{v}}{dt} = -\nabla p + \mu \nabla^2 \mathrm{v} + \rho \mathrm{F}$.

– When the acceleration terms in the Navier–Stokes equation are neglected, i.e. when $\rho \frac{d\mathrm{v}}{dt} = 0$, we get $0 = -\nabla p + \mu \nabla^2 \mathrm{v} + \rho \mathrm{F}$, which is called the Stokes flow equation. The equation is occasionally referred to as the creeping flow equation, as the term $[\rho \mathrm{v} \cdot \nabla \mathrm{v}]$, which is a quadratic function of the velocity, can be disregarded when the flow is exceptionally sluggish. In certain flows, such as the Hagen–Poiseuille tube flow, the term "restriction to slow flow" is not applicable and is eliminated. The Stokes flow equation is a crucial component of lubrication theory, which examines the movement of particles in suspension, the flow of fluids through porous materials and the locomotion of microorganisms. There exists an extensive body of literature pertaining to this particular issue.

– When the effects of viscous forces are disregarded ($\mu \nabla^2 \mathrm{v} = 0$), the equation of motion transforms into what is referred to as the Euler equation for an "inviscid" fluid. While it is true that there are no fluids that can be considered to be completely "inviscid", there are numerous flow situations where the effects of viscosity are quite insignificant. Instances include the aerodynamic flow around aircraft wings (except the immediate vicinity of the solid boundary), the hydrodynamic flow of rivers around the upstream surfaces of bridge abutments, certain issues in compressible gas dynamics and the flow of ocean currents. The Bernoulli equation, which describes the flow of inviscid fluids under steady conditions, is widely recognized as one of the most renowned equations in classical fluid dynamics. The equation can be readily obtained from Euler's equation (Bird et al. 2002). The next part provides examples of how the Bernoulli equation is used to real-world challenges.

6.1.4. *Definitions in fluid flow*

6.1.4.1. *Bernoulli equation*

Bernoulli expresses the conservation of energy for liquids. A fluid possesses pressure energy (dynamic pressure) and potential energy (Figure 6.1):

$$p_1 + \rho g z_1 + \frac{1}{2}\rho v_1^2 = p_2 + \rho g z_2 + \frac{1}{2}\rho v_2^2 \qquad [6.27]$$

where z is the height coordinate.

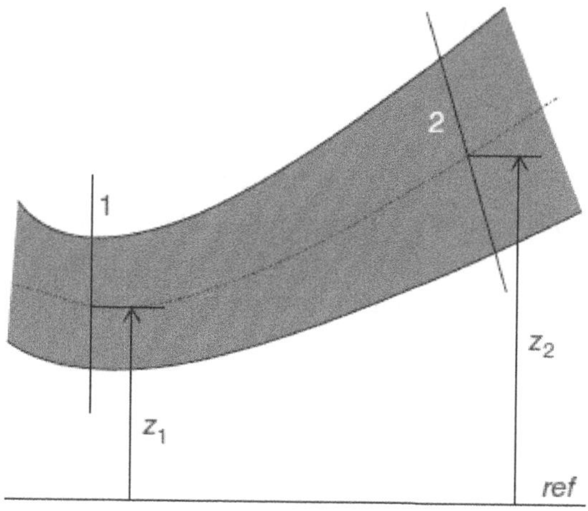

Figure 6.1. *Bernoulli law (adopted after permission from Borremans (2019))*

In the above equation, symbol F was identified as the gravitational acceleration constant g.

6.1.4.2. *Static and dynamic pressure*

Consider a horizontal pipe. The law of Bernoulli is applied to points 1 and 2.

$$p_1 + \frac{1}{2}\rho v_1^2 = p_2 + \frac{1}{2}\rho v_2^2$$

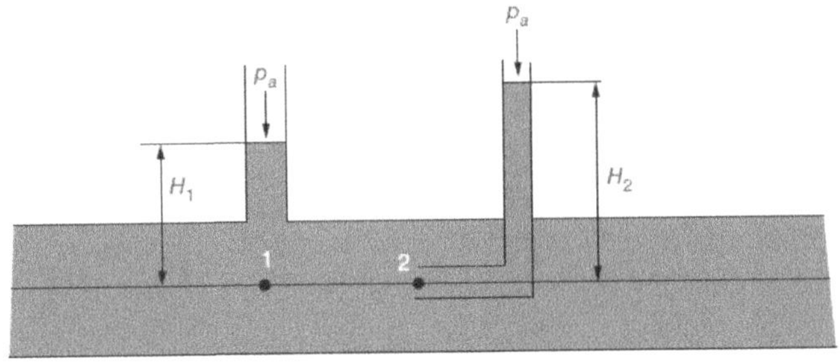

Figure 6.2. *Static and dynamic pressure (adopted after permission from Borremans (2019))*

Point 2 is a stagnation point: v2 = 0 So: $p_1 + \frac{1}{2}\rho v_1^2 = p_2$

from which: $p_2 > p_1$

In vertical sense, there is no movement and we can apply hydrostatics: $p_1 = p_a + \rho g H_1$; $p_2 = p_a + \rho g H_2$ from which $p_1 = p_2 + \rho g(H_2 - H_1)$.

Look now at a tube with a narrowed passage (a venturi) (Figure 6.3).

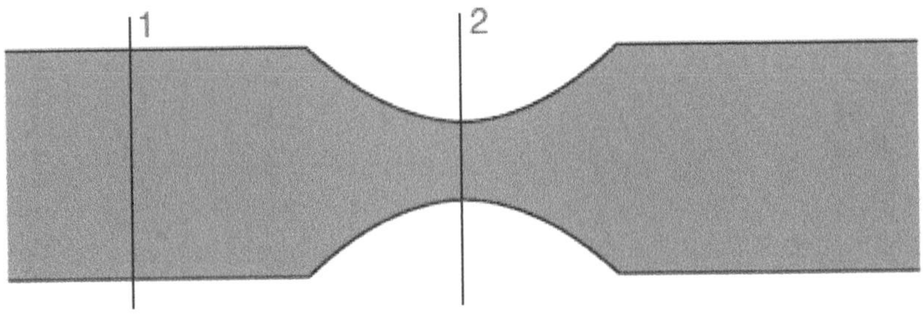

Figure 6.3. *Venturi (adopted after permission from Borremans (2019))*

$Q_v = v_1 \cdot A_1 = v_2 \cdot A_2$ with: $A_1 > A_2$ $v_1 < v_2$

In the throat, the liquid velocity will be higher than in other regions, resulting in greater dynamic pressure. Applying Bernoulli's principle leads to:

$$p_1 + \frac{1}{2}\rho v_1^2 = p_2 + \frac{1}{2}\rho v_2^2$$

so that $p1 > p2$

It is observed that static pressure can be converted into dynamic pressure, and vice versa. This is why, in many applications of applied mechanics, the total pressure of a fluid is often described as the sum of static and dynamic pressure. The total pressure thus represents the "total energy content" of the fluid.

6.1.4.3. *Extension of Bernoulli's law*

Consider a pipeline with friction loss (Figure 1.4). The classic law of Bernoulli, written as $p_1 + \frac{1}{2}\rho v_1^2 = p_2 + \frac{1}{2}\rho v_2^2$, is not valid anymore because: $p_1 + \frac{1}{2}\rho v_1^2 > p_2 + \frac{1}{2}\rho v_2^2$.

Note the pressure loss by friction p_f, then:

$$p_1 + \frac{1}{2}\rho v_1^2 = p_2 + \frac{1}{2}\rho v_2^2 + p_f \quad [6.28]$$

In general, this becomes:

$$p_1 + \rho g z_1 + \frac{1}{2}\rho v_1^2 = p_2 + \rho g z_2 + \frac{1}{2}\rho v_2^2 + p_f \quad [6.29]$$

Dividing by ρ, g leads to the law of Bernoulli being seen in the unit meter:

$$\frac{p_1}{\rho g} + z_1 + \frac{1}{2}\frac{v_1^2}{g} = \frac{p_2}{\rho g} + z_2 + + \frac{1}{2}\frac{v_2^2}{g} + H_f$$

with the friction loss in meters

$$H_f = \frac{p_f}{rg} \quad [6.30]$$

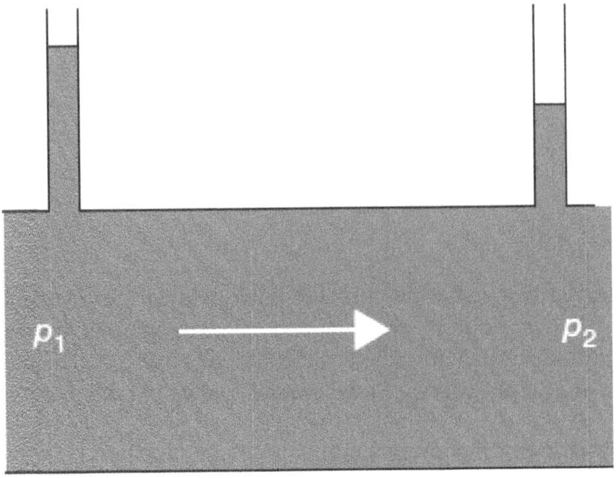

Figure 6.4. *Friction loss (adopted after permission from Borremans (2019))*

6.1.4.4. *Laminar and turbulent flow*

Examine a container in which a pipeline is attached (Figure 6.5). A valve is installed on the pipeline to regulate the flow and thus control the velocity of the liquid. A pigment is introduced through an aperture. At low velocities, the stream exhibits a great degree of regularity, with the colored portions flowing in parallel strata. This is a smooth and orderly flow. At increased speeds, the color particles follow a nonuniform trajectory and start to blend with the liquid, resulting in the exchange of energy between the layers. This is a flow characterized by turbulence.

The distinction between laminar and turbulent flow is determined by the value of the Reynolds number (Re), which is defined as:

$$\mathrm{Re} = \frac{\rho w D}{\mu}$$

where D is the pipe diameter and w is the mean velocity in the tube. The Reynolds number is dimensionless.

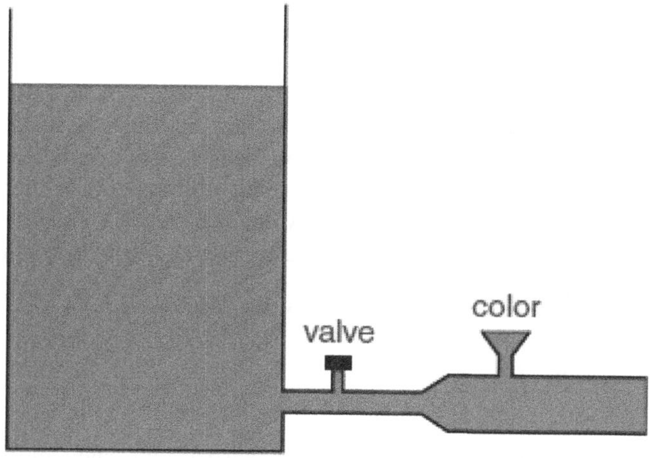

Figure 6.5. *Types of flows (adopted after permission from Borremans (2019))*

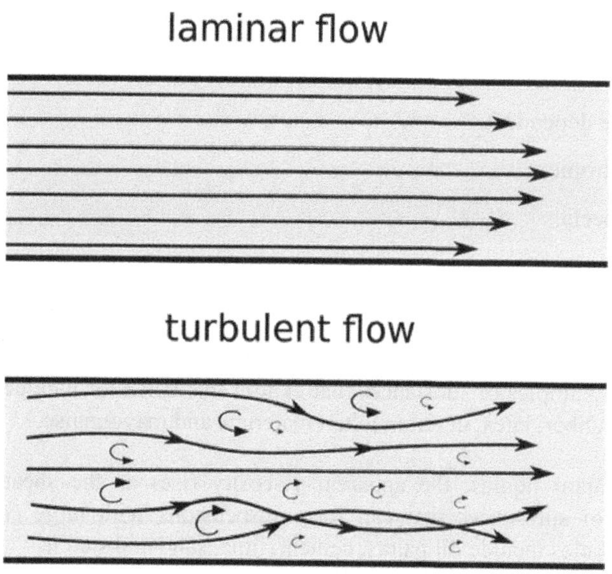

Figure 6.6. *Laminar versus turbulent flow in a pipe (adopted from The Constructor (n.d.))*

6.1.4.5. *Non-Newtonian viscosity and compressibility*

The kinematic viscosity, closely associated with shear viscosity, is defined as the ratio of viscosity to density. Fluids that meet the criteria for the Newtonian viscosity are referred to be Newtonian fluids. Examples of substances include water, solutions including water, liquids with low molecular weight, and oils and distillates derived from oil. Black liquor, liquid resin and resinous sebacic acid exhibit Newtonian fluid behavior (Nesbitt 2006). Liquids that exhibit different flow behaviors than Newtonian fluids are referred to as non-Newtonian liquids. To simplify matters, we can express the shear tensor in relation to the apparent viscosity μ1, which is similar to how Newton's law is written

$$\overset{0}{P^{\nu}} = -2\mu_1 \overset{0}{V}$$

The majority of high molecular liquids, suspensions and emulsions exhibit non-Newtonian characteristics. Nesbitt (2006) categorizes non-Newtonian liquids into three primary types:

i) Non-time-dependent:

pseudo-plastic;

dilatant;

plastic.

ii) Time-dependent:

thixotropic;

rheopectic;

irreversible.

iii) Visco-elastic.

Pseudo-plastic liquids exhibit a drop in the apparent viscosity as the shear rate increases. Examples of substances that exhibit this behavior include high molecular solutions, rubber, latex, certain molten materials and mayonnaise.

For dilatant liquids, the apparent viscosity rises as the shear rate increases. Examples of substances that can form suspensions with large concentrations of minute particles include oil paints, cement, lime, sand and starch.

Plastic liquids necessitate a specific minimum shear stress, known as the yield stress, to commence flowing. As the shear rate increases, the apparent viscosity drops from a value that is effectively infinite. Examples of substances include toothpaste, ointments, grease, margarine, paper pulp, printers ink and emulsions.

Time-dependent liquids exhibit a change in the apparent viscosity that is influenced by both the shear rate and the duration of flow. Thixotropic liquids see a decrease in the apparent viscosity when they start to flow. Once the flow stops, the liquid reverts back to its initial viscosity after a specific duration. Examples: pigment, viscous edibles, etc.

Rheopectic liquids exhibit a phenomenon where their viscosity increases when subjected to mechanical forces and returns to their normal viscosity when the forces are removed. Examples: specific solutions containing gypsum particles that are temporarily dispersed in a liquid. Irreversible liquids do not regain their initial viscosity immediately or sometimes only after a significant amount of time following the elimination of the external impact. These fluids necessitate cautious pumping. Examples of food items that undergo coagulation include cheese, yogurt and marmalade.

Viscoelastic liquids are composed of substances that display both elastic and viscous characteristics. Viscoelastic liquids experience both elastic and viscous deformations. When the flow stops, there is a partial reversal of the elastic deformation. Examples of materials include asphalt, liquid nylon, rubber and polymer solutions.

All fluids are compressible. Any values used must relate to the operating conditions. Classically, compressibility is expressed in terms of the bulk modulus mb defined by the relationship:

$$\text{Compressibility} = 1/m_b$$

$$m_b = -V_f \frac{\Delta p}{\Delta V_f}$$

Symbol V_f represents the volume of the fluid.

6.2. Pumps, compressors and flowmeters

6.2.1. *Pumps*

Pumps are used in all areas where liquid conveyance is required. Occasionally, the liquid being pumped is in a clean state, while other times it serves as a channel for transporting heat energy or suspended solid particles. There is a vast range of pump designs available due to the diverse operational needs. In an effort to present a comprehensive overview of the pumps now in use, they have been classified into four primary categories. Each pump type currently in use should be examined separately to aid in the selection of the suitable kind (Nesbitt 2006):

- rotodynamic pumps;
- special rotodynamic pumps;
- positive displacement pumps;
- other pumps.

Rotodynamic machines aim to maintain a constant energy state but cannot fully achieve this due to varying inefficiencies and manufacturing limitations. Positive displacement machines seek to operate as "continuous flow" devices but face challenges in doing so due to the properties of the liquid and manufacturing constraints.

Rotodynamic pumps feature one or more impellers with vanes that rotate within a pump casing. As the liquid flows around the vanes, the resulting forces are used to convert energy. These pumps are categorized as either radial or axial, depending on the primary direction in which the liquid moves through the impellers. Radial pumps are commonly referred to as "centrifugal pumps", while axial pumps are known as "propeller pumps". Mixed-flow pumps, which serve as intermediate types between purely radial and purely axial designs, are also available.

Certain characteristics of specialized rotodynamic pumps are closely related to those of radial rotodynamic pumps. However, these specialized pumps do not fully qualify as rotodynamic because their performance does not align with predictions from rotodynamic equations. While these pumps convert the kinetic energy of the liquid into static head, their actual performance must be validated through testing.

Peripheral pumps use impellers with vanes to increase the kinetic energy of the liquid, while Pitot tube pumps employ a rotating drum for the same purpose. Both types of pumps offer a significant advantage by providing a higher differential head increase compared to centrifugal pumps operating at the same speed. Positive displacement pumps, on the other hand, work by trapping volumes of liquid and moving them in the pumping direction or compressing and discharging them at the outlet.

The piston pump is the oldest and most well-known type of a positive displacement pump. Each stroke of the piston expels a specific volume of liquid from the cylinder. A plunger pump is a modified version of the piston pump, which also operates using a reciprocating motion.

The second sub-category includes rotating or rotary positive displacement pumps. Gear pumps are an example, as they transport liquid through the gaps between the gear teeth and maintain pressure differences through the seals formed at the contact surfaces of the meshing gears. Other examples of rotary displacement pumps include screw pumps, vane pumps and hose pumps.

Ejectors and air-lift pumps exemplify the broader category known as "other pump types". In a jet pump, the static pressure in the nozzle decreases as the velocity of the motive fluid increases, drawing the liquid into the system to mix with the driving fluid. Within the venturi (or diffuser) of the jet pump, the fluid flow speed decreases and static pressure increases. Jet pumps are versatile, capable of handling both liquids and gases, making them useful for priming other pumps. Ejectors, also referred to as jet pumps, are typically powered by steam or compressed air. Air-lift pumps function by reducing the effective density of the liquid by introducing significant amounts of gas, usually in the form of compressed air. When surface tension is adequate, the liquid can foam. The atmospheric or gas pressure acting on the liquid's surface causes the lower-density mixture to rise above the stationary liquid.

Performance is highly unpredictable and typically necessitates testing and fine-tuning in order to attain the intended outcomes. Pump performance, particularly that of rotodynamic pumps, is typically represented as a curve. The primary focus lies in examining the correlation between the amount of liquid moved during a specific timeframe, known as the volume flow or flow rate, and the increase in pressure or elevation achieved by the pump. The several primary groups exhibit significantly distinct characteristics in this regard, as depicted in Figure 3.1. The volume flow of positive displacement pumps is mostly unaffected by changes in pressure difference, resulting in a performance curve that is nearly vertical. The performance curve of the centrifugal pump demonstrates a positive correlation between pressure rise and a decrease in volume flow. Peripheral pumps have a characteristic curve that falls between the curves of these two categories of pumps.

The flow rate of a pump is determined by the ratio of the stated rise in the pump pressure to the back-pressure, which represents the resistance in the system that is necessary to move the flow forward. The piping system possesses a distinctive curve that represents the relationship between flow, differential pressure or head. This curve must be determined by the designer of the system. The pump can only function at the point where the pump characteristic meets with the system characteristic. By altering the pump speed, we can modify the pump characteristic. The system characteristic can be altered by manipulating the valves to either open or close them, or by adjusting the levels in the tanks. If all variables remain unchanged, the flow rate of the pump will remain constant. However, the pump may experience wear and the piping may become partially obstructed as a result of deposits or corrosion byproducts. Operating circumstances typically undergo changes throughout time.

When the pressure of the liquid in a pump's suction is lowered to match the vapor pressure of the liquid, the liquid undergoes evaporation. Cavitation, also known as the process of vapor bubbles transitioning back into a liquid form under greater pressure conditions, leads to the occurrence of highly concentrated pressure spikes in localized places. Cavitation is undesirable as it diminishes both the efficiency of the pump and the lifespan of the pump. The concept of NPSH (net positive suction head) is used to quantify the danger of cavitation. It refers to the minimum pressure or head required above the vapor pressure of the liquid to prevent or minimize the occurrence of cavitation. NPSH, which stands for net positive suction head, is a notion that is relevant to the occurrence of cavitation in rotodynamic pumps. NPSH stands for net positive suction head, which refers to the measurement of the pressure available at the suction side of a pump to prevent cavitation

$$\text{NPSH} = (\text{suction pressure} - \text{vapor pressure})/\rho g$$

Suction pressure is the pressure at the inlet of a pump.

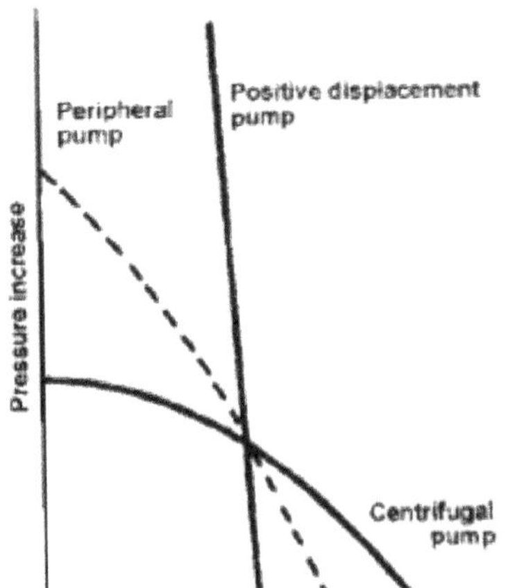

Figure 6.7. *Pump curves for different pump types (adopted with permission from Nesbitt (2006))*

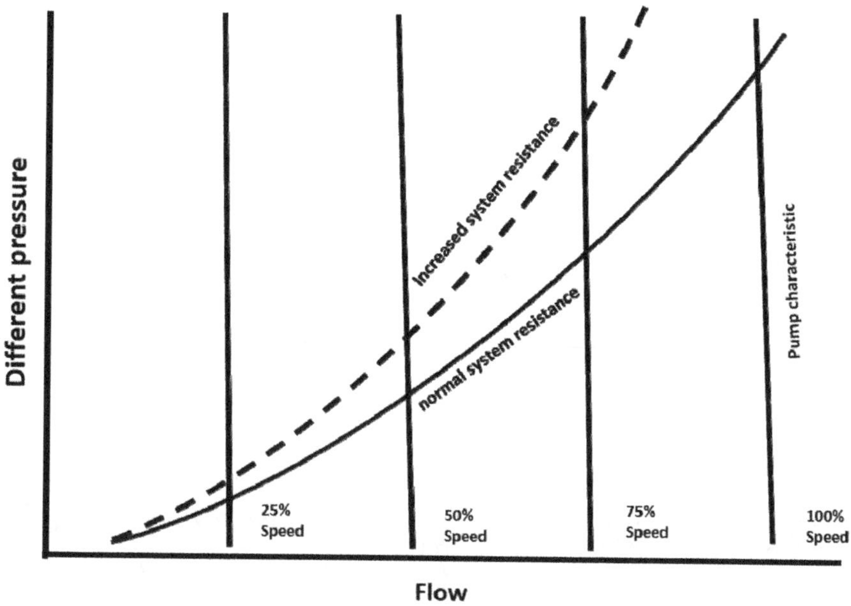

Figure 6.8. *Positive displacement pump and system characteristics (adopted with permission from Nesbitt (2006))*

The interested reader about pump details and design has to resort to Nesbitt (2006) and Karassik et al. (2008).

6.2.2. *Compressors*

The differentiation between fans, blowers and compressors lacks precise delineation. Various types of fans are categorized as having exceptionally low compression ratios. Fans are employed to produce static pressures of up to a maximum of 30 inches of water column. Blowers generally function with compression ratios that do not surpass 1.4 per stage. The maximum compression ratio in machines with numerous stages is 4:1. Compressors are commonly defined as devices that elevate the pressure of a substance.

Figure 6.9 depicts the primary classifications of compressors and their respective explanations. Cam, diaphragm and diffusion compressors are excluded because of their specific use and relatively small sizes.

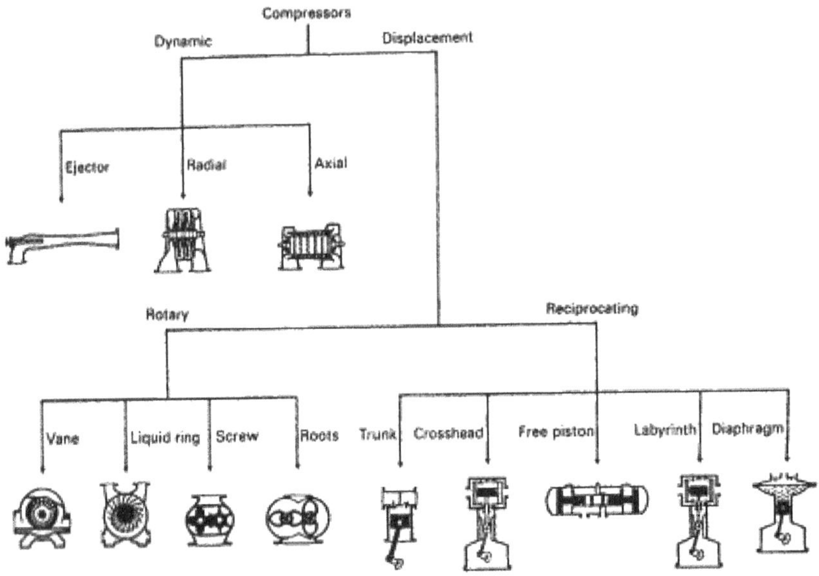

Figure 6.9. *Types of compressors for industrial use*

Positive displacement units are mechanisms that capture and compress consecutive amounts of gas within a confined space, leading to a rise in pressure. Rotary positive displacement compressors are mechanisms that accomplish compression and displacement by means of the direct motion of rotating parts.

Sliding vane compressors belong to the category of rotary positive displacement devices. The system comprises of a rotor that is positioned off-center in a cylindrical casing, and axial vanes that move smoothly in a radial direction within the rotor. The gas trapped between the vanes is squeezed and displaced from its initial location.

Liquid piston compressors are a variant of rotary positive displacement machines that employ water or another liquid as the piston to compress and displace the gas being handled. Two-impeller straight-lobe compressors belong to the category of a rotary positive displacement machinery. The system employs a pair of straight-lobed impellers that collaborate to gather gas and convey it from the intake to the discharge. There is no internal compression.

Helical or spiral lobe compressors belong to the category of rotary positive displacement devices. They are composed of two rotors that mesh together, with each rotor having a spiral shape. These rotors compress and move the gas. Dynamic compressors are a category of rotary continuous-flow machines that employ a rapidly rotating component to enhance the velocity of the gas as it flows through. The acceleration leads to a transformation of the gas's kinetic energy into pressure. This phenomenon occurs both in the rotating component and in the stationary diffusers or blades.

Centrifugal compressors are dynamic devices that use one or more rotating impellers, typically enclosed on the sides, to increase the velocity of gas. The primary gas flow in these compressors is radial. In contrast, axial compressors are mechanical devices that use a bladed rotor to accelerate gas along an axial flow direction. Mixed flow compressors feature an impeller design that combines attributes of both centrifugal and axial compressors.

Gas compression can be achieved through four main methods, divided into two intermittent methods and two continuous flow methods. These terms are descriptive and not related to thermodynamics or work classification:

1. Intermittent method 1: collect consecutive quantities of gas in a confined space, reduce the volume (which increases pressure) and then release the compressed gas from the enclosure.

2. Intermittent method 2: capture incremental quantities of gas in a container, transfer it to the exit point without changing its volume, compress the gas by reversing the flow from the discharge system and finally expel the compressed gas from the container.

3. Continuous flow method 1: use the mechanical force generated by rapidly rotating impellers or bladed rotors to compress the gas, thereby increasing its speed and pressure (velocity is converted into pressure within stationary diffusers or blades).

4. Continuous flow method 2: introduce the gas into a rapidly flowing stream of the same or a different gas (often steam), and convert the resulting high speed of the mixture into pressure using a diffuser.

Compressors employing methods 1 and 2 are classified as positive displacement compressors and fall into the intermittent category. Those using method 3 are referred to as dynamic compressors, while compressors using method 4 are known as ejectors and typically operate at intake pressures below atmospheric pressure.

Modern process facilities are equipped with various large compressors, which can be categorized as either positive displacement or dynamic machines based on specific requirements. In the positive displacement category, the focus is primarily on reciprocating compressors and twin-screw machines, with reciprocating compressors being the older technology.

Rotating compressors operate using the positive displacement principle and employ rotating motion to compress gas. Numerous variations and designs exist within this category, each using distinct mechanical principles and configurations. While rotary compressors are commonly used in petroleum and chemical plants, they are not as prevalent as reciprocating and centrifugal compressors. The rotary compressor segment mainly consists of units designed for moderate discharge pressures and low power consumption. Although some rotary compressors require less maintenance compared to reciprocating compressors, it is important to note that others may have higher maintenance demands due to factors such as service conditions and design features.

Currently, twin-screw rotary compressors with capacities of up to roughly 100,000 m^3/hr are being used successfully and have a proven track record. Increasingly, machines with power inputs over 7,000 horsepower and a proven track record of continuous operation in demanding coke gas settings are becoming increasingly prevalent. Figure 6.10 depicts the typical scope of uses for each compressor.

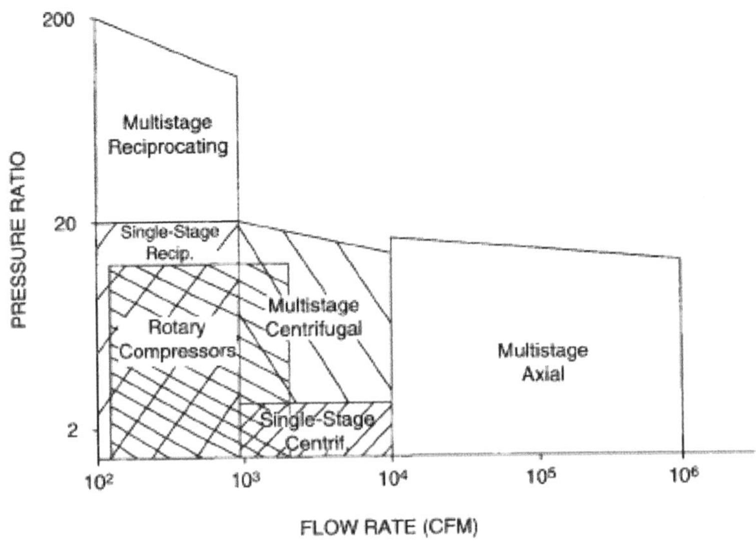

Figure 6.10. *Typical application range of compressors*

For a detailed presentation, including calculations, see Bloch and Hoefner (1996), Bloch (2005), Brown (2005), Bloch and Godse (2006) and Giampaolo (2023).

6.2.3. *Flowmeters*

This section closely follows the KLM Practical Engineering Guidelines for Processing Plant Solutions (KLM Technology Group 2024). Measuring fluid flow can be challenging, as it involves quantifying the mass or volume of the material as it passes through a pipe or conduit. Complications may arise due to the complex nature of fluid flow dynamics. Flow measurements depend on various physical characteristics that are often difficult to define.

This design guideline outlines the methodology for selecting the appropriate flow measurement instrument based on specific applications. It also assists engineers in determining the appropriate dimensions for various flow measuring devices, such as orifice plates, venturi tubes, pitot tubes, nozzles, rotameters, vane meters and positive displacement flowmeters.

The definition section provides clear explanations of all key parameters used in the guideline, enabling readers to understand the significance of these terms. This design guideline delves into the process of selecting different types of flow measurement instruments, including differential pressure flowmeters, variable area flowmeters and positive displacement meters.

The importance of industrial flow measurement has grown over the past 50 years, not only due to its widespread use for accounting purposes – such as the transfer of fluids from suppliers to customers – but also because of its application in production processes. Industries that rely on flow measurement include food and beverage, oil and gas, medical, petrochemical, power generation and water distribution.

In custody transfer metering, recognizing that flow measurement directly correlates to monetary value shifts our perspective significantly. The goal is to convert custody transfer measurements into financial figures with zero error. Quantities for custody transfer are considered absolute when invoiced, and the primary aim of this measurement is to minimize errors so that a mutually agreed-upon quantity can be established for custody exchange. The acceptable tolerance for control measurement is ± 2%, while operational measurement should not exceed ± 5%. In contrast, the target for custody transfer metering is ± 0%.

Flow measurement involves quantifying the volume of a fluid, such as a liquid or vapor, that flows through a pipe, duct or open channel. Flow can be expressed in terms of the rate at which a fluid moves, either by volume (e.g. gallons per minute, cubic meters per minute, cubic feet per minute) or by mass (e.g. kilograms per hour, pounds per hour). It can also be described as the total volume or amount of fluid that moves within a specified time period.

Fluid flow measurement can be categorized into various types, each requiring careful consideration of factors such as accuracy requirements, cost implications and the use of flow information to achieve desired outcomes. Typically, flowmeters measure flow indirectly by assessing a related characteristic, such as the differential pressure across a flow restriction or the velocity of fluid within a pipe. Flow measurement devices rely on various fundamental physical principles to operate effectively.

There are various types of flowmeters available on the market, which can be classified as follows:

i) Differential pressure flowmeters:

- orifice plate;
- venturi tube;
- pitot tube;
- nozzle.

ii) Variable area flowmeters:

- rotameter;
- movable vane meter;
- weir;
- flume.

iii) Positive displacement flowmeters:

- tri-rotor PD meter;
- birotor PD meter;
- piston-type PD meter;
- oval gear PD meter;
- nutating disk-type PD meter;
- sliding-vane-type PD meter;
- roots PD meter;
- CVM meter;
- diaphragm meter.

iv) Turbine flowmeters.

v) Electromagnetic flowmeters.

vi) Ultrasonic flowmeters:

- Doppler;
- transit-time.

vii) Coriolis (mass) flowmeters.

viii) Vortex shedding meters.

Selecting the right flowmeter for a specific application can be challenging due to the wide variety of options available on the market. This process requires a comprehensive evaluation of several factors, including total cost, fluid state, flow conditions, Reynolds number, density, rangeability, mechanical installation constraints and accuracy requirements. Design engineers must establish criteria based on mass, volume (operating standard) or energy. Additionally, it is essential to provide the pressure and temperature of the fluid to the flowmeter and transmitter to account for any variations during the process. An analyzer that measures a specific gravity or density may also be necessary to address fluctuations in stream composition.

The term "precision accuracy" is commonly used in flow measurement, though it is often misunderstood or misapplied. Unfortunately, it has become a marketing tool exploited by both suppliers and consumers of metering equipment. The supplier with the highest accuracy claims may win the bid, even when users may require precision levels beyond what any available measuring device can provide. Historically, "accuracy" is referred to a meter's ability to measure flow correctly, quantified as the ratio of the indicated measurement to the true measurement.

The opposite of uncertainty reflects the maximum potential error within a specified confidence level. Precision is vital for flowmeters due to its role in obtaining accurate measurements, particularly in custody transfers where financial implications are significant. The accuracy of a flowmeter is defined as the maximum discrepancy between the meter's reading and the actual flow rate or total flow. Flowmeter performance is evaluated by comparing its indicated flow rate to a highly accurate value obtained through rigorous calibration procedures. ISO 5725 addresses accuracy in this context.

There are two ways to represent precision, each with distinct meanings:

1. Percentage of the measured value or actual reading: this method determines accuracy based on the indicated tolerance, expressed as a percentage of the measured value or actual reading at various flow rates.

2. Percentage of full scale deflection (FSD): this represents measurement inaccuracy as a percentage of the maximum flow rate that the flowmeter can measure. Generally, the error expressed as a percentage of FSD is smaller than that expressed as a percentage of the actual reading.

The flow range refers to the difference between the lowest and highest flow rates at which a meter can accurately measure and meet specified performance criteria. Rangeability is the ability of a flowmeter to measure a broad spectrum of flow rates, while "meter turndown" denotes the ratio of the maximum flow rate to the minimum flow rate. The specific rangeability of a flowmeter is a crucial consideration during the selection process, with further details on the relative rangeability of different flowmeters discussed in the following section.

The discharge coefficient adjusts the theoretical flow rate equation to account for the effects of the velocity profile, as influenced by the Reynolds number. Actual testing by various organizations has established specific discharge coefficients for different flowmeter shapes, which are critical in determining the configuration of the flow channel. The orifice flow rate is significantly influenced by several factors, including the orifice diameter, pipe diameter, fluid velocity, fluid density and fluid viscosity.

The Reynolds number is a dimensionless quantity in fluid mechanics that characterizes fluid flow over a surface. It is valuable for predicting how a meter will respond to changes in fluid properties, applicable to both gases and liquids. Given the impracticality of conducting extensive studies on every meter for every fluid type, understanding the correlations between different fluid properties is advantageous. A detailed presentation on this subject was provided by Baker (2000).

6.3. Pressure drop in designing equipment

6.3.1. *Pressure drop in single-phase flow*

6.3.1.1. *General pressure drop equation*

The energy requirements for fluid flow through a heat exchanger are directly proportional to the heat transfer coefficient. With adequate pressure, it is possible to achieve high values of the heat transfer coefficient (h), which can lead to smaller transfer areas. When designing equipment, it is essential to compare capital and operating costs to identify the most cost-effective solution.

Performance evaluation should not be limited to calculating the pressure drop within the tubes or around the tube bundle, as this only accounts for a fraction of the flow losses within the system. The calculations must include the effects of connections, extensions, flow restrictions, valves, bends, elbows and baffles located outside the tube bundle.

Data relevant to this topic are presented in the following pages. In subsequent chapters, the term "fluid" will refer to both compressible gases, which exhibit significant density variations, and liquids, which experience minimal density changes and low compressibility. The application of similarity principles is only relevant to substances that conform to Newton's law, meaning their viscosity remains constant regardless of shear.

The subsequent equation for calculating pressure drop is applicable to many flow patterns, such as laminar or turbulent:

$$\Delta p = fa\frac{\rho w^2}{2} \qquad [6.31]$$

with f as the drag coefficient and a as a factor that depends on the nature of the flow problem.

Thus, for flow through pipes and ducts,

$$a = F(l/d_i) \text{ (l = length of pipe)} \qquad [6.32]$$

for flow through valves and pipeline fittings,

$$a = 1; \qquad [6.33]$$

for crossflow over tube bundles,

$$a = F(n, l/d_i) \text{ (n = number of tubes)} \qquad [6.34]$$

The average values for the density ρ and the velocity w in the direction of flow should be inserted into equation [6.31]. If the pressure drop leads to a significant change in density, the calculation should be carried out in multiple steps. This approach is commonly used in practice for a non-isothermal flow as well.

6.3.1.2. *Pressure drop in flow through pipes*

This section follows closely Perry's Handbook (Green and Southard 2019) and VDI Heat Atlas (VDI 2010).

The pressure drop in the pipe flow is determined by

$$\Delta p = f\frac{l}{D}\frac{\rho w^2}{2} \qquad [6.35]$$

The drag coefficient, represented as f, is influenced by the Reynolds number, which characterizes the flow within the tube. The fluid parameters μ (viscosity) and ρ (density), along with the velocity w, correspond to the mean pressure (p'+p'')/2 and the mean temperature $(T_1+T_2)/2$ in the tube. If the flow direction changes significantly, it is advisable to calculate the pressure drop for each segment. Flow is considered laminar when the Reynolds number (Re) is below approximately 2320, while turbulent flow is likely when Re exceeds this threshold.

In the range of Reynolds numbers between approximately 2320 and 8000, flow in smooth-walled tubes can remain laminar if the inlet flow is relatively gentle and the tube intake is adequately polished. As the pipe's surface roughness increases, the transition from laminar to turbulent flow occurs at lower Reynolds numbers, although it never falls below the minimum value of 2320.

The Hagen–Poiseuille law is particularly applicable to smooth tubes made of materials such as glass, brass or copper. It also applies to rough tubes with a roughness coefficient (ϵ) of up to 0.07, which includes slightly rusted steel or cast iron pipes

$$\Delta p = \frac{32\mu w l}{D^2} \text{ or } f = 64/\text{Re} \qquad [6.36]$$

The surface texture of the tube wall is a crucial determinant. Accurate resistance rules can only be applied to surfaces that are technically smooth, such as glass, drawn brass and copper tubes. Blasius states that the given formula is applicable when the Reynolds number is approximately between 3,000 and 100,000

$$f = \frac{0.3164}{\sqrt[4]{\text{Re}}} \qquad [6.37]$$

The Hermann equation is suitable for the Reynolds number range of approximately 2×10^4–2×10^6

$$f = 0.00540 + \frac{0.3964}{\text{Re}^{0.3}} \qquad [6.38]$$

For Reynolds numbers greater than 10^6 (Re > 106), the value of f can be determined using the Prandtl and von Karman equation:

$$\frac{1}{\sqrt{f}} = -0.8 + 2\log\left(\text{Re } \sqrt{f}\right) \quad [6.39]$$

Rough surfaces present significantly greater resistance to flow. In turbulent flow, this resistance is primarily dependent on the surface roughness. However, in the transition zone, it also depends on the Reynolds number (Re). Despite numerous observations, a universal relationship between pressure drop and roughness has not been established, leaving a degree of uncertainty in calculating pressure drops in rough pipes. The following guidelines aim to improve accuracy and are accompanied by brief explanations to ensure proper application.

Roughness can be classified into two main forms:

– extended undulations;

– slight protuberances.

The latter can be either rounded and uniformly distributed or sharp and angular with a random distribution. At low Reynolds numbers, the fluid's boundary layer is typically thick enough to cover all projections. While the pressure drop, in this scenario, is somewhat higher than in a smooth tube, the relationship to Re follows similar principles.

As the Reynolds number increases, the boundary layer becomes thinner, exposing the protuberances in the order of their size, which gradually increases turbulence until the relationship becomes quadratic. Consequently, the shape of the resistance curves derived from measurements provides insight into the roughness.

Additionally, changes in the shape of these curves over extended periods can indicate the thickness and nature of any deposits. The roughness of a tube's inner surface is influenced by its material and generally progresses from undulatory to protuberant in the following sequence: vitreous, bituminized, drawn, sandblasted, galvanized, briefly stored, rusted, fouled after prolonged operation and covered with scale.

Regardless of the numerous forms that roughness can assume and how it is distributed, it can be defined in general terms by a "relative roughness", which is given by

$$\varepsilon = \frac{r}{D}$$

where r is the average height of all the projections. If flow is completely governed by roughness, the quadratic relationship for the resistance applies, and f is independent of Re.

The Prandtl and Karman equation for this case is

$$\frac{1}{\sqrt{f}} = 2\log\left(\frac{1}{\varepsilon}\right) + 1.14 \qquad [6.40]$$

The relationship to both Re and ε is allowed for in the Colebrook and White equation for the transition between flow over smooth and flow over rough surfaces, i.e.

$$\frac{1}{\sqrt{f}} = -2\log\left[\frac{2.51}{\text{Re}\ \sqrt{f}} + \frac{\varepsilon}{3.71}\right] \qquad [6.41]$$

For practical convenience, [6.40] and [6.41] are presented graphically in Figure 6.11.

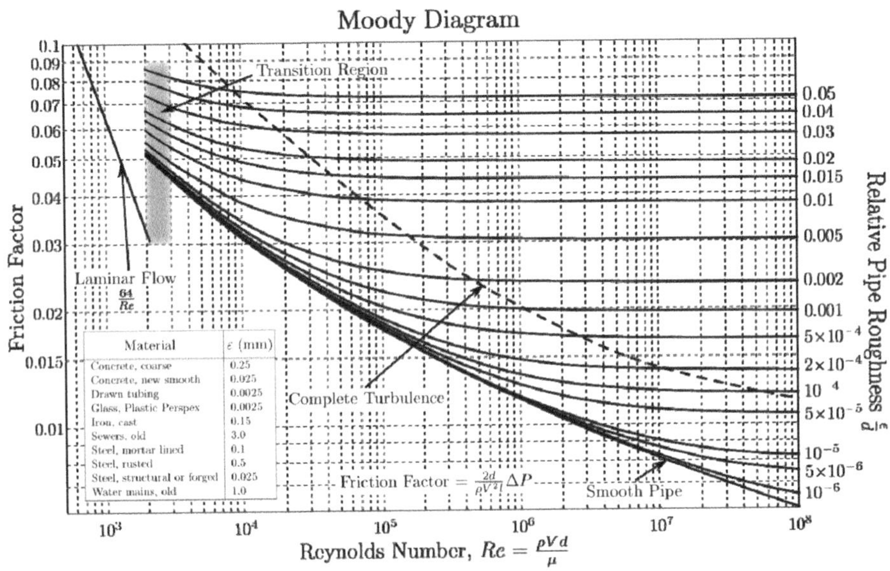

Figure 6.11. *Friction factor, f for rough tubes as a function of the relative roughness r/D and the Reynolds number Re (Moody Diagram, by Beck and Collins, see: en.wikipedia.org/wiki/File:Moody_EN.svg, CC BY-SA 4.0). For a color version of this figure, see www.iste.co.uk/kumararya/chemicalengineering1.zip*

6.3.1.3. *Pressure drop in flow through pipes of noncircular cross-section*

The use of hydraulic diameter (D_h) enables the application of the Blasius formula to any given cross-section. The hydraulic diameter is given by

$$D_h = 4A / Per \qquad [6.42]$$

where A is the cross-sectional area and Per is the length of the participating periphery.

6.3.1.4. *Pressure drop in flow through pipes of changing cross-section*

When fluid enters a narrow passage or moves from the top of a heat exchanger into a group of tubes, it experiences a decrease in pressure, in addition to what is predicted by the Bernoulli equation

$$\Delta p = f_{in} \frac{\rho w^2}{2} \qquad [6.43]$$

The increase in pressure is due to the constriction of the fluid flow at the entrance and the subsequent dissipation of vortices after traveling a distance of approximately 8–10 times the inner diameter of the tube, D. Before achieving the final smooth flow profile, it is necessary to travel down a significantly longer attenuation path, referred to as l_A. If the flow is laminar,

$$l_A \approx 0.13 \mathrm{Re} \ D \qquad [6.44]$$

If the flow is turbulent,

$$l_A \approx 0.0575 \mathrm{Re} \ D \qquad [6.45]$$

Initially, the flow velocity is the same over the entire cross-section. The decrease in velocity within the length l_A is due to the slowdown produced by the increased friction and the kinetic energy required to establish the flow profile at the tube walls. The term f_{in} encompasses all of these effects. If the fluid mass at the tube intake is not moving or moving at a very slow speed, energy needs to be supplied to increase its velocity to the average velocity w, and the total pressure decrease is

$$\Delta p = (1 + f_{in}) \frac{\rho w^2}{2} \qquad [6.46]$$

The topic of pressure drop in the fluid flow due to changes in the cross-sectional area encompasses the following aspects:

- narrowed tubes;
- inflow in tube bundles;
- outflow from vessels;
- continuous reduction in the cross-sectional area;
- abrupt and continuous widening of cross-section;
- orifice, nozzle and venturi meters;
- tees and oblique junctions;
- changes in flow direction, such as bends and elbows;
- valves, including regulating, throttling, cocks, and butterfly valves.

A comprehensive discussion of these topics can be found in the VDI Heat Atlas (VDI 2010).

6.3.1.5. *Pressure drop in one phase flow chemical industry equipment*

Strongly related to the pressure drop in pipes is the pressure drop in heat transfer equipment as heat exchangers. A detailed overview in this area could be found in Kern's *Process Heat Transfer* book (Flynn et al. 2019).

Another significant domain of the single-phase flow pertains to the flow within packed beds. Fixed beds are composed of individual particles that allow the passage of a fluid, whether it be a liquid or gas, while retaining solid accumulations. Packing materials, in contrast, are often permeated by a two-phase mixture of liquids and gases. Contrary to individual components of packing materials such as Raschig rings and saddle bodies, which typically have a high surface area per unit volume and low-pressure drop, the particles in fixed beds are compact in shape and are sometimes referred to as "spherical". For a comprehensive understanding of this subject, see Molerus (1993).

A gas or vapor that passes through a mass transfer column, such as an absorption column or distillation column, has a substantial decrease in pressure, often known as pressure loss. The study conducted by Stichlmair and Fair (1998) provides a comprehensive analysis of the dry pressure loss in tray columns.

6.3.2. Two-phase gas–liquid flow in chemical industry

The calculation of pressure drop in a gas–liquid flow differs greatly from that of a single-phase flow. The main reason for this is the interaction between the two phases, which may be explained by incorporating the void fraction into the equations of mass, momentum and energy conservation.

Areas where this methodology is used encompass the assessment of pressure decreases, the examination of expansion properties and the procedures involving phase transition. The cross-sectional distributions of the phases fluctuate depending on the flow conditions due to their interaction. Moreover, the arrangement of the phases in different sections is also affected by the tube's orientation.

The subsequent text identifies and explains the attributes of the common types of adiabatic tube flows, as denoted and depicted in Figure 6.2(A–E and a–e, correspondingly).

– For the upward movement of fluid through vertical tubes:

(A) Bubble flow: this is distinguished by a substantial number of bubbles that are distributed in a nearly uniform manner across the liquid. However, the liquid phase does adhere to the tube wall.

(B) Slug flow: this refers to a specific type of flow pattern in which gas and liquid phases move in distinct slugs or pockets. Slug flow is characterized by the formation of exceptionally enormous bubbles that can extend to several times the diameter of the tube. Additionally, there are diminutive bubbles present.

(C) Chaotic flow (or churn flow): this refers to a type of fluid flow characterized by a disordered and turbulent motion. In the turbulent flow, there are macroscopic and microscopic bubbles that appear to be randomly dispersed.

(D) Wispy annular flow: this term refers to a certain type of flow pattern. In this particular flow pattern, the liquid is primarily distributed in the vicinity of the tube wall. The gaseous phase is found within the central region of the tube, intermingled with clusters of droplets that have the ability to coalesce into elongated structures.

(E) Annular flow: this refers to the flow pattern in which a thin film of liquid surrounds a gas core, with little or no liquid present in the center of the pipe. The liquid is predominantly dispersed along the tube wall, with only a small number of droplets suspended in the gas phase.

– For horizontal flow:

(a) Bubble flow: this refers to the movement of gas bubbles through a liquid or fluid. The gas phase exists as little bubbles that are primarily dispersed in the upper portion of the tube due to the force of gravity.

(b) Stratified flow: this refers to the movement of fluids in distinct layers or strata, with each layer having different properties such as density or temperature. The gaseous phase is situated above the liquid phase, characterized by the absence of any disturbances at the boundary between the two phases.

(c) Wavy flow: this flow exhibits a wavy pattern. The gaseous phase is situated above the liquid phase. Waves are observed at the boundary between phases in this scenario.

(d) Slug flow: this refers to the intermittent movement of a two-phase mixture, often gas and liquid, in which large gas bubbles or slugs are separated by liquid slugs. The liquid and gas phases are distinct. The waves are sufficiently forceful to occasionally completely fill the whole cross section of the tube. In these regions, there is a heightened prevalence of bubbles in the liquid phase and droplets in the gas phase.

(e) Annular flow: this refers to the flow pattern in which a thin film of liquid surrounds a core of gas or vapor. The gas is flowing within the central region of the tube. The liquid phase wets the tube wall. The gas phase contains droplets. While the entire tube wall is covered with liquid, most of the liquid flows near the bottom of the tube.

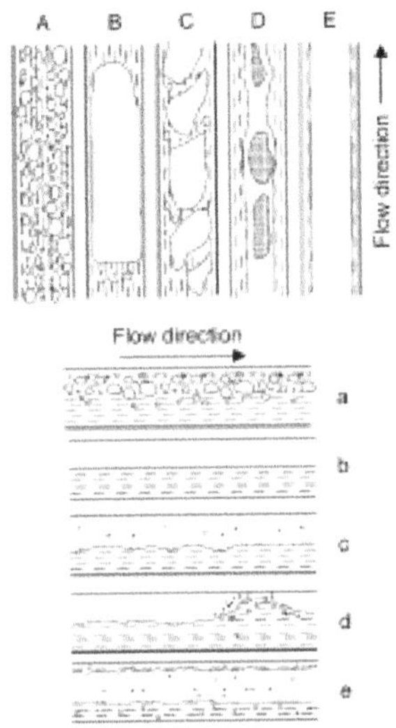

Figure 6.12. *Types of flow patterns (adopted after permission from VDI Heat Atlas (VDI 2010))*

Studying two-phase flow phenomena is essential for determining critical mass flux. When pressure is held constant at the inlet of a nozzle, orifice or duct, and gradually decreased at the exit, the mass flow rate increases. However, once the pressure falls below a certain threshold, further increases in mass flow rate cease. This maximum mass flow rate achievable under these conditions is known as the critical mass flux. This outcome is influenced by several factors, including the intake pressure P_0, temperature T_0, void fraction, gas mass flow fraction and other physical parameters at the inlet.

Another significant aspect of two-phase phenomena involves the simultaneous flow of gases or vapors and liquids in opposite directions. This is often observed in film columns, such as condensers with vertical tube bundles, where liquid phases flow as thin films. Specialized mechanisms are frequently employed to distribute the liquid phases at the entrance of vertically oriented tubes, ensuring that liquid films flow downward along the inner surfaces. The direction of the gas phase flow is dictated by the mass and heat flux, which establish concentration and temperature gradients. In the countercurrent flow, the gas phase moves upward, while in the concurrent flow, it moves downward. Technical applications of this principle include the condensation of vapors in vertically positioned tubes or tube bundles and the selective absorption of components from gaseous mixtures into a flowing liquid layer. In both cases, the opposing flow of phases maximizes concentration and temperature gradients along the liquid films' surfaces.

Packed bed systems also demonstrate significant characteristics of a two-phase flow. Columns filled with packing materials serve as both heat exchangers and mass transport devices. Two common types of packing are arranged-type packings, which include fabric and stacked elements, and dumped-type packing elements. In separation processes such as distillation and absorption, gas or vapor and liquid flow countercurrently. The design of these columns necessitates calculations of gas pressure drop under both conditions – without downward liquid flow and with it – as well as assessments of maximum throughput. This chapter specifically addresses the pressure drop and flooding point of both arranged-type and dumped-type packings, focusing on scenarios where gas or vapor flows upward in countercurrent to the liquid.

Common packing elements include Raschig rings, Pall rings, Berl, Intalox, Novolox and Torus saddles, with dimensions ranging from 5 to 200 mm. These components are typically made from materials such as stoneware, porcelain, glass, stainless steel and plastic, often featuring grill or lattice constructions for packing. For information on pressure drop in packed beds, reference to the VDI Heat Atlas (VDI 2010) is recommended.

6.3.3. *Particle motion in fluids*

The movement of particles – such as solids, droplets and bubbles – within fluids plays a crucial role in various applications across chemical engineering, biotechnology, process engineering and energy conversion. The forces acting on these particles often inform process design and optimization. Examples of such processes include pneumatic conveying, fluidized beds, sedimentation, particle dispersion in stirred tanks, aerodynamic classification of powders and particle separation in cyclones.

"Dispersed flows" refer to two-phase flows where individual particles are uniformly distributed within a continuous phase, whether gaseous or liquid. In diluted dispersed systems, particle transport is primarily governed by fluid dynamic forces acting on the particle surface, as well as external forces such as gravity, buoyancy and electrical influences. In contrast, dense distributed two-phase flows, such as fluidized beds, highlight the significance of fluid dynamic interactions and collisions between particles.

Many straightforward design concepts assume a quasi-steady particle motion. For instance, these methods are used to determine the settling rate of particles in a liquid, which is foundational for sizing sedimentation tanks or centrifuges. They also calculate the particle size required for effective separation in cyclones, based on a balance of forces.

However, when designing processes that predominantly involve turbulent flows, it is essential to solve the complete equations of motion for particles to ascertain critical factors such as particle residence time and residence time distribution. In such chaotic flows, we must consider turbulent eddies along with fluid mechanics and external influences. Analyzing the residence time distribution of particles is vital for optimizing operational parameters in various multiphase chemical reactors, including stirred vessels, bubble columns and loop reactors.

A thorough understanding of particle motion and the forces involved is essential for accurately modeling dispersed multiphase flows using numerical methods. Two primary approaches used in technical and industrial multiphase flows are the two-fluid (or Euler/Euler) method and the Euler/Lagrange method, which is combined with an appropriate turbulence model (Crowe et al. 1997). For additional insights into the numerical computation of the dispersed multiphase flow, refer to Michaelides (2006).

The Lagrangian approach can be employed to analyze particle movement in fluids, involving the resolution of a system of ordinary differential equations that trace the particle's trajectory. This method allows for the determination of changes in particle position and both linear and angular velocity components, necessitating the consideration of all relevant forces acting on the particle. The equation of motion for small spherical particles in a highly viscous still fluid, applicable at low particle Reynolds numbers (known as Stokes flow), can be attributed to the pioneering work of Boussinesq (1885), Basset (1888) and Oseen (1927).

Consequently, this equation of motion is commonly referred to as the BBO equation. Various studies have expanded the BBO equation to accommodate turbulent flows, with Maxey and Riley (1983) providing a detailed derivation of the motion equation for small spherical particles in a nonuniform flow. When factoring in particle rotation, calculating particle trajectories involves solving three ordinary differential equations while neglecting heat and mass transfer phenomena. The vector form of these differential equations used to determine the particle's position and its linear and angular velocities is provided below:

$$\frac{d\vec{x}_p}{dt} = \vec{u}_p,$$

$$m_p \frac{d\vec{u}_p}{dt} = \sum \vec{F}_i,$$

$$I_p \frac{d\vec{\omega}_p}{dt} = \vec{T},$$

where $m_p = \pi/6\rho_p D_p^3$ is the particle mass, $I_p = 0.1 m_p D_p^2$ is the moment of inertia for a sphere, $\vec{F}_i$ represents the different relevant forces acting on the particle and $\vec{T}$ stands for the torque acting on a rotating particle due to the viscous interaction with the fluid.

Analytical solutions for the various forces and torque are only applicable to small particle Reynolds numbers, namely in the Stokes regime. An expansion to higher Reynolds numbers typically involves including empirical correlations, which are obtained from experiments or, more recently, via direct numerical simulations (DNS). Significant applications encompass fluidized beds, cyclones used for the precipitation of solid particles and pneumatic conveying systems.

6.4. Computational fluid dynamics

Currently, a multitude of issues related to the movement of fluids in machinery can be efficiently addressed by the application of numerical techniques, frequently using computer technology via computational fluid dynamics (CFD). CFD use numerical analysis and data structures to analyze and solve issues pertaining to fluid flow. Computers are employed to perform the calculations required for simulating the unhindered motion of fluids, as well as the interaction between fluids (liquids and gases) and surfaces defined by boundary conditions. Artificial intelligence is currently being used to accelerate the advancement of CFD. One of the most robust numerical methods to solve fluid mechanics is the Galerkin finite element method (GFEM). In what follows, a tutorial example of CFD as the GFEM analysis of flow past a surface-mounted obstacle (flow over a step) is given. Finally, the recent advance in the field is summarized.

6.4.1. *A tutorial example: flow over a step*

The flow over a surface-mounted obstacle is the result of the combined flow over a step that faces forward and a step that faces backward. This flow serves as an excellent prototype for various applications in industry, such as the analysis of air pollution, determining optimal locations for power plants, studying wind effects on tall buildings, analyzing turbomachines, investigating car aerodynamics, studying the cooling of electronic equipment and gas-cooled reactors, analyzing meteorology and wind energy applications, studying the flow over road embankments and designing fluid-handling devices (Fragos et al. 1997). This section closely adheres to the research conducted by Fragos et al. (1997).

Figure 6.13 displays the computational domain for the flow over a barrier that is installed on a surface. A fluid with constant density and viscosity, following Newton's laws of motion, moves towards a closed wind tunnel with a uniform velocity u0. As the fluid enters the wind tunnel, it slows down along the wall due to the noslip boundary condition. It then moves towards the surface-mounted obstruction, which is a step. Flow separation and reattachment occur in the area of the obstacle, causing the fluid to eventually leave the computational domain at a position where these flow phenomena diminish.

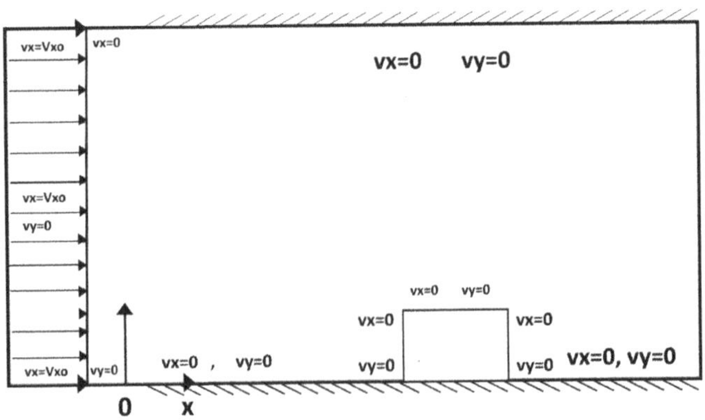

Figure 6.13. *Computational domain for flow over a surface-mounted obstacle (adopted after permission from Fragos et al. (1997)).*

The fluid flow is characterized by laminar behavior, with constant temperature and density, and occurring in a two-dimensional space. The flow is also at a steady state, meaning that it does not change over time. The governing equations for this flow are the dimensionless Navier–Stokes equations

$$\nabla \cdot \mathbf{v} = 0 \qquad [6.47]$$

$$\mathbf{v} \cdot \nabla \cdot \mathbf{v} = -\nabla p + \frac{1}{\mathrm{Re}} \nabla^2 \mathbf{v} \qquad [6.48]$$

Equations [6.47] and [6.48] denote the principles of mass conservation (continuity equation) and momentum, respectively. The velocity vector in the fluid is represented by $v = (v_x, v_y)$, where v_x and v_y are its components in the x- and y-directions, respectively. p represents the pressure, and Re is the Reynolds number, which is calculated using the formula $\rho u_0 H/\mu$. In this formula, u_0 represents the uniform approaching velocity of the fluid from the wind tunnel, H represents the height of the step, ρ is the density and μ stands for the viscosity of the fluid. The pressure, p, has been scaled to a dimensionless value by with the magnitude ρu_0^2. Figure 6.13 illustrates the boundary conditions for this flow. At the outlet, open boundary conditions were considered (Papanastasiou et al. 1992).

The subsequent stage involves discretizing the computing domain. Figure 6.14 displays a sample computational mesh.

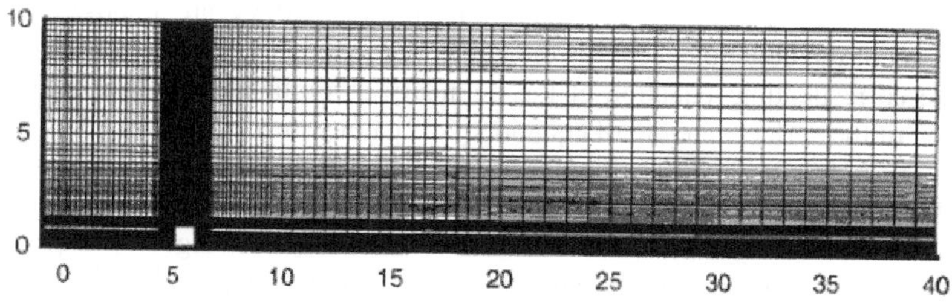

Figure 6.14. *Representative computational mesh for flow over a step (adopted after permission from Fragos et al. (1997))*

The velocities v_x and v_y, as well as the pressure p, in the governing equations [6.47] and [6.48] are expanded using Galerkin basis functions as follows:

$$v_x = \sum_{i=1}^{9} v_{xi}\phi^i \quad , \quad v_y = \sum_{i=1}^{9} v_{yi}\phi^i \quad , \quad p = \sum_{i=1}^{4} p_i\psi^i \qquad [6.49]$$

The basis functions φi are biquadratic, while the basis functions ψi are bilinear. The governing equations, when integrated with the basis functions, produced the following residuals for continuity (R_C^i) and momentum R_M^i :

$$R_C^i = \int_V \nabla \cdot \mathbf{v}\psi^i dV \qquad [6.50]$$

$$R_M^i = \int_V \left[\mathbf{v}.\cdot\nabla\mathbf{v} - \nabla\cdot\left(-p\mathbf{I} + \frac{1}{\mathrm{Re}}\mathbf{T}_1\right)\right]\phi^i dV \qquad [6.51]$$

Here, I is the unit matrix $\mathbf{T}_1 = \nabla\mathbf{v} + (\nabla\mathbf{v})^T$ and $\nabla^2\mathbf{v} = \nabla\cdot\mathbf{T}_1$. By applying the divergence theorem in order to decrease the order of differentiation, equation [6.51] reduces to

$$R_M^i = \int_V \left[(\mathbf{v}.\cdot\nabla\mathbf{v})\phi^i - \nabla\cdot\left(-p\mathbf{I} + \frac{1}{\mathrm{Re}}\mathbf{T}_1\right)\cdot\nabla\phi^i \right]^i dV - \int_S \mathbf{n}\cdot\left(-p\mathbf{I}\frac{1}{\mathrm{Re}}\mathbf{T}_1\right)\phi^i dS \qquad [6.52]$$

Given that essential boundary conditions for vx and vy are imposed on all boundaries of the domain except for the outflow, equation [6.52] will be substituted with the corresponding equation at the boundary as outlined in Figure 6.13. The inclusion of the surface integral of equation [6.52] along the outflow boundary of the domain is essentially the imposition of the free boundary condition (Papanastasiou et al. 1992). The residuals are computed numerically using a nine-point Gaussian integration method. The problem leads to a set of nonlinear algebraic equations, which are solved using the Newton–Raphson iterative approach. The method follows the scheme: $\mathbf{q}^{(n+1)} = \mathbf{q} - \mathbf{J}^{-1}\mathbf{R}(\mathbf{q}^{(n)})$, where the vector of unknowns, represented by q = [vx_1, vy_1, p1 … vx_n, vy_n, pn] is updated using the Jacobian matrix $\mathbf{J} = \partial\mathbf{R}/\partial\mathbf{q}$ of the residuals R with respect to the nodal unknowns q. At each iteration, the linear equations' banded matrix is solved using a frontal solver (Hood 1974). The zeroth-order continuation method has been employed as an initial estimate to progress from one Reynolds (Re) solution to another. In all computer simulations, the Newton approach converged quadratically within four to six iterations.

The model predictions were found in an excellent agreement with available experimental data.

Figure 6.15 depicts the comparison of model prediction with available experimental data.

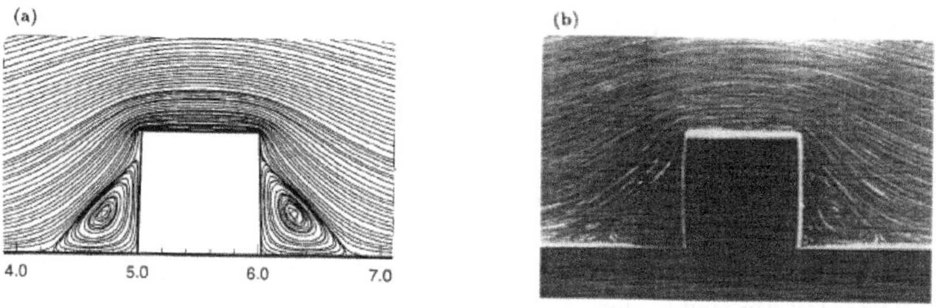

Figure 6.15. *Separated Stokes flow around a surface-mounted obstacle: (a) numerical prediction of this work; (b) experimental data from Taneda (1979) (adopted after permission from Fragos et al. (1997))*

Details about the application of GFEM in CFD could be found in Gresho and Sani (1998) and Zienkiewicz and Taylor (2000). Alternative methods in CFD include the finite volume method (Moukalled et al. 2016; Maliska 2023) and spectral methods (Canuto 1988).

6.4.2. *Recent advancements in CFD*

Given the growing demand for high heat transfer in modern industry, it is crucial to find effective ways to enhance the thermal performance of heat exchangers. The concentric pipe heat exchanger is commonly used in heat transfer applications because of its uncomplicated design and easy production. Technological advancements have prompted various investigations on concentric pipe heat exchangers in recent decades. The literature review conducted by Li et al. (2022) provides a comprehensive analysis of heat transfer enhancement and flow characteristics in concentric pipe heat exchangers, specifically focusing on the double and triple concentric pipe configurations.

Prior studies on heat transfer enhancement in pipe flow indicate that enhancing thermal performance typically results in an associated rise in pressure drop. This factor should be duly considered when applying these findings in actual scenarios. To address this predicament, drag reduction (DR) technology shows great promise and has garnered increasing attention in related research. The work by

Gong et al. (2021) categorizes several DR technologies and provides a comprehensive assessment of their development. The initial section of the study centers on the investigations carried out on additive defect recognition, whereas the subsequent section focuses on defect recognition through surface alteration. The former category is subdivided into polymer DR and surfactant DR, based on the various additives used. The latter is categorized into two groups: DR by implementing superhydrophobic surfaces and DR by implementing slippery surfaces. An in-depth analysis has been conducted on the advancement of the aforementioned technologies, and their impact on heat transfer, as well as their relevance to current and possible future applications, has been summarized.

The process of fluid discharge with intricate rheological properties from industrial tanks, pipes, dispensers or packaging containers through an opening is a multifaceted phenomenon commonly encountered in industry. If not adequately characterized, it can lead to slow discharge and a substantial amount of leftover fluid in the container. The flow rate is significantly influenced by the shape of the opening and the characteristics of the fluid's flow behavior. An essay by Teoman et al. (2022) provides a comprehensive overview of the latest advancements in the flow of intricate fluids via orifices.

Centrifugal pumps are extensively employed in many industries to transport fluids containing particles through pipelines when there is no requirement for high head and flow rate. Slurry erosion and cavitation are intricate and incompletely comprehended processes that arise in centrifugal pumps. Nevertheless, it is possible to mitigate these undesired occurrences to some degree. Accurate design and execution of experiments are necessary to provide reliable predictions about slurry erosion and cavitation. CFD approach is useful in situations when there are restrictions to testing equipment, as it supports analytical solutions and tests. The review paper by Noon et al. (2021) outlines the diverse methods employed by researchers to mitigate slurry erosion and cavitation. Ananya and Patel (2024) conducted a thorough examination of the flow characteristics and erosion wear of several types of equipment used for handling multi-phase flows. El-Emam et al. (2022) investigated computational techniques for analyzing erosion wear in centrifugal pumps.

Mixers are extensively used in chemical and hydrometallurgical procedures. One of the major obstacles in the field of hydrometallurgy process is the agitation caused by the impeller inside mixers. Solvent extraction is a method used to separate components in a solution based on their distribution between two liquids that do not mix together. Dynamic mixers are considered highly effective containers for mixing in processes involving mass, momentum, heat transfer and chemical reactions. One of the benefits of these kinds of containers is their exceptional efficiency. The study conducted by Parvizi et al. (2016) provides an overview of the current conceptual and technological advancements in several dynamic mixers.

In the highly competitive global industrial market, enhancing the capabilities of current manufacturing equipment, such as industrial reheating furnaces (IRFs), leads to improved profitability through increased productivity, enhanced quality and superior energy efficiency. The review by Zhao et al. (2021) provides an overview of the latest advancements and current state of technology in IRFs. The recent applications and developments of zone modeling and CFD simulations in IRFs have received significant attention. This focus specifically emphasizes the integration of slap heating processes characteristics, flow and radiative heat transfer analyses, control strategy and furnace optimization.

Olaleye et al. (2022) conducted a CFD modeling investigation on a vertical gas–liquid separator. In their study, Shah et al. (2016) conducted a comprehensive analysis of the CFD modeling of fluid catalytic cracking riser.

CFD is a practical method for analyzing and resolving issues with many types of equipment used in the pharmaceutical business or healthcare. Given that unit operations handle a significant volume of fluids, even small improvements in performance and efficiency can lead to increased revenues and decreased expenses. The review conducted by Singh et al. (2024) specifically examines the various applications of CFD within the pharmaceutical industry.

The dry powder inhaler (DPI) is a pulmonary formulation that is both patient-friendly and effective, surpassing standard and alternative pulmonary dosage forms in specific illness situations. The creation of DPI (Dry Powder Inhaler) products, however, is more intricate compared to other types of medication, especially in terms of device design and the incorporation of the drug formulation. The review conducted by Dhoble et al. (2024) specifically examines the capabilities of DPI devices in delivering drugs to the lungs, with a particular focus on the design of the devices and the creation of drug formulations.

Twin-screw granulation is now a crucial procedure in powder processing industries and the pharmaceutical sector for producing granules with precise qualities. The review by Arthur and Rahmanian (2024) examines the key elements of the twin-screw granulation process, encompassing the fundamental principles, equipment design, process parameters and simulation approaches.

Szpicer et al. (2023) conducted a comprehensive analysis of the utilization of CFD simulations in the food business. Ajani et al. (2021) discussed the latest developments in using CFD to describe cooling processes and systems in the agrifood business at many scales. In their study, Manoharan and Radhakrishnan (2022) investigated the computational aspects of cold plasma dynamics and its possible utilization in the field of food processing. Malekjani and Jafari (2018) examined the application of CFD in simulating food drying processes.

The placement and dimensions of energy stations (ESs) and pipeline networks (PNs) have a substantial impact on the perception of the economic viability of a regional integrated energy system (RIES). The research by Wang et al. (2021) introduces a strategy for synergistically planning the location and sizing of energy storage systems (ESs) and power networks (PNs). This chapter provides a comprehensive analysis of the fundamental framework of ESs and PNs in RIES, while explicitly identifying the key aspects that need to be taken into account in the planning model.

Agricultural drainage is crucial on a global scale for both food production and soil conservation. It also helps protect investments in agricultural output and irrigation facilities. It has the potential to enhance agricultural output and soil fertility, particularly in areas with inadequate drainage and prolonged water saturation. Both the materials used for subsurface drainage and the procedures employed in their installation have a lengthy historical lineage that can be traced back to prehistoric times. Over time, there has been continual development of new subsurface drainage materials, installation techniques and modernized equipment to leverage technological advancements resulting from research and development. Additionally, improvements have been made in the planning and management of the implementation process. Modern advancements in materials and installation techniques can provide solutions to previously unresolved difficulties, but they can also introduce new challenges. The study conducted by Yannopoulos et al. (2020) examines the progression of fundamental materials used in subsurface drainage and the corresponding procedures employed for their installation, tracing their historical development and adaptation. Additionally, the research explores potential future directions in the design and implementation of drainage systems.

The primary challenge in the 21st century is the phenomenon of global warming, which poses a significant threat to humanity. The building sector, which accounts for 40% of worldwide energy consumption and greenhouse gas (GHG) emissions, plays a crucial role in contributing to this problem. The significance of cooling systems should not be overlooked, since they, along with ventilation and heating systems, collectively contribute about 60% of the energy consumption in buildings. Passive cooling systems have the potential to be a favorable option for decreasing energy usage. Windcatcher is one of the oldest passive cooling systems that is still in use today. Windcatchers can provide sufficient ventilation in buildings by managing pressure differences and using the buoyancy effect. Since most of the previous windcatcher studies assessed the design characteristics, the work of Jomehzadeh et al. (2017) focused on the indoor air quality (IAQ) and thermal comfort aspects.

This review provides a comprehensive analysis and comparison of the many theoretical and experimental techniques used by researchers in diverse case studies to evaluate the IAQ and thermal comfort. Del Hoyo Arce et al. (2018) conducted a comprehensive analysis of the efficient modeling of district heating and cooling networks.

6.5. Conclusions

In this chapter, the fundamental laws regarding flow in chemical industry equipment were derived by using sound principles of continuity mechanics and nonequilibrium thermodynamics. The practical implementation of fluid mechanics based on the Bernoulli equation was also considered. Critical equipment for chemical processes such as pumps, compressors and flowmeters was reviewed. Moreover, the pressure drop implementation in equipment such as fluid flow in pipes, heat exchangers, beds, cyclones, etc., including the multi-phase flow was presented. Finally, CFD tutorial calculations based on the Galerkin finite element method are presented and the recent advance in the area is presented. Emphasis throughout this work was given to give key references for the interested reader. We believe that this work is useful not only in academia but also in industry.

6.6. References

Acrivos, A., Leal, L.G., Snowden, D.D., Pan, F. (1968). Further experiments on steady separated flows past bluff objects. *Journal of Fluid Mechanics*, 34, 25–48.

Ajani C.K., Zhu Z., Sun D.-W. (2021). Recent advances in multiscale CFD modelling of cooling processes and systems for the agrifood industry. *Critical Reviews in Food Science and Nutrition*, 61(15), 2455–2470. doi: 10.1080/10408398.2020.1809992.

Ananya, L. and Patel, V.K. (2024). A comprehensive review on the flow characteristics and erosion wear of the various multi-phase flow handling equipment. *Particulate Science and Technology*, 42(7), 1266–1281. doi: 10.1080/02726351.2024.2347832.

Arthur, T.B. and Rahmanian, N. (2024). Process simulation of twin-screw granulation: A review. *Pharmaceutics*, 16(6), 706. doi: 10.3390/pharmaceutics16060706.

Baker, R.C. (2000). *Flow Measurement Handbook*. Cambridge University Press, Cambridge.

Basset, A.B. (1888). On the motion of a sphere in a viscous liquid. *Philosophical Transactions of the Royal Society A*, 179, 43–69

Bird, R.B., Stewart, W.E., Lightfoot, E.N. (2002). *Transport Phenomena*, 2nd edition. John Wiley & Sons, New York.

Bloch, H.P. (2005). *A Practical Guide to Compressor Technology*. John Wiley & Sons, New York. doi: 10.1002/0471929786.

Bloch, H.P. and Godse, A. (2006). *Compressors and Modern Process Applications*. John Wiley & Sons, New York. doi: 10.1002/0470047208.

Bloch, H.P. and Hoefner, J.F. (1996). Reciprocating compressors and their applications. In *Reciprocating Compressors*, Bloch, H.P. and Hoefner, J.J. (eds). Butterworth–Heinemann, Oxford.

Borremans, M. (2019). Piston compressors. In *Pumps and Compressors*. John Wiley & Sons, New York.

Boussinesq, J.V. (1885) Sur la résistance d'une sphère solide. *Comptes rendus hebdomadaires des séances de l'Académie des sciences*, 100, 935.

Brown, R.N. (2005). *Compressors*, 3rd edition. Gulf Professional Publishing, Houston.

Canuto, C., Yousuff Hussaini, M., Quarteroni, A., Zang, T.A. (1988). *Spectral Methods in Fluid Dynamics*. Springer, Berlin, Heidelberg. doi: 10.1007/978-3-642-84108-8.

Crowe, C.T., Schwarzkopf, J.D., Sommerfeld, M., Tsuji, Y. (1997). *Multiphase Flows with Droplets and Particles*. CRC Press, Boca Raton.

Demirel, Y. (2007). *Nonequilibrium Thermodynamics, Transport and Rate Processes in Physical, Chemical and Biological Systems*, 2nd edition. Elsevier, Amsterdam.

Dhoble, S., Kapse, A., Ghegade, V., Chogale, M., Ghodake, V., Patravale, V., Vora, L.K. (2024). Design, development, and technical considerations for dry powder inhaler devices, *Drug Discovery Today*, 29(5), 103954, doi: 10.1016/j.drudis.2024.103954.

El-Emam, M.A., Zhou, L., Yasser, E., Bai, L., Shi, W. (2022). Computational methods of erosion wear in centrifugal pump: A state-of-the-art review. *Archives of Computational Methods in Engineering*, 29(6), 3789–3814. doi: 10.1007/s11831-022-09714-x.

Flynn, M., Akashige, T., Theodore, L. (2019). *Kern's Process Heat Transfer*. John Wiley & Sons New York.

Fragos, V.P., Psychoudaki, S.P., Malamataris, N.A. (1997). Computer-aided analysis of flow past a surface-mounted obstacle. *International Journal for Numerical Methods in Fluids*, 25(5), 495–512. doi: 10.1002/(sici)1097-0363.

Giampaolo, T. (2023). *Compressor Handbook: Principles and Practice*, 2nd edition. River Publishers, Aalborg.

Gong, W., Shen, J., Dai, W., Li, K., Gong, M. (2021). Research and applications of drag reduction in thermal equipment: A review. *International Journal of Heat and Mass Transfer*, 172, 121152. doi: 10.1016/j.ijheatmasstransfer.2021.121152.

Green, D.W. and Southard, M.Z. (eds) (2019). *Perry's Chemical Engineers' Handbook*, 9th edition. McGraw Hill, New York [Online]. Available at: https://www.accessengineeringlibrary.com/content/book/9780071834087.

Gresho, P.M. and Sani, R.L. (1998). *Incompressible Flow and The Finite Element Method*. John Wiley & Sons, New York.

de Groot, S.R. and Mazur, P. (1984). *Non-Equilibrium Thermodynamics*, 2nd edition. Dover Publications, New York.

Hood, P. (1974). Frontal solution program for unsymmetric matrices. *International Journal for Numerical Methods in Engineering*, 10, 379–399.

del Hoyo Arce, I., Herrero López, S., López Perez, S., Rämä, M., Klobut, K., Febres, J.A. (2018). Models for fast modelling of district heating and cooling networks. *Renewable and Sustainable Energy Reviews*, 82, 1863–1873. doi: 10.1016/j.rser.2017.06.109.

Jomehzadeh, F., Nejat, P., Calautit, J.K., Yusof, M.B.M., Zaki, S.A., Hughes, B.R., Yazid, M.N.A.W.M. (2017). A review on windcatcher for passive cooling and natural ventilation in buildings. Part 1: Indoor air quality and thermal comfort assessment. *Renewable and Sustainable Energy Reviews*, 70, 736–756. doi: 10.1016/j.rser.2016.11.254.

Jou, D., Lebon, G., Casas-Vazquez, J. (2010). *Extended Irreversible Thermodynamics*, 3rd edition. Springer, Berlin.

Karassik, I.J., Messina, J.P., Cooper, P., Heald, C.C. (eds) (2008). *Pump Handbook*, 4th edition. McGraw Hill, New York.

KLM Technology Group (2024). Practical engineering guidelines for processing plant solutions [Online]. Available at: www.klmtechgroup.com [Accessed July 2024].

Kondepudi, D. and Prigogine, I. (2003). Thermodynamics nonequilibrium. In *Encyclopedia of Applied Physics*, 2nd edition, Trigg, G.L. (ed.). Wiley-VCH, Weinheim.

Kuiken, G.D.C. (1994). *Thermodynamics of Irreversible Processes – Applications to Diffusion and Rheology*, 1st edition. John Wiley & Sons, New York.

Li, H., Wang, Y., Han, Y., Li, W., Yang, L., Guo, J., Liu, Y., Zhang, J., Zhang, M., Jiang, F. (2022). A comprehensive review of heat transfer enhancement and flow characteristics in the concentric pipe heat exchanger. *Powder Technology*, 397, 117037. doi: 10.1016/j.powtec.2021.117037.

Malekjani, N. and Jafari, S.M. (2018). Simulation of food drying processes by Computational Fluid Dynamics (CFD); recent advances and approaches. *Trends in Food Science and Technology*, 78, 206–223. doi: 10.1016/j.tifs.2018.06.006.

Maliska, C.R. (2023). *The Finite Volume Method*. Springer, Cham. doi: 10.1007/978-3-031-18235-8.

Manoharan, D. and Radhakrishnan, M. (2022). Computational cold plasma dynamics and its potential application in food processing. *Reviews in Chemical Engineering*, 38(8), 1089–1105. doi:10.1515/revce-2021-0005.

Maxey, M.R. and Riley, J.J. (1983). Equation of motion for a small rigid sphere in a nonuniform flow. *Physics of Fluids*, 26(4), 883–889.

Michaelides, E.E. (2006). *Particles, Bubbles and Drops: Their Motion, Heat and Mass Transfer*. World Scientific Publishing, Singapore.

Molerus, O. (1993). *Principles of Flow in Disperse Systems*. Chapman & Hall, London.

Moukalled, F., Mangani, L., Darwish, M. (2016). *The Finite Volume Method in Computational Fluid Dynamics*. Springer, Cham. doi: 10.1007/978-3-319-16874-6.

Nesbitt, B. (2006). *Handbook of Pumps and Pumping: Pumping Manual International*. Elsevier, Amsterdam.

Noon, A.A., Jabbar, A.U., Koten, H., Kim, M.-H., Ahmed, H.W., Mueed, U., Shoukat, A.A., Anwar, B. (2021). Strive to reduce slurry erosion and cavitation in pumps through flow modifications, design optimization and some other techniques: Long term impact on process industry. *Materials*, 14(3), 521, 1–3. doi: 10.3390/ma14030521.

Olaleye, O.E., Van Ombele, B., Li, X., Adeshina, O.A. (2022). CFD modeling analysis of a vertical gas liquid separator. *Journal of Petroleum Science and Engineering*, 216, 110733. doi: 10.1016/j.petrol.2022.110733.

Oseen, C.W. (1927). *Hydromechanik*. Akademische Verlagsgem, Leipzig.

Papanastasiou, T.C., Malamataris, N., Ellwood, K. (1992). A new outflow boundary condition. *International Journal for Numerical Methods in Fluids*, 14(5), 587–608.

Parvizi, S., Keshavarz Alamdari, E., Hashemabadi, S.H., Kavousi, M., Sattari, A. (2016). Investigating factors affecting on the efficiency of dynamic mixers. *Mineral Processing and Extractive Metallurgy Review*, 37(5), 342–368. doi: 10.1080/08827508.2016.1218868.

Rastogi, R.P. (2007). *Introduction to Non-equilibrium Physical Chemistry*, 1st edition. Elsevier, Amsterdam.

Shah, M.T., Utikar, R.P., Pareek, V.K., Evans, G.M., Joshi, J.B. (2016). Computational fluid dynamic modelling of fluid catalytic cracking riser: A review. *Chemical Engineering Research and Design*, 111, 403–448. doi: 10.1016/j.cherd.2016.04.017.

Singh, V., Shah, K., Garg, A., Dewangan, H.K. (2024). Computational fluid dynamics: Insights and applications in the pharmaceutical field. *Letters in Drug Design and Discovery*, 21(3), 440–450. doi: 10.2174/157018082066 6221117142108.

Stichlmair, J. and Fair, J.R. (1998). *Distillation – Principles and Practice*. Wiley-VCH, New York.

Szpicer, A., Bińkowska, W., Wojtasik-Kalinowska, I., Salih, S.M., Półtorak, A. (2023). Application of computational fluid dynamics simulations in food industry. *European Food Research and Technology*, 249(6), 1411–1430.

Taneda, S. (1979). Visualization of separating Stokes flows. *Journal of the Physical Society of Japan*, 46, 1935–1942 .

Teoman, B., Potanin, A., Armenante, P.M. (2022). The discharge of complex fluids through an orifice: A review. *Chemical Engineering Research and Design*, 179, 346–364. doi: 10.1016/j.cherd.2022.01.023.

The Constructor (n.d.). Laminar flow and turbulent flow [Online]. Available at: https://theconstructor.org/fluid-mechanics/laminar-turbulent-flow/559432/.

VDI (ed.) (2010). *VDI Heat Atlas*. Springer, Berlin, Heidelberg.

Verros, G.D. (2007). On the validity of the Onsager reciprocal relations in multi-component diffusion. *Physics Letters A*, 365(1–2), 34–38.

Wang, Y., Wang, J., Gao, M., Zhang, D., Liu, Y., Tan, Z., Zhu, J. (2021). Cost-based siting and sizing of energy stations and pipeline networks in integrated energy system. *Energy Conversion and Management*, 235, 113958, doi: 10.1016/j.enconman.2021.113958.

Yannopoulos, S.I., Grismer, M.E., Bali, K.M., Angelakis, A.N. (2020). Evolution of the materials and methods used for subsurface drainage of agricultural lands from antiquity to the present. *Water (Switzerland)*, 12(6), 1767. doi: 10.3390/w12061767.

Zhao, J., Ma, L., Zayed, M.E., Elsheikh, A.H., Li, W., Yan, Q., Wang, J. (2021). Industrial reheating furnaces: A review of energy efficiency assessments, waste heat recovery potentials, heating process characteristics and perspectives for steel industry. *Process Safety and Environmental Protection*, 147, 1209–1228. doi: 10.1016/j.psep.2021.01.045.

Zienkiewicz, O.C. and Taylor, R.L. (2000). *The Finite Element Method*, 5th edition. Butterworth–Heinemann, Oxford.

7

Conduction, Convection and Radiation

The three fundamental concepts behind heat transfer methods are convection, radiation and conduction. Understanding the behavior of heat in various physical systems requires understanding the complicated interplay of these three processes.

Conduction is the mechanism through which heat is transported through a solid material because of molecular interactions. It delves into thermal conductivity, Fourier's law and thermal resistance.

The second heat transfer mechanism, convection through fluids, is described in natural and forced convection environments. The significance of fluid properties in the convective transfer of heat is explained, as well as boundary layers and the Nusselt number.

The third process, radiation, is thoroughly explored, focusing on electromagnetic waves' emission, absorption and transmission. In this context, the Stefan–Boltzmann law, Kirchhoff's law and the concept of emissivity are discussed.

In addition, this chapter discusses practical applications and examples of each heat transfer mechanism, emphasizing its importance in engineering, physics and everyday life. This in-depth look at conduction, convection and radiation will give readers a solid basis for understanding and implementing heat transfer principles in various disciplines.

7.1. Fundamentals of heat transfer mechanisms

Heat: heat is energy that flows from a hot object to a cold object.

A body with a high temperature has more thermal energy than one with a low temperature.

Chapter written by Thangallapalli SRINIVAS, G. SRINIVAS, B.K. PUROHIT and A.V. RAGHAVENDRA RAO.

Heat transfers energy from a hot body to a cold one (Diller and Ryan 1998).

1 cal = 4.184 Joule (J)	1 Therm = 10^5 Btu
1 Btu = 1055.056 Joule (J)	1 Celsius heat unit (Chu) = 1.8 Btu

Table 7.1. *Heat Transfer Rate Units*

Heat transfer: when two objects with different temperatures touch, heat is transferred until they reach thermal equilibrium. The temperature gradient catalyzes thermal energy transfer. The scientific field of heat transfer studies heat energy transfer rates (Novikov 1958; Curzon and Ahlborn 1975; Callen 1985; Bejan 1996).

Thermodynamics	Amount of energy transfer equilibrium state condition
Heat transfer	Rate of energy transfer, time to reach a specific temperature Non-equilibrium state condition

Table 7.2. *Definitions of thermodynamics and heat transfer*

7.1.1. *Introduction to conduction, convection and radiation*

Mechanisms of heat flow: heat flow to a cold receiver from a hot source may be due to three mechanisms (Flik et al. 1992), as shown in Figure 7.1.

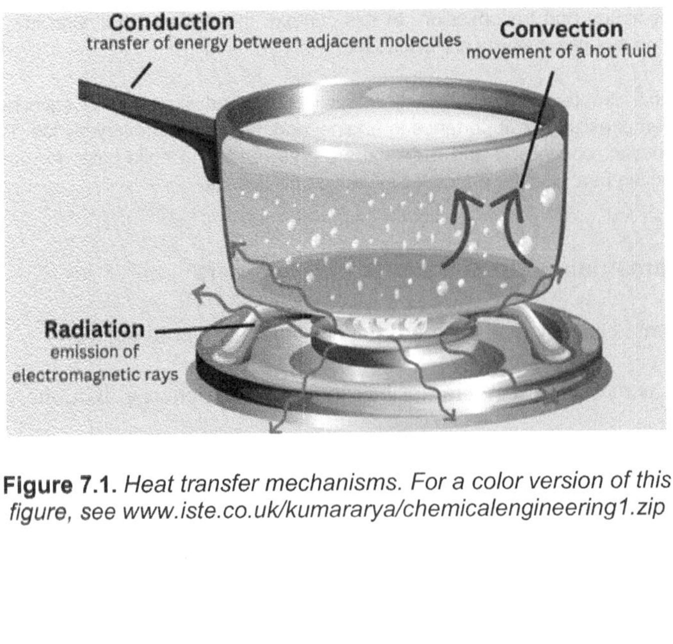

Figure 7.1. *Heat transfer mechanisms. For a color version of this figure, see www.iste.co.uk/kumararya/chemicalengineering1.zip*

– Conduction.

– Convection.

– Radiation.

Without molecule motion, heat is transferred between substances in physical contact.

Lattice vibration and free electron movement cause conduction heat transfer at rest. Conduction occurs when liquid and gas molecules collide, as shown in Figure 7.2. It always vibrates substance molecules. Heat energizes molecules and speeds up vibration. This causes them to collide with nearby molecules. Collisions transfer energy and heat to neighboring molecules.

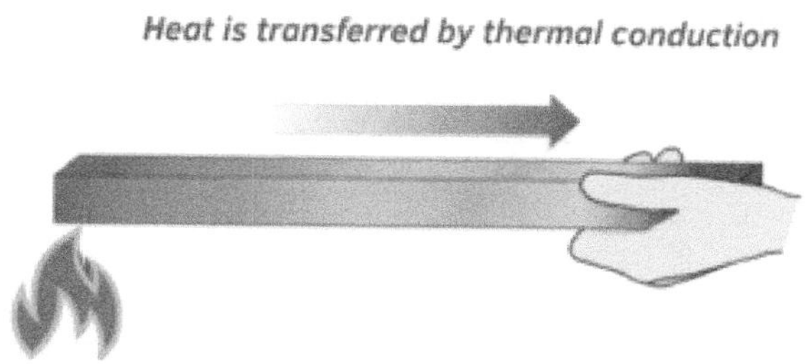

Figure 7.2. *The conduction mode of the heat transfer mechanism. For a color version of this figure, see www.iste.co.uk/kumararya/chemicalengineering1.zip*

EXAMPLE.– heat flow across the metal sheet of the boiler, brick wall of a furnace, heat exchangers metal walls, etc.

Fourier's law of heat conduction:

ASSUMPTION.– one-dimensional (1D) heat flow through an isothermal surface. Heat flow is at a steady state, with constant thermal conductivity and with no accumulation of heat (Chen 1996):

> The area of heat transfer perpendicular to the heat flow and temperature gradient determines the rate of heat transfer through a homogeneous and isothermal surface.

$$Q = -kA\frac{dT}{dx}$$

where Q is the rate of heat transfer (J/s or W); A is the heat transfer area (m^2); $\frac{dT}{dx}$ is the temperature gradient (K/m) and k is the thermal conductivity (W/(m.K)).

Here, $\frac{dT}{dx}$ is negative. The temperature decreases with increasing distance (x). A negative sign in the equation means heat flow is positive. This negative sign indicates a temperature drop in the heat flow direction.

Heat flux is the rate at which heat is transferred per unit area. Its unit is W/m^2.

$$\text{Heat flux (q)} = \frac{Rate\ of\ heat\ transfer\ (Q)}{Area\ (A)}$$

$$\text{Conduction rate of heat transfer, } Q = -kA\frac{dT}{dx}$$

$$Heat\ flux\ (q) = \frac{Q}{A} = -k\frac{dT}{dx}$$

Summary:

– Conduction can occur in gases, solids and liquids, while convection can occur only in fluids (liquid and gas) and radiation in a vacuum.

– Both conduction and convection of heat transfer occur only in the presence of a medium, and for vacuum heat transfer (radiation), no medium is required.

7.1.2. *Relationship to thermodynamics*

Heat transfer and thermodynamics are intimately linked. Here is a breakdown of their relationship with some reference materials.

7.1.2.1. *Heat transfer as a mechanism in thermodynamics*

Temperature differences cause heat transfer. The study of thermodynamics focuses on heat, energy (like work) and matter.

The Zeroth Law of Thermodynamics defines temperature and equilibrium. If two systems are not isolated, they will reach thermal equilibrium in thermal contact (Moran and Shapiro 2004). The First Law of Thermodynamics states the conservation of energy. A closed system's change in internal energy is equal to the heat that is transferred into it (positive value) minus the work that the system does on its surroundings (Cengel and Boles 2018). This law emphasizes heat transfer as a means of energy transformation.

7.1.2.2. *Heat transfer equations in thermodynamics*

The equation for the first law is $\Delta U = Q - W$, where ΔU is the internal energy change, Q is the heat transfer and W is the system work. This equation uses heat transfer (Q) as a quantifiable term.

Specific thermodynamic processes (isobaric, isothermal, adiabatic) involve specific relationships between heat transfer, work and other thermodynamic properties such as pressure and volume (Cengel and Boles 2018). Heat transfer calculations become part of analysing these processes.

7.2. Conduction

7.2.1. *Fourier's heat conduction law*

In heat transfer, Fourier's law describes heat conduction. The negative temperature gradient and the area normal to the heat flow direction determine heat transfer through a material. Here is a detailed breakdown of the equation with a book reference (Flik et al. 1992):

The equation: the mathematical expression for Fourier's Law in one dimension is:

$$q = -k\,(dT/dx)$$

where:

– q (W/m2) represents the heat flux;

– k (W/m·K) represents the material thermal conductivity;

– dT/dx represents the temperature gradient (K/m).

7.2.2. *The matter's thermal properties*

7.2.2.1. *Thermal conductivity*

The thermal conductivity constant is Fourier's law of heat conduction's proportionality constant. A substance's thermal conductivity measures its heat transfer (Klemens 1969; McLaughlin 1969; Foust 1972).

$$k = \frac{Q}{A\,(^{\Delta T}/_{\Delta x})}$$

Thermal conductivity measures how well heat is transferred by conduction through a medium, taking into account its area, thickness and temperature.

Unit and dimension of thermal conductivity (k)

$$Unit\ of\ k = \frac{Q}{A\,(^{\Delta T}/_{\Delta x})} = \frac{(^{J}/_{s})}{m^2.(^{K}/_{m})} = \frac{W}{m.K};\qquad Dimension\ of\ k = \frac{W}{m^2.(^{K}/_{m})} = \frac{kg.m}{s^3.K} = \frac{M.L}{T^3.\theta} = MLT^{-3}\theta^{-1}$$

Temperature dependency of thermal conductivity (Bergman et al. 2011)

Greater value of k	High heat transfer rate (Q)	Good conductor/carrier of heat
The lower value of k	Low heat transfer rate (Q)	Good insulator of heat

Table 7.3. *Best conductors of heat: the higher thermal conductivity value*

Silver = 420; Copper = 395; Gold = 302; Aluminium = 210 $\frac{W}{m.K}$.

Best insulators of heat: the materials with thermal conductivity less than 0.2 $\frac{W}{m.K}$ are called insulators. These materials – such as asbestos, glass wool and cork – reduce the rate at which heat is transferred.

Thermal conductivity (k) changes with respect to temperature. k = f (temperature).

– value of thermal conductivity in increasing order:

Gases < Water < Ice < bricks < low carbon stell < Aluminium < Copper

– for metals, the value of thermal conductivity generally changes with respect to temperature. For plane metal wall: $k = k_0(1 + \beta T)$.

For spherical and cylindrical metal:

$$k = k_0(1 + \alpha T + \beta T^2)$$

where k is at temperature T (K), thermal conductivity; k_0 is at temperature 0°C (273.15 K), thermal conductivity; T is the absolute temperature in Kelvin (K); α and β are the coefficient of thermal conductivity.

	k Value	**Range of k values**	**When the value of temperature (T)**	**The value of thermal conductivity (k)**
Solid	Higher	10 to 400 $\frac{W}{m.K}$	T increases (↑)	k decreases (↓)
			T decreases (↓)	k increases (↑)
			Except for Uranium	
Liquid	Low	0.1 to 1 $\frac{W}{m.K}$	T increases (↑)	k decreases (↓)
			T decreases (↓)	k increases (↑)
			Except for Water	
Gas	Lower	0.01 to 0.1 $\frac{W}{m.K}$	T increases (↑)	k increases (↑)

Table 7.4. *Effect of temperature on thermal conductivities of solids, liquids and gases*

EXAMPLE.–

A plane wall metal has a thermal conductivity of 10 W/(m. K) at 0°C and a coefficient of thermal conductivity of 0.8. Calculate the thermal conductivity of the metal at 30°C.

– k_0 = thermal conductivity at 0°C (273.15 K) = 10 W/(m. K);

– T = absolute temperature in Kelvin 30°C = 273.15 + 30 = 303.15 K;

– β = coefficient of thermal conductivity = 0.8.

For metals, thermal conductivity at a temperature of 303.15 K: $k = k_0(1 + \beta T)$

$$k = 10 \times (1 + 0.5 \times 303.15) = 2435.2\ W/(m.K)$$

7.2.3. Heat conduction through solids and composite materials

Heat conduction through a single-layer wall (plane wall) in a steady state

Let us consider a solid medium with having a thickness of x, a cross-sectional area of A and having a constant thermal conductivity (k). This type of single-layer solid medium (as shown in Figure 7.3) is also known as a plane wall (Geankoplis 1993).

ASSUMPTION.– 1D heat flow through an isothermal surface. Heat flow is at a steady state, with constant thermal conductivity and with no accumulation of heat.

If both sides of the plane wall are kept at T1 and T2, with T1 > T2, then heat will flow from hot to cold.

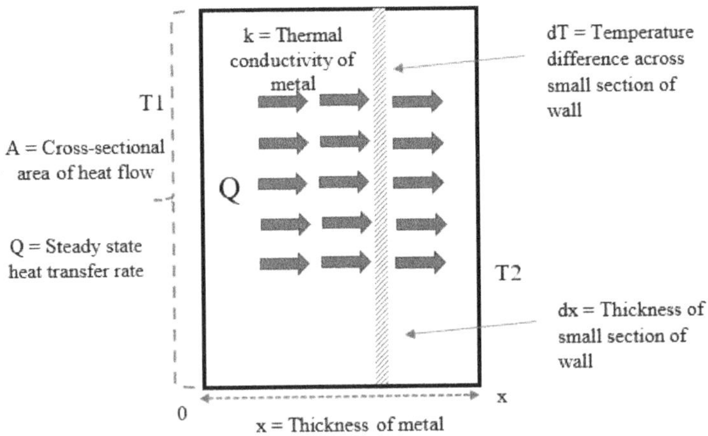

Figure 7.3. *Heat conduction through the plain wall*

Let us consider a very thin element of plane wall with thickness "dx" and the temperature difference across this thin wall is "dT". Then, according to Fourier's law, the rate of heat flow through this thin wall will be

$$Q = -kA\frac{dT}{dx}$$

$$Qdx = -kAdT$$

Integrating both sides of the above equation (to calculate the overall rate of heat flow throughout the overall thickness),

$$\int Qdx = \int -kAdT$$

$$Q\int_{0}^{x} dx = -kA\int_{T1}^{T2} dT$$

$$Q\int_{0}^{x} x^{0}dx = -kA\int_{T1}^{T2} T^{0}dT \left[x^{0} = 1;\ T^{0} = 1;\ \int x^{n}dx = \frac{x^{n+1}}{n+1}\right]$$

$$Q\, x\}_{0}^{x} = -kA\, T\}_{T1}^{T2}$$

$$Q\,(x-0) = -kA\,(T2 - T1)$$

$$Q\,x = kA\,(T1 - T2) = kA\,\Delta T\ [\Delta T = Overall\ temperature\ difference]$$

$$Q = kA\frac{\Delta T}{x} = \frac{\Delta T}{\left(x/_{kA}\right)} = \frac{\Delta T}{R}$$

$$Q = \frac{\Delta T}{\left(x/_{kA}\right)} = \frac{\Delta T}{R}$$

$$Rate\ of\ heat\ transfer\ (Q) = \frac{Driving\ force\ (\Delta T)}{Thermal\ resistance\ (R)}$$

where R is the thermal resistance of medium = $x/_{kA}$; unit = $K/_{W}$.

The inverse of thermal resistance is referred to as conductance.

$$C = 1/_{R} = kA/_{x};$$

– Q (J/s or W) is the steady-state heat flow per unit time;

– k (W/(m. K)) is the material thermal conductivity;

– A (m^2) is the heat transfer area;

– x (m) is the wall thickness;

– ΔT (K) is the temperature difference between hot and cold surface.

$$Q = kA\frac{\Delta T}{x} = \frac{\Delta T}{\left({}^{x}/_{kA}\right)} = \frac{\Delta T}{R}$$

EXAMPLE:–

Question: the heat loss rate through a red brick wall with 0.70 W/(m.K) thermal conductivity. The wall measures 5 m long, 4 m high and 250 mm thick. If the wall surface is kept at 100, 30°C, again, under identical conditions, calculate heat loss when the wall is 100 mm and 500 mm thick.

Answer:

Heat transfer area = A = 5×4 m^2 = 20 m^2; wall thickness = Δx = x = 0.25 m.

Thermal conductivity of wall = k = 0.70 W/(m.K); wall surface temperatures = T1 = 100°C = 373 K and T2 = 30°C = 303 K; temperature difference = ΔT = T1 - T2 = 373 – 303 = 70 K and heat loss rate in 0.25 m thick wall

$$Q = -kA\frac{dT}{dx} = kA\frac{\Delta T}{\Delta x} = kA\frac{T1 - T2}{x} = 0.7 \times 20 \times \frac{70 \ \ W.m^2.K}{0.25 \ \ m.K.m}$$
$$= 3{,}920 \ W$$

Again, the rate of heat loss for 100 mm and 500 mm thick walls are:

Rate of heat loss in 0.1 m thick wall, $Q = kA\frac{T1-T2}{x} = 0.7 \times 20 \times \frac{70 \ W.m^2.K}{0.1 \ m.K.m} = 9{,}800 \ W$.

Rate of heat loss in 0.5 m thick wall, $Q = kA\frac{T1-T2}{x} = 0.7 \times 20 \times \frac{70 \ W.m^2.K}{0.5 \ m.K.m} = 1{,}960 \ W$.

If at the same conditions, the thickness of the wall will be varied to 0.1 m and 0.5 m.

The heat transfer rate will be observed as 9,800 W and 1,960 W.

Wall thickness (x)	Heat loss rate (Q)
100 mm	9,800 W
250 mm	1,960 W
500 mm	1,960 W

Table 7.5. *Effect of wall thickness on heat transfer rate*

Multi-layered composite wall heat conduction under steady-state conditions.

A composite wall is formed by the combination of layers of different materials in parallel or series or a combination of both (Bergman et al. 2011).

Let us consider three different plane walls with names 1, 2 and 3 with having the same cross-sectional area (A) (shown in Figure 7.4). Let us assume that these three different plane walls have different thermal conductivities (k1, k2 and k3) and different thicknesses (x1, x2 and x3), respectively.

ASSUMPTION.– heat transfer in one dimension via an isothermal medium. Heat flow is at a steady state (Q), with constant thermal conductivity. There is no accumulation and generation of heat.

Fourier's law estimates the first layer's heat transfer (Q1) to be

$$Q1 = -kA\frac{dT}{dx} = \frac{\Delta T1}{R1} = \frac{\Delta T1}{\left({}^{x1}/_{k1A}\right)} \; or \; \Delta T1 = Q1R1$$

Similarly, for the wall's second and third layers, heat transfer rate will be

$$Q2 = -kA\frac{dT}{dx} = \frac{\Delta T2}{R2} = \frac{\Delta T2}{\left({}^{x2}/_{k2A}\right)} \; or \; \Delta T2 = Q2R2$$

$$Q3 = -kA\frac{dT}{dx} = \frac{\Delta T3}{R3} = \frac{\Delta T3}{\left({}^{x3}/_{k3A}\right)} \; or \; \Delta T3 = Q3R3$$

where the thermal resistances for the first, second and third plane wall are R1, R2 and R3, respectively. Now, let a composite wall medium be formed by the combination of these three layers in series.

$$Overall\ rate\ of\ heat\ transfer\ (Q) = \frac{Overall\ temperature\ difference\ (\Delta T)}{overall\ thermal\ resistance\ (R)}$$

$$\Delta T = QR$$

Overall or effective thermal resistance of composite wall $= R = \frac{x}{kA}$;

– x = total thickness; A = heat transfer area of composite wall;

– ΔT = T1 – T2 = overall temperature drop due to the composite layer;

– k = overall or effective thermal conductivity of composite wall.

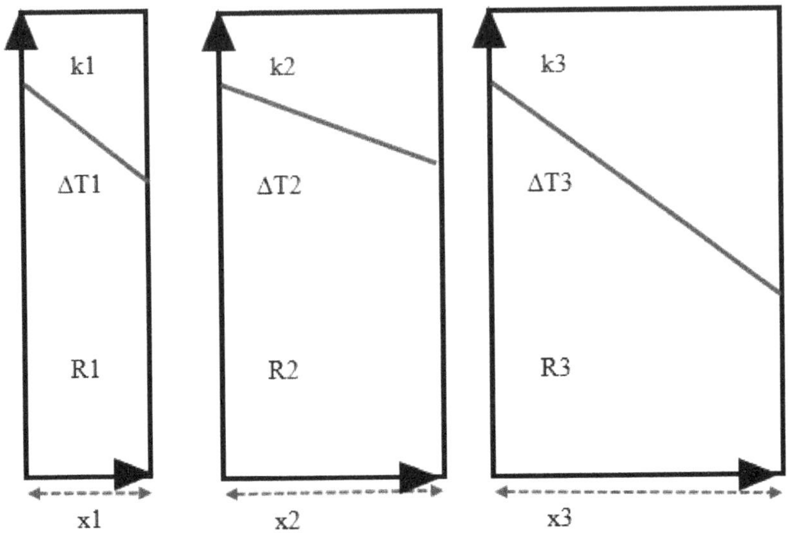

Figure 7.4. *Heat conduction under steady-state conditions through multiple-layer wall*

The composite wall is made by combining the three plane walls in series (as shown in Figure 7.4). Let T3 and T4 be the surface temperatures on both sides of the middle layer wall. T1 > T2 > T3 > T4 is the steady-state heat transfer temperature profile. Let ΔT1, ΔT2 and ΔT3 represent the observed temperature drop for each layer.

Then, the overall temperature drop (ΔT = T1 – T2) due to the composite layer will be equal to the sum of temperature drop observed by individual plane walls.

$$\Delta T = \Delta T1 + \Delta T2 + \Delta T3$$

$$(T1 - T2) = (T1 - T3) + (T3 - T4) + (T4 - T2)$$

$$\Delta T = QR = Q1R1 + Q2R2 + Q3R3$$

The assumption is made that the heat is transferring in a steady state through the walls. Then, the rate of heat transfer will be the same throughout the composite wall.

$$Q1 = Q2 = Q3 = Q \text{ (Steady − state heat transfer condition)}$$

Now, $\Delta T = QR = QR1 + QR2 + QR3 = Q(R1 + R2 + R3)$

$$\Delta T = QR = Q(R1 + R2 + R3)$$

Here, the overall resistance $= R = R1 + R2 + R3$
$= \left(\frac{x1}{k1A} + \frac{x2}{k2A} + \frac{x3}{k3A}\right)$

Then, heat transfer rate through composite wall will be

$$Q = \frac{\Delta T}{R} = \frac{\Delta T}{(R1 + R2 + R3)} = \frac{\Delta T}{\left(\frac{x1}{k1A} + \frac{x2}{k2A} + \frac{x3}{k3A}\right)}$$

NOTE.– when individual walls are connected in series, then the overall thermal resistance for the composite wall will be the sum of individual wall thermal resistances.

$$R = R1 + R2 + R3 \text{ or } \frac{x}{kA} = \frac{x1}{k1A1} + \frac{x2}{k2A2} + \frac{x3}{k3A3}$$

When individual walls are connected in parallel, then the overall thermal resistance for composite wall will be the sum of the reciprocal of individual wall thermal resistances.

$$\frac{1}{R} = \frac{1}{R1} + \frac{1}{R2} + \frac{1}{R3} \text{ or } \frac{kA}{x} = \frac{k1A1}{x1} + \frac{k2A2}{x2} + \frac{k3A3}{x3}$$

Here, x is the total length and k is the overall thermal conductivity of the composite wall.

Here, x1, x2, x3 are the individual thickness, k1, k2, k3 are the individual thermal conductivity and A1, A2, A3 are the individual heat transfer area for walls, respectively, in which a composite wall is made.

NOTE.– the heat transfer is constant under steady state.

$$Q1 = Q2 = Q3 = Q$$

$$\frac{\Delta T1}{R1} = \frac{\Delta T2}{R2} = \frac{\Delta T3}{R3} = \frac{\Delta T}{R} = Q \text{ or}$$

$$\frac{T1 - T3}{R1} = \frac{T3 - T4}{R2} = \frac{T4 - T2}{R3} = \frac{T1 - T2}{R} = Q$$

Steady-state heat conduction through cylinder (single layer)

Let us consider a thick-walled hollow cylinder with a length of "L" as shown in Figure 7.5 (cross-sectional view). Let the inside and outside radii of the hollow cylinder be "ri" and "ro", respectively. Let k be the thermal conductivity of the material in which the cylinder is made. Inside the cylinder (radius "ri") and outside (radius "ro"), the temperatures are "Ti" and "To".

Assume that Ti > To. Then, heat will flow from the higher temperature surface or inner side (at radius ri) to the lower temperature surface or outer side (at radius ro) of the cylinder.

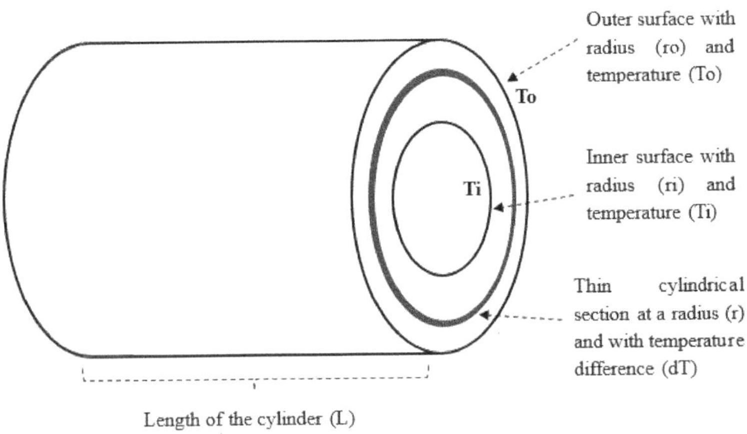

Figure 7.5. *Steady-state heat conduction through cylinder*

ASSUMPTION.– 1D heat flow through an isothermal surface. Heat flow is at a steady state, with constant thermal conductivity. There is no accumulation and generation of heat.

A very thin cylindrical element of thickness "dr" inside the cylinder at radius "r" is considered. Following Fourier's law, heat flow through this thin wall cylinder at radius "r" will be

$$Rate\ of\ heat\ transfer = \frac{Driving\ force}{Thermal\ resistance}$$

$$Thermal\ resistance\ (R) = \frac{x}{k\ A_{lm}} = \frac{(ro - ri)}{k\ 2\pi r_{lm} L} = \frac{(ro - ri)}{k\ 2\pi \frac{(ro - ri)}{\left(\ln \frac{ro}{ri}\right)} L}$$

$$= \frac{\left(\ln \frac{ro}{ri}\right)}{k\ 2\pi L}$$

Steady-state heat conduction through cylinder (composite layer)

Let us consider a composite layer of cylinder that is formed by the combination of a series of three layers of different metals, as shown in Figure 7.6 (say 1, 2 and 3). Let the thermal conductivities of them be k1, k2 and k3, respectively. The different layers have different thickness as x1, x2 and x3, respectively. Let the inside and outside radii of these cylinders be (r1, r2), (r2, r3) and (r3, r4), respectively (Geankoplis 1993).

Then, the thickness of cylinders will be, x1 = r2 - r1; x2 = r3 – r2; x3 = r4 – r3.

Let T1 > T2 on both sides of the composite wall. From T1 to T2, heat will flow. Both sides of the middle layer wall have surface temperatures T3 and T4. The cross-sectional view of steady-state heat transfer temperature profile is (T1 > T2 > T3 > T4).

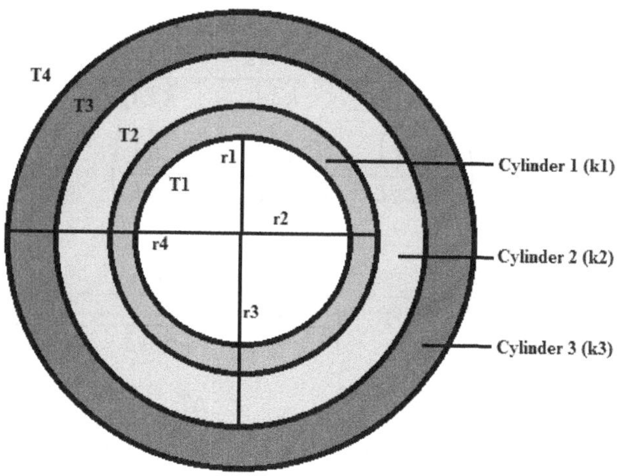

Figure 7.6. *Steady-state heat conduction through cylinder (composite layer). For a color version of this figure, see www.iste.co.uk/kumararya/chemicalengineering1.zip*

$$Overall\ rate\ of\ heat\ transfer\ (Q) = \frac{Overall\ temperature\ difference\ (\Delta T)}{overall\ thermal\ resistance\ (R)}$$

Overall temperature drop (ΔT) in the composite layers will be, $\Delta T = QR = \Delta T1 + \Delta T2 + \Delta T3$

Let ΔT1, ΔT2 and ΔT3 be the respective observed temperature drop for the individual layers.

$$\Delta T = (T1 - T2) = (T1 - T3) + (T3 - T4) + (T4 - T2)$$

NOTE.– the heat transfer is constant under steady-state conditions,

$$\frac{\Delta T1}{R1} = \frac{\Delta T2}{R2} = \frac{\Delta T3}{R3} = \frac{\Delta T}{R} = Q$$

where R1, R2 and R3 denote the thermal resistances of the initial, secondary and third cylindrical walls, respectively.

$$\Delta T = QR = QR1 + QR2 + QR3 == Q(R1 + R2 + R3)$$

$$Here, the\ overall\ resistance =$$

$$R = R1 + R2 + R3 = \left(\frac{x1}{k1A_{lm,1}} + \frac{x2}{k2A_{lm,2}} + \frac{x3}{k3A_{lm,3}}\right)$$
$$= \left(\frac{\left(\ln\frac{r2}{r1}\right)}{k1\ 2\pi L} + \frac{\left(\ln\frac{r3}{r2}\right)}{k2\ 2\pi L} + \frac{\left(\ln\frac{r4}{r3}\right)}{k3\ 2\pi L}\right)$$

$$Then, Q = \frac{\Delta T}{(R1 + R2 + R3)} = \frac{\Delta T}{\left(\frac{x1}{k1A_{lm,1}} + \frac{x2}{k2A_{lm,2}} + \frac{x3}{k3A_{lm,3}}\right)}$$

$$or, Q = \frac{\Delta T}{(R1 + R2 + R3)} = \frac{\Delta T}{\left(\frac{\left(\ln\frac{r2}{r1}\right)}{k1\ 2\pi L} + \frac{\left(\ln\frac{r3}{r2}\right)}{k2\ 2\pi L} + \frac{\left(\ln\frac{r4}{r3}\right)}{k3\ 2\pi L}\right)}$$

Here, L is the length of cylinder and k is the overall thermal conductivity of the composite cylinders.

Here, x1, x2, x3 are the individual thickness, k1, k2, k3 are the individual thermal conductivity and $A_{lm}1$, $A_{lm}2$, $A_{lm}3$ are the individual logarithmic heat transfer area for each cylinder, respectively, in which a composite cylinder is made.

Steady-state heat conduction through sphere (single layer)

Figure 7.7 (cross-sectional) shows a hollow sphere with thick walls. “ri” and “ro” represent the hollow sphere’s inside and outside radii. Sphere material thermal conductivity is k (Zeng et al. 1995).

The temperatures are “Ti” and “To,” respectively, inside and outside the sphere (radius “ri” and “ro”). Assume that Ti > To. Then, heat will flow from the higher temperature surface or inner side (at radius ri) to the lower temperature surface or outer side (at radius ro) of the sphere.

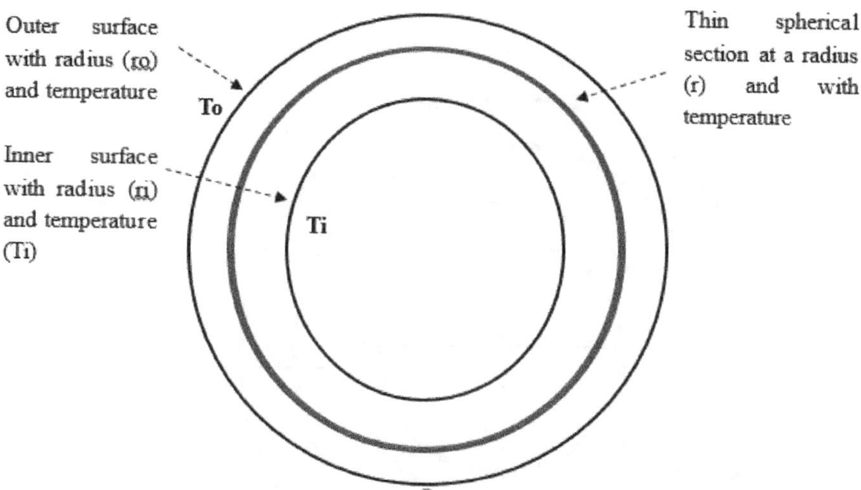

Figure 7.7. *Steady-state heat conduction through sphere (single layer)*

ASSUMPTION.– 1D heat flow through an isothermal surface. Heat flow is at a steady state, with constant thermal conductivity. There is no accumulation and generation of heat.

Let us consider a thin spherical element of thickness "dr" inside the sphere at radius "r". Fourier's law states that heat flow through this thin wall sphere at radius "r" is

$$\left[A_{gm} = 4\pi r_{gm}^2 = (4\pi)(ri.ro) = (4\pi ri)(4\pi ro) = geomatric\ mean\ area\right]$$

$$Q = \frac{k\,A_{gm}}{x}\,\Delta T\ [ro - ri = x = thickness\ of\ sphere]$$

$$Q = \frac{\Delta T}{\left(x/k\,A_{gm}\right)} = \frac{\Delta T}{R}\ or\ Rate\ of\ heat\ transfer$$

$$= \frac{Driving\ force}{Thermal\ resistance}$$

$$Thermal\ resistance\ (R) = \frac{x}{k\,A_{gm}} = \frac{(ro - ri)}{k\,4\pi r_{gm}^2} = \frac{(ro - ri)}{k\,4\pi ri.ro}$$

Steady-state heat conduction through sphere (composite layer)

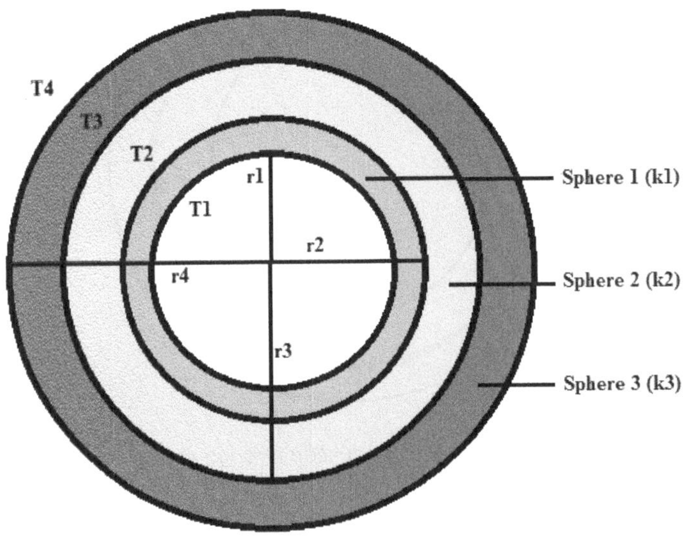

Figure 7.8. *Steady-state heat conduction through cylinder (composite layer). For a color version of this figure, see www.iste.co.uk/kumararya/chemicalengineering1.zip*

Let us consider a composite layer of sphere that is formed by the combination of a series of three layers of different metals (say 1, 2 and 3), as in Figure 7.8. Let the thermal conductivities of them be k1, k2 and k3, respectively. The different layers have different thickness as x1, x2 and x3, respectively. Let the inside and outside radius of these cylinders be (r1, r2), (r2, r3) and (r3, r4), respectively (Zeng et al. 1995).

$$Q = \frac{\Delta T}{(R1 + R2 + R3)} = \frac{\Delta T}{\left(\frac{(r2 - r1)}{k1\ 4\pi r2r1} + \frac{(r3 - r2)}{k2\ 4\pi r3r2} + \frac{(r4 - r2)}{k3\ 2\pi r4r2}\right)}$$

7.3. Convection heat transfer

Convection heat transfer is brought on by intermolecular forces and bulk molecule transport. The transfer of thermal energy from solid surfaces to bulk fluid media or between gases and liquids is known as convection. Conduction and circulation are involved in convection heat transfer. Heat is transferred by conduction between a solid surface and a fluid in the absence of fluid flow. Molecules with high thermal energy move more quickly through the fluid. Thus, convective heat transfer is enhanced by bulk fluid motion. Convection heat transfer processes include boiling and condensation (McCabe et al. 1993; Dutta 2023). Example of heat transfer by convection are heating water in cooking pans or through electric rods, cooling of an object in a room, mixing hot and cold fluid, heating/cooling a fluid flowing in pipe, etc.

7.3.1. *Classification of convection heat transfer*

Heat is transferred when a fluid moves or mixes with its hot and cold components. Convectional heat transfer can occur naturally or by forced convection (shown in Figure 7.9). Consider heating of water in a metal container using an external heat source from the bottom. The water layer contacted to the bottom of the vessel will be at higher temperature than other available water in the container. Fluid at a higher temperature will have less density compared to cold fluid. Thus, the hot fluid layer will move upward and the cold fluids at the top will move downward, due to difference in density. The cold water that sinks to the bottom is then heated up again and will move upward. This produces a pattern of water circulation that uses fluid molecules to transfer heat from the vessel's bottom to its top. When a system mixes or transfers heat because of a buoyancy force or density difference, this is known as natural free convection. Stirrers that are external can combine hot and cold liquid. There is forced convection. As external stirrers hasten mixing, forced convection heats more quickly than natural convection.

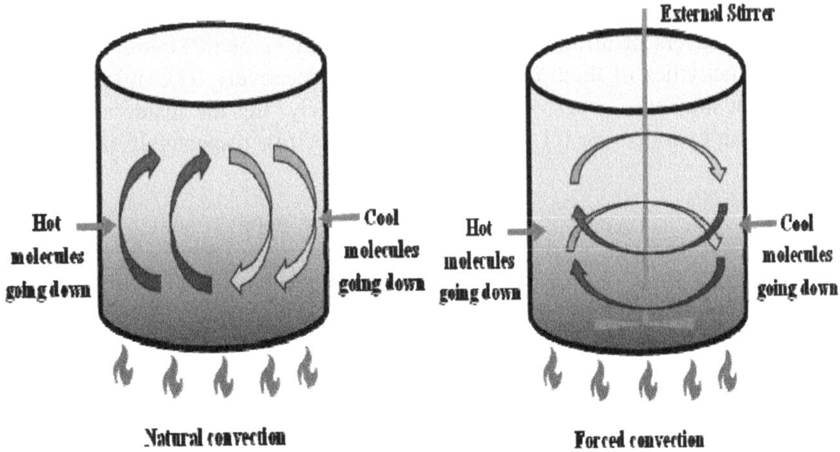

Figure 7.9. *Convective heat transfer mechanism. For a color version of this figure, see www.iste.co.uk/kumararya/chemicalengineering1.zip*

7.3.2. *Newton's law of cooling*

Isaac Newton's law of cooling calculates convective heat transfer speed, which is affected by medium temperature, heat transfer or contact area and fluid velocity (McCabe et al. 1993). Based on this law, "the rate of heat transfer by convection is directly proportional to the area of heat transfer taken perpendicular to the direction of heat flow and the temperature difference."

Heat transfer rate (Q) α, heat transfer area (A), temperature difference (ΔT)

$$Q \alpha A \Delta T$$

Newton's law of cooling describes convection heat transfer. Consider 1D heat flow in an isothermal medium. Heat flow is at a steady state, with a constant thermal conductivity. There is no accumulation and generation of heat in the medium.

$$Q = h\, A\, \Delta T = \frac{\Delta T}{\left(1/_{h\,A}\right)} = \frac{\Delta T}{R} \text{ or } h = \frac{Q}{A\, \Delta T}$$

The constant of proportionality in Newton's law of cooling is termed as heat transfer coefficient (h) and R is the resistance to heat transfer by convection. Heat transfer coefficient is defined as the rate of heat transfer through a medium by convection having the unit area and having unit temperature differences. The heat transfer coefficient, sometimes referred to as film coefficient as heat transfer at a phase boundary, is observed occurring through a thin stagnant film adhering to the interface. The temperature gradient is observed at this film layer and beyond this film fluid is assumed to be well mixed (Dutta 2023).

$$\text{Unit of h} = \frac{Q}{A\,\Delta T} = \frac{W}{m^2.K};\ \text{Dimension of h} = \frac{W}{m^2.K} = \frac{(N.m/s)}{m^2.K} = \frac{kg}{s^3.K} = M\Theta^{-3}T^{-1}$$

Convective heat transfer studies desire to identify the flow field, fluid temperature distribution and heat transfer coefficient. However, heat transfer coefficient determination complicates heat transfer rate calculations.

7.3.3. *Local and average convection coefficients*

The local convection coefficient, h(x), is the heat transfer rate per unit area, per unit temperature difference, at a specific solid surface location (x). It describes surface heat transfer rate variation due to boundary layer development, surface geometry and fluid properties (Bergman et al. 2011). One value for the entire heat transfer surface is the average convection coefficient (h_avg). The local convection coefficient is averaged over the surface area to calculate it. A represents the total surface area, and ∫ represents the surface-wide integration.

$$h_{avg} = \frac{1}{A}\int h(x)dA;$$

Individual and overall heat transfer coefficient: three series resistances will cause heat flow from bulk hot fluid to bulk cold fluid (Figure 7.10).

– resistance offered by hot fluid film (convection);

– resistance offered by metal wall (conduction);

– resistance offered by cold fluid film (convection).

h_i is the inside fluid heat transfer coefficient and A_i is the tube's inner side heat transfer area; convection from bulk hot fluid to the inner surface of metal cylinder will transfer heat inside the pipe.

$$Q_i = h_i\,A_i\,(T1 - T3) = \frac{(T1 - T3)}{\left(\frac{1}{h_i\,A_i}\right)} = \frac{\Delta T1}{R_i}\ \mathit{or}\ \Delta T1 = Q_iR_i = \frac{Q_i}{h_i\,A_i}$$

Here, $R_i = \frac{1}{h_i A_i}$ = resistance offered by hot fluid film.

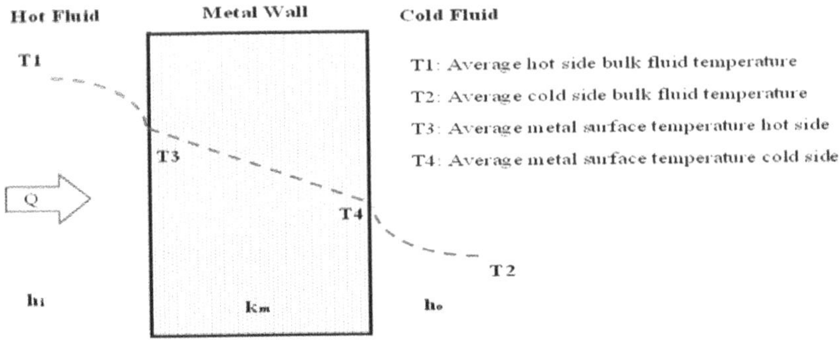

Figure 7.10. *Heat transfer from inner to the outer side of circular pipe through metal wall*

The metal pipe's rate of heat transfer by conduction from hot to cold side of the metal cylinder is (k_m is metal wall thermal conductivity and A_m is tube heat transfer area):

$$Q_m = k_m A_m \frac{(T3 - T4)}{x_m} = \frac{(T3 - T4)}{\left(\frac{x_m}{k_m A_m}\right)} = \frac{\Delta T2}{R_m} \text{ or } \Delta T2 = Q_m R_m = \frac{Q_m x_m}{k_m A_m}$$

$$= \frac{Q_m}{\left(k_m A_m / x_m\right)}$$

Here, $R_m = \frac{x_m}{k_m A_m}$ = resistance offered by metal wall.

From the outer surface of the metal cylinder to the bulk cold fluid, convection will transfer heat outside the pipe (h_o is the outer side fluid heat transfer coefficient and A_o is the tube's outer side heat transfer area).

$$Q_o = h_o A_o (T4 - T2) = \frac{(T4 - T2)}{\left(\frac{1}{h_o A_o}\right)} = \frac{\Delta T3}{R_0} \text{ or } \Delta T3 = Q_o R_o = \frac{Q_o}{h_o A_o}$$

Here, $R_o = \frac{1}{h_o A_o}$ = resistance offered by cold fluid film.

ASSUMPTION.– under steady-state condition of heat flow, the overall heat transfer from hot fluid to cold fluid will be Q ($Q = Q_o = Q_m = Q_i$). Again, if all have the same heat transfer area, A_i = A_m = A_o = A. The total temperature difference (T1-T2 or ΔT) between hot and cold fluids is ΔT = ΔT1 + ΔT2 + ΔT3.

$$\Delta T = Q_i R_i + Q_m R_m + Q_o R_o = Q(R_i + R_m + R_o)$$

$$Q = \frac{\Delta T}{(R_i + R_m + R_o)} = \frac{\Delta T}{\left(\frac{1}{h_i A_i} + \frac{x_m}{k_m A_m} + \frac{1}{h_o A_o}\right)} = \frac{\Delta T}{R} = \frac{\Delta T}{\left(\frac{1}{UA}\right)} = UA\Delta T$$

Overall heat flow resistance (R) relates to the overall heat transfer coefficient (U) and individual fluid heat transfer coefficients (h_i and h_o). Overall heat transfer rate from bulk hot fluid to bulk cold fluid through conduction and convection is calculated using this expression.

7.3.4. *Boundary layers, Nusselt number and Prandtl number*

Concept of boundary layer: due to viscosity, fluid velocity near the solid surface up to a certain thickness is zero (no-slip condition) (McCabe et al. 1993). Film is a thin layer of stagnant fluid near a solid surface with zero velocity. Cold side fluid is called “cold side film”, and hot side fluid is called “hot fluid film” when fluid velocity is zero. This no-slip condition increases flow velocity away from the solid surface wall until it reaches free stream velocity.

The flow of fluid along a surface creates a thin boundary layer. The intensity of the boundary layer depends on the fluid’s velocity, viscosity and surface configuration. Faster, less viscous and more streamlined fluids have a thinner boundary layer. According to Prandtl’s hypothesis, shear stress and fluid friction decrease near a contact surface or boundary layer as the Reynolds number increases. In addition, boundary layer pressure remains constant. The pressure inside the boundary layer equals the pressure outside it in inviscid flow. Varying velocity and temperature profiles define boundary layers.

The hydrodynamic boundary layer (HBL) is the fluid thickness over a solid plate with velocity gradient and shear stress. The fluid velocity is 0.99 times the free stream velocity at the thickness perpendicular to the solid body. The fluid thickness over the solid plate with thermal gradient is the thermal boundary layer (TBL). Different temperatures create an imaginary layer called the TBL. Prandtl’s number (Pr) relates to HBL and TBL thicknesses. For liquids, Pr > 1, solids, air, gases = 1 and liquid metals < 1.

$$(\mathrm{Pr})^{\frac{1}{3}} = \frac{\text{Thickness of hydrodynamic boundary layer (HBL)}}{\text{Thickness of thermal boundary layer (TBL)}}$$

The five series resistances will cause heat flow from bulk hot fluid to bulk cold fluid in the above heat transfer rate expression. Two additional resistances affect heat transfer due to the solid's thin fluid film. Heat flow from inside to outside will have resistances (1) from the hot fluid to the layer of hot fluid film, (2) from the hot fluid film layer to the metal pipe, (3) from the metal pipe layer (from hot side to cold side), (4) from the metal pipe to the cold fluid film layer and (5) from the cold fluid film layer to the cold fluid.

Due to turbulence or mixing, hot and cold fluids typically have uniform temperatures. This means the hot and cold sides (flowing region) have uniform temperatures. Hot and cold fluid film boundary layers and metal layers govern heat transfer resistance. Due to zero fluid transport in both film layers, heat transfer is conduction only.

Calculation of heat transfer coefficient: heat transfer coefficient "h" depends on fluid properties, surface size and flow velocity or type. The convective heat transfer coefficient is assessed using dimensional analysis. Dimensional analysis uses mathematics to correlate the number of variables or influencing parameters of a physical event to a single effect equation. In dimensional analysis, Buckingham's π-theorem establishes a mathematical relationship between the convective heat transfer coefficient and its influencing variables. Dimensional analysis was used by Nusselt to derive equations for free and forced convection heat transfer coefficients (Welty et al. 2020).

Dimensional analysis for calculation of heat transfer by a forced convection: in forced convection, the heat transfer coefficient (h) depends on factors such as fluid density (ρ), viscosity (μ), thermal conductivity (k), specific heat (Cp), surface dimensions (L) and fluid velocity (u) (Welty et al. 2020).

$$h = f(\rho, \mu, \mathrm{k}, C_P, \mathrm{L}, \mathrm{u}) \text{ or } f(h, \rho, \mu, \mathrm{k}, C_P, \mathrm{L}, \mathrm{u}) = 0$$

Then, the expression with seven number of variables $\mathrm{f}(\mathrm{h}, \mathrm{L}, \mathrm{C_P}, \rho, \mu, \mathrm{k}, \mathrm{u}) = 0$ can be written with these three dimensionless terms. $\mathrm{f}(\pi 1, \pi 2, \pi 3) = 0$. Here, the total number of variables = n = 7; number of fundamental dimensions required to express all the variables = m = 4 (M, L, T, Θ). Here, m = number of repeating variables in π expressions = 4; repeating variables are assigned in π terms according to the order of fluid properties, thermal properties, flow properties, geometric properties.

Then, number of dimensionless or π terms = n – m = 7 – 4 = 3

$$\pi 1 = \mu^a * k^b * u^c * L^d * h;\ \pi 2 = \mu^a * k^b * u^c * L^d * \rho;\ \pi 3 = \mu^a * k^b * u^c * L^d * C_P$$

By applying Buckingham's π-theorem, assigning respective dimensions to the variable in each of three π expressions, and equating the powers of fundamental quantity M, L, T, on both sides of each π of expression, we have

$$\pi 1 = \frac{h * L}{k} = \text{Nusselt's Number (Nu)}$$

$$\pi 2 = \frac{\rho * u * L}{\mu} = \text{Reynold's Number (Re)}$$

$$\pi 3 = \frac{\mu * C_P}{k} = \text{Prandtl's Number (Pr)}$$

The dimensionless Nusselt's number (Nu) describes heat transfer between a moving fluid (convection) and a solid body (conduction). Larger Nu values increase convection heat transfer over conduction.

$$Nusselt's\ number\ (Nu) = \frac{\text{Convective heat transfer rate}}{\text{Conductive heat transfer rate}} = \frac{h * A * \Delta T}{k * A * \frac{\Delta T}{L}} = \frac{h * L}{k}$$

– Reynold's number (Re) is dimensionless and is used to determine the flow behaviour of fluid. It is the ratio between inertial force to the viscous force.

$$Reynold's\ number\ (Re) = \frac{\text{Inertial force}}{\text{Viscous force}} = \frac{\rho * u * L}{\mu} = \frac{u * L}{\gamma}$$

Kinematic or momentum viscosity of fluid = $\gamma = {}^{\mu}/_{\rho}$.

For flow through flat plates, Re < $5*10^5$ (laminar flow); Re > $5*10^5$ (turbulent flow).

For flow through circular pipes, Re < 2,100 (laminar flow); 2,100 < Re < 4,000 (transient region); Re > 4,000 (turbulent flow).

– Prandtl's number (Pr) integrates heat transfer from momentum and thermal energy diffusion processes in fluids.

$$\text{Prandtl's number (Pr)} = \frac{\text{Momentum diffusivity through the fluid } (\gamma)}{\text{Thermal diffusivity through the fluid } (\alpha)}$$

$$= \frac{(\mu/\rho)}{(k/\rho C_P)} = \frac{\mu * C_P}{k}$$

Due to fluid velocity (momentum) profile, heat diffuses quickly when Pr is small because the TBL thickness is higher. More Pr numbers mean more momentum diffusivity through fluids. More mixing or convection heat transfer increases the Nu number.

Now, putting these values of $\pi 1, \pi 2, \pi 3$ in the main dimensionless expression.

$$f(\pi 1, \pi 2, \pi 3) = 0 \text{ or } f\left(\frac{h * L}{k}, \frac{\rho * u * L}{\mu}, \frac{\mu * C_P}{k}\right) = 0 \text{ or } f(Nu, Re, Pr) = 0$$

$$Nu = f(Re, Pr) = f(Pe)$$

Here, the Pecklet number (Pe) is a dimensionless number and is used for force convection.

$$\text{Pecklet Number (Pe)} = \mathit{Reynold\ Number\ (Re) * Prandtl\ Number (Pr)}$$

$$Pe = Re * Pr = \frac{\rho * \vartheta * L}{\mu} * \frac{\mu * C_P}{k} = \vartheta * L * \frac{\rho * C_P}{k} = \frac{L * \vartheta}{\alpha}$$

Forced convection heat transfer coefficient (h) is calculated by correlating Nu, Re and Pr. The above correlation data for forced convection systems can be expressed with reasonable accuracy by a simple power-law relation.

$$Nu = f(Re, Pr) = C\ (Re)^a\ (Pr)^b$$

where the constant C and exponents "a" and "b" are determined through experiments.

Different empirical equations used for calculation of heat transfer coefficient (h), based on the given system, in forced convection are (Cengel 2000; Welty et al. 2020; Dutta 2023):

1) *When flow is over a flat plate:*

If fluid flow is laminar: Blassius equation ($Re < 5*10^5$; $0.5 \leq Pr \leq 50$)

$$\mathrm{Nu} = \frac{\mathrm{h} * \mathrm{L}}{\mathrm{k}} = 0.664\ (\mathrm{Re})^{0.5}\ (\mathrm{Pr})^{0.33}$$

If flow of fluid is turbulent: Dittus–Boelter equation ($Re > 5*10^5$; $0.6 \leq Pr \leq 60$)

$$\mathrm{Nu} = \frac{\mathrm{h} * \mathrm{L}}{\mathrm{k}} = 0.037\ (\mathrm{Re})^{0.8}\ (\mathrm{Pr})^{\mathrm{n}}\ [\ \mathrm{L} = \text{length of plate};\ \mathrm{n} = 0.4, \text{for heating and } \mathrm{n} = 0.3]$$

If flow of fluid is at critical: Reynold's number ($Re = 5*10^5$; $0.6 \leq Pr \leq 60$)

$$\mathrm{Nu} = \frac{\mathrm{h} * \mathrm{L}}{\mathrm{k}} = (0.037\ (\mathrm{Re})^{0.8} - 871)\ (\mathrm{Pr})^{\mathrm{n}}$$

If flow of liquid metals such as mercury: (higher thermal conductivity, $Pr < 0.05$)

$$\mathrm{Nu} = \frac{\mathrm{h} * \mathrm{L}}{\mathrm{k}} = 0.056\ (\mathrm{Re})^{0.5}\ (\mathrm{Pr})^{0.5}$$

For all fluids, the common and single correlation of Nu at all Re and Pr are

$$\mathrm{Nu} = \frac{\mathrm{h} * \mathrm{L}}{\mathrm{k}} = \frac{0.3387\ (\mathrm{Re})^{0.5}\ (\mathrm{Pr})^{0.33}}{\left[1 + \left(0.0468/\mathrm{Pr}\right)^{0.66}\right]^{0.25}}$$

2) *For flow through a cylindrical pipe:*

If flow of fluid is laminar: Sieder–Tate equation

For less viscous fluid ($Re < 2{,}100$ and $\mathrm{Re.Pr.}\left(\frac{D}{L}\right) > 12$);

$$\mathrm{Nu} = \frac{\mathrm{h} * \mathrm{D}}{\mathrm{k}} = 1.62(\mathrm{Re})^{0.33}\ (\mathrm{Pr})^{0.33} \left(\frac{\mathrm{D}}{\mathrm{L}}\right)^{0.33}$$

For high viscous fluid ($Re < 2{,}100$; $0.48 < Pr < 16{,}700$; $\mathrm{Re.Pr.}\left(\frac{D}{L}\right) > 10^8$);

$$\mathrm{Nu} = \frac{\mathrm{h} * \mathrm{D}}{\mathrm{k}} = 1.86\ (\mathrm{Re})^{0.33}\ (\mathrm{Pr})^{0.33} \left(\frac{\mathrm{D}}{\mathrm{L}}\right)^{0.33} \left(\frac{\mu}{\mu_{\mathrm{w}}}\right)^{0.14}$$

The Sieder–Tate correction is $(\mu/\mu_w)^{0.14}$. Here, μ and μ_w represent the fluid's viscosity at bulk and wall temperatures, respectively.

If fluid flow is turbulent:

For less viscous fluid ($Re \geq 10{,}000$; $0.7 \leq Pr \leq 160$ and $\left(\frac{D}{L}\right) < 0.1$); Dittus–Boelter equation

$$Nu = \frac{h * L}{k} = 0.023\,(Re)^{0.8}\,(Pr)^{n}$$

[D = diameter of the pipe; L = length of pipe]
[n = 0.3 for cooling and 0.4 for heating]

For high viscous fluid ($Re \geq 10{,}000$; $0.7 \leq Pr \leq 160$ and $\left(\frac{D}{L}\right) < 0.1$); Sieder–Tate equation

$$Nu = \frac{h * D}{k} = 0.027\,(Re)^{0.8}\,(Pr)^{0.33}\left(\frac{\mu}{\mu_W}\right)^{0.14}$$

If flow of fluid (Gas) is turbulent: ($Pr \approx 0.74$) $Nu = \frac{h*L}{k} = 0.02\,(Re.Pr)^{0.8}$

For airflow at 1 atm pressure, $Nu = \frac{h*L}{k} = \frac{3.5\,u^{0.8}}{D^{0.2}}$

If the flow of fluid is transient ($2{,}100 < Re < 4{,}000$):

$$Nu = \frac{h * L}{k} = 0.116\,[[(Re)^{0.66} - 125](Pr)^{0.33}\left[1 + \left(\frac{D}{L}\right)^{0.14}\right]\left(\frac{\mu}{\mu_W}\right)^{0.14}$$

For the flow of liquid metals ($Re.Pr > 100$; $\left(\frac{L}{D}\right) < 60$): $Nu = \frac{h*L}{k} = 5 + 0.025\,(Re.Pr)^{0.8}$

3) *For flow across a cylindrical pipe:* ($40 < Re < 105$; $0.65 < Pr < 300$)

$$Nu = \frac{h * D}{k} = [0.4(Re)^{0.5} + 0.06(Re)^{0.66}]\,(Pr)^{0.4}\left(\frac{\mu}{\mu_W}\right)^{0.25}$$

4) *For flow across a sphere:*

$$Nu = \frac{h * D}{k} = 2 + [0.4(Re)^{0.5} + 0.06(Re)^{0.66}]\,(Pr)^{0.4}\left(\frac{\mu}{\mu_W}\right)^{0.25}$$

Dimensional analysis for heat transfer by a natural convection: in natural convection, the convective heat transfer coefficient (h) is affected by the same factors as forced convection, except for fluid velocity (u). An external source causes forced convection, while temperature variations drive free convection through density differences between fluid layers (Welty et al. 2020).

Factors such as fluid density, viscosity, thermal conductivity, specific heat, surface dimensions, temperature difference, volumetric expansion coefficient and gravity acceleration influence the heat transfer coefficient (h) in free convection. $h = f(\rho, \mu, k, C_P, L, \Delta T, \beta, g)$ or $f(h, \rho, \mu, k, C_P, L, \Delta T, \beta, g) = 0$.

However, in the context of free convection, the combination of $(\Delta T\ \beta\ g)$ is considered as a single component because the velocity of fluid particles depends on these specific quantities. Hence, the given expression can be formulated as $f(h, \rho, \mu, k, C_P, L, (\Delta T\ \beta\ g)) = 0$. Then, a relationship between these variables can be determined by using dimensional analysis. Here, the total number of variables = n = 7; number of fundamental dimensions required to express the variables = m = 4 (M, L, T, Θ: repeating variables in π expressions).

Then, the number of dimensionless or π terms = n – m = 7 – 4 = 3

$$\pi 1 = \mu^a * k^b * \rho^c * L^d * h;\ \pi 2 = \mu^a * k^b * \rho^c * L^d * (\Delta T\ \beta\ g);$$
$$\pi 3 = \mu^a * k^b * \rho^c * L^d * C_P$$

Then, the expression with seven number of variables $f(h, L, C_P, \rho, \mu, k, (\Delta T\ \beta\ g)) = 0$ can be written with these three dimensionless terms. $f(\pi 1, \pi 2, \pi 3) = 0$.

By applying Buckingham's π-theorem, assigning respective dimensions to the variable in each of three π expressions, and equating the powers of fundamental quantity M, L, T, ανδ Θ on both sides of each π of expression, we have

$$\pi 1 = \frac{h * L}{k} == \text{Nusselt's number (Nu)}$$

$$\pi 2 = \frac{\rho^2 * (\Delta T\ \beta\ g) * L^3}{\mu^2} = \text{Grashoff's number (Gr)}$$

$$\pi 3 = \frac{\mu * C_P}{k} = \text{Prandtl's number (Pr)}$$

Grashoff's number is a dimensionless number and is used for free convection.

$$\text{Grashoff's number (Gr)} = \frac{\text{Inertial force} * \text{buoyancy force}}{\text{Viscous force}}$$
$$= \frac{\rho^2 * (\Delta T\ \beta\ g) * L^3}{\mu^2}$$

Now, putting these values of $\pi1, \pi2, \pi3$ in the main dimensionless expression, we have

$$f(\pi1, \pi2, \pi3) = 0 \text{ or } f\left(\frac{h * L}{k}, \frac{\rho^2 * (\Delta T\ \beta\ g) * L^3}{\mu^2}, \frac{\mu * C_P}{k}\right)$$
$$= 0 \text{ or } f(Nu, Gr, Pr) = 0$$

$$Nu = f(Gr, Pr) = f(Ra)$$

Here, the Rayleigh number (Ra) is a dimensionless number and is used for free convection.

($10^4 < Ra < 10^9$, Laminar flow and $Ra > 10^9$, turbulent flow)

$$\text{Rayleigh number (Ra)} = \text{Grashoff number (Gr)} * \text{Prandtl number (Pr)}$$

$$Ra = Gr * Pr = \frac{\rho^2 * (\Delta T\ \beta\ g) * L^3}{\mu^2} * \frac{\mu * C_P}{k} = (\Delta T\ \beta\ g) * L^3 * \frac{\rho * C_P}{k} * \frac{\rho}{\mu}$$
$$= \frac{(\Delta T\ \beta\ g) * L^3}{\alpha * \gamma}$$

For natural or free convection systems, the heat transfer coefficient (h) is calculated by correlating Nu, Gr and Pr. The experimental data of the above correlation natural or free convection system is usually expressed with reasonable accuracy by a simple power-law relation.

$$Nu = f(Gr, Pr) = C\ (Gr)^a\ (Pr)^b$$

where experiments determine C, a and b. Depending on the system, empirical equations for heat transfer coefficient (h) in natural or free convection include (Geankoplis 1993; McCabe et al. 1993; Zeng et al. 1995; Bergman et al. 2011; Welty et al. 2020; Dutta 2023):

1) *For flow over a vertical surface:*

If the flow is laminar ($10^4 < Ra < 10^9$) $Nu = \frac{h*L}{k} = 0.59\ (Gr)^{0.25}\ (Pr)^{0.25}$

If the flow is turbulent ($10^9 < Ra < 10^{12}$) $Nu = \frac{h*L}{k} = 0.13\ (Gr)^{0.33}\ (Pr)^{0.33} = 0.13\ (Ra)^{0.33}$

For all fluids, at entire range of Ra number

$$Nu = \frac{h*L}{k} = \left(0.825 + \frac{0.387\ (Ra)^{1/6}}{\left[1 + \left(0.492/_{Pr}\right)^{9/16}\right]^{8/27}}\right)^2$$

2) *For flow within a horizontal plate:* (hot solid surface is above cold fluid)

If the flow is laminar ($10^4 < Ra < 10^7$) $Nu = \frac{h*L}{k} = 0.54\ (Gr)^{0.25}\ (Pr)^{0.25} = 0.54\ (Ra)^{0.25}$

If the flow is turbulent ($10^7 < Ra < 10^{11}$) $Nu = \frac{h*L}{k} = 0.15\ (Gr)^{0.33}\ (Pr)^{0.33} = 0.15\ (Ra)^{0.33}$

3) *For flow within a horizontal plate:* (hot solid surface is below cold fluid)

If the flow is in range ($10^5 < Ra < 10^{10}$) $Nu = \frac{h*L}{k} = 0.27\ (Gr)^{0.25}\ (Pr)^{0.25} = 0.27\ (Ra)^{0.25}$

4) *For flow over a horizontal cylinder:* if the flow is in range ($Ra \leq 10^{12}$)

$$Nu = \frac{h*L}{k} = \left(0.6 + \frac{0.387\ (Ra)^{1/6}}{\left[1 + \left(0.559/_{Pr}\right)^{9/16}\right]^{8/27}}\right)^2$$

5) *For flow over a sphere:* if the flow is in range ($Ra \leq 10^{11}$; $Pr \geq 0.7$)

$$Nu = \frac{h*L}{k} = 2 + \frac{0.589\ (Ra)^{1/4}}{\left[1 + \left(0.469/_{Pr}\right)^{9/16}\right]^{4/9}}$$

7.4. Radiation heat transfer

Thermal radiation transmits energy via electromagnetic waves. Radiation behaves like visible light. Linear light can reflect on a surface. Radiation (Figure 7.11) is the only major manner of energy/heat transfer in a medium or vacuum (Bergman et al. 2011).

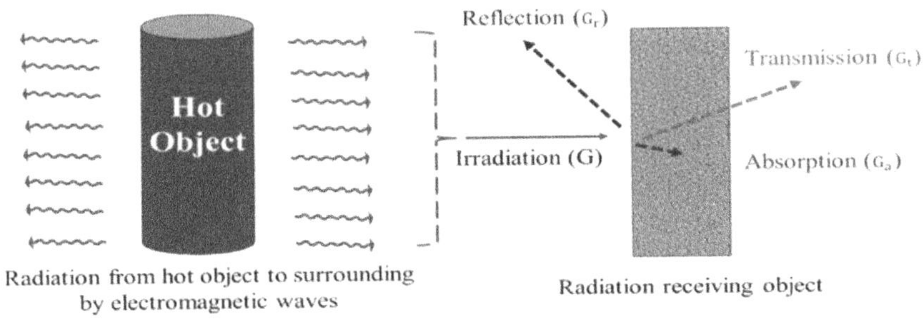

Figure 7.11. *Radiative heat transfer (Bergman et al. 2011). For a color version of this figure, see www.iste.co.uk/kumararya/chemicalengineering1.zip*

Thermal radiation refers to the transmission of energy between two systems by the emission of photons resulting from a difference in temperature. Thermal radiation propagates by the emission of discrete units known as photons or quanta, which vary in wavelength. Every quantum possesses energy, denoted as E = h*γ. Let h be the value of Plank's constant = 6.625 * 10^{-34} J.s, and γ represent the frequency of the radiation, measured in cycles per second or Hertz. The photons that carry thermal energy propagate as rays. As these rays come into contact with the receiving surface, they are either absorbed, reflected or transmitted. The conversion of the absorbed wave motion into thermal energy occurs within the body (Holman 1986; McCabe et al. 1993; Cengel 2000; Welty et al. 2020; Dutta 2023). Instances of heat transfer through radiation are the transfer of heat from the sun to the ground; the dissipation of heat from an uninsulated steam pipe; the utilization of solar energy in solar heaters; and the warming of a chilly room with a radiant electric heater. Heat transfer is in furnaces, combustion chambers, nuclear explosions and space applications.

7.4.1. *Fundamental concepts*

Total emissive power (E) is the amount of radiant energy a body emits per unit surface area and time period. The units are J/s.m^2 or W/m^2. A body's monochromatic (spectral) emissive power (Eλ) measures the amount of radiant energy released per unit surface area and time, for a specific wavelength (λ). It uses $W/(m^2·\mu m)$ units. Integration of the monochromatic emissive power yields the total emissive power, E. Total emissive power (E)= $\int_0^\infty E_\lambda \, d\lambda$.

The total energy density (u) of a body is its radiant energy output across all wavelengths per volume. It uses J/m^3. For a specific wavelength (λ), the monochromatic (spectral) energy density (uλ) measures the amount of radiant energy emitted per unit volume by a body. It uses $J/(m^3·\mu m)$ units. Integration of the monochromatic energy density yields the total energy density, u. Total emissive power (u) = $\int_0^\infty u_\lambda \, d\lambda$.

Irradiation: radiant energy of all wavelengths incident on a body per unit surface area per unit time. G indicates it. W/m^2 units.

A body can absorb, reflect and transmit thermal radiation from another body.

– Absorption: when light interacts with the atoms or molecules within a material, some of the light's energy can be absorbed. This absorption can excite electrons in the material to higher energy levels. The absorbed energy is then converted into other forms, often heat, through various mechanisms like vibrational relaxation. The extent of absorption depends on the material's properties and the wavelength of the light.

– Reflection: when light encounters a surface, it can interact with the electrons at the surface and be reflected back into the surrounding medium. The reflection can be specular (mirror-like) for smooth surfaces, where the reflected ray follows the law of reflection. Rough surfaces cause diffuse reflection, where the light is scattered in various directions.

– Transmission: light can also pass through a material if it is partially transparent. The transmitted light may be refracted (bent) at the interface between the material and the surrounding medium due to the difference in refractive indices. In addition, some of the light may be absorbed within the material, reducing the intensity of the transmitted ray.

If a body receives G thermal radiation, then, G_a and G_r are the body's energy absorption and reflection. G_t is the body energy transmission. Then, according to conservation of energy principle,

Heat reflected + Heat absorbed + Heat transmitted = Incident radiation

$$G = G_a + G_r + G_t \text{ or } 1 = \frac{G_a}{G} + \frac{G_r}{G} + \frac{G_t}{G} \; or$$

$$1 = \alpha + \rho + \tau$$

– Absorptivity (α): a body's absorptivity is the fraction of incident radiation it absorbs. Use the letter "α" to represent it.

– Reflectivity (ρ): reflectivity is the percentage of incident radiation reflected by a body, denoted by "ρ".

– Transmissivity (τ): the fraction of the incident radiation on a body that is transmitted through the body is known as the transmissivity and may be denoted by the letter "τ".

Based on these parameters, the objects can be classified under different names:

– Opaque Body: if the incident energy will not pass through the body and will only be absorbed and reflected. These materials are called as opaque body. Here, $\tau = 0$, $\alpha + \rho = 1$ and represents a non-transparent surface.

– Transparent Body: if the entire incident energy passes through the body, it is referred to as a perfectly transparent body. Here, $\tau = 1$, $\alpha = \rho = 0$ and represents a perfectly transparent surface. Example: rock salt, quartz and fluorite.

– White Body: a white body is defined as one that reflects all incident energy that falls on it. Here, $\rho = 1$, $\alpha = \tau = 0$ and represents a perfect reflector.

– Non-reflecting Surface: if all the incident energy fall on a body will not be reflected by the surface of the body. Here, $\rho = 0$, $\alpha + \tau = 1$ and represents a non-reflecting surface.

– Gray Body: a gray body is defined as one with uniform radiative fractions (α, ρ, τ) across the entire wavelength.

– *Black body concept:* radiation is completely absorbed by a black body. The values of α, ρ and τ are 1, 0 and 0, respectively, representing a perfectly absorbing surface. It is regarded as an optimal radiation receiver. The body does not need to have a dark surface in order to absorb all radiation. A black body is a theoretical construct that does not exist in reality. All surfaces reflect radiation to some degree, meaning that there are no perfect black bodies. Even the darkest surfaces found in nature still have a reflectance of approximately 1%. While completely black bodies do not occur naturally, several materials can come close to achieving this. Lampblack and platinum black are most closely resembling a black body, capable of absorbing approximately 96% of the incoming radiation (Kern 1950; Gavhane 2015).

Every substance releases radiation, with the characteristics and amount of radiation depending on the temperature and qualities of the material making up the emitting object. It may be demonstrated that materials that effectively absorb a specific wavelength at a given temperature also emit that same wavelength effectively. Black bodies emit all wavelengths, making them the most efficient heat emitters or radiators. Black bodies emit the most energy at a given temperature. A fully black body has one absorptivity and emissivity (McCabe et al. 1993).

7.4.2. Radiation heat fluxes – different laws of thermal radiation

1) *Stefan–Boltzmann's law:* a perfect black body emits radiation heat energy (heat flux) when no heat energy is received from surroundings (Cengel 2000; Welty et al. 2020; Dutta 2023). "A black body's total emissive power is directly proportional to its fourth power of absolute temperature."

$$\text{Emissive power of a black body } (E_b) \ \alpha \ \text{Temperature}^4 \ (T^4)$$

$$E_b = \sigma T^4 = \frac{Q}{A} = q \text{ (thermal radiation heat flux)}$$

where Q is the heat transfer per unit time (J/s or W); A is the heat transfer area (m^2); T is the absolute temperature (K); σ is the constant of proportionality = Stefan–Boltzmann's constant ($5.67 * 10^{-8}$ $W/(m^2.K^4)$); Dimension of σ: $[M\Theta^{-3}T^{-4}]$.

Emissivity (ε): This refers to the capacity of a body's surface to emit heat radiation. Emissivity compares a body's total emissive power (E) to a black body's (Eb) at the same temperature. Temperature and radiation wavelength affect the relationship. A black body has emissivity (ε) equal to one, while a non-black body has ε less than one.

$$\epsilon = \frac{E}{E_b}$$

Emissive power or radiant energy lost per unit time per unit area from a non-black body at "T1" is $E_1 = \sigma T_1^4 = \frac{Q}{A} = q =$ heat flux.

Suppose a non-black body with emissivity "ϵ_1" at temperature "T_1" emits radiant energy per unit area equal to $E_1 = \epsilon_1 * E_b = \epsilon_1 \sigma (T_1)^4$.

If the non-black body "1" having an emissivity "ϵ_1" at temperature "T_1" is surrounded by another non-black body "2" having an emissivity "ϵ_2" at temperature "T_2". Consider emissive power of body "1" is more than body "2" (E1 > E2) and both the bodies placed together. The net radiation energy or heat transferred from body 1 to body 2 will be

$$E_1 - E_2 = \epsilon_1\sigma\,(T_1)^4 - \epsilon_2\sigma\,(T_2)^4 = \frac{Q}{A} = q \text{ (thermal radiation heat flux)}$$

If the non-black body "1" having an emissivity "ϵ_1" at temperature "T_1" is surrounded by another black body "2" having an emissivity "$\epsilon_2 = 1$" at temperature "T_b". The net radiation energy or heat transferred from body 1 to body 2 will be

$$E_1 - E_b = \epsilon_1\sigma\,(T_1)^4 - \sigma\,(T_b)^4 = \frac{Q}{A} = q \text{ (thermal radiation heat flux)}$$

2) *Planck's law:* this rule establishes a correlation relating a black body's monochromatic emissive power, its absolute temperature and its corresponding wavelength.

$$E_{b,\lambda} = \frac{2\pi hc^2\lambda^{-5}}{e^{hc/k\lambda T} - 1} = \frac{C_1\lambda^{-5}}{e^{C_2/\lambda T} - 1}$$

Here, Eb, λ is the black body's monochromatic emissive power (W/(m^2·μm)), h is the Planck's constant, k is the Boltzmann constant, c is the speed of light, T is the absolute temperature (K), and λ is the radiation wavelength. With C1 and C2 constants, the above equation can be written. C1 = 3.472 * 10–16 W·m^2 • C2 = 0.01439 m·K.

3) *Wien's displacement law:* the wavelength in which monochromatic emissive power peaks (λmax) is inversely proportional to absolute temperature (T), where λ_{max} is in micrometres and T is in Kelvins.

$$T \propto \frac{1}{\lambda_{max}} \text{ or } T = \frac{C}{\lambda_{max}} \text{ or } T\lambda_{max} = C = 2{,}890$$

4) *Kirchhoff's law:* this states that all bodies have the same total emissive power–absorptivity ratio at thermal equilibrium.

– Case 1: two non-black bodies in thermal equilibrium with each other or with their surroundings. Per Kirchhoff's law.

$$\frac{E_1}{\alpha_1} = \frac{E_2}{\alpha_2}$$

– Case 2: when a black body or a non-black body is thermally balanced with its surroundings. According to Kirchhoff's law.

$$\frac{E_1}{\alpha_1} = \frac{E_b}{\alpha_b} \text{ or } \frac{E_1}{\alpha_1} = \frac{E_b}{1} \text{ or } \frac{E_1}{E_b} = \alpha_1 \text{ or } \in = \alpha_1$$

NOTE:– emissivity and absorptivity are equal when a black body is thermally balanced.

7.4.3. *View factor*

Angle, geometrical, centrifugal and shape are view factors. How much radiation is diffused from one surface and reflected directly from another. Two non-black bodies (body "1" and "2") have surface area (A1 and A2), temperature (T1 and T2) and emissivity (∈_1 and ∈_2). Consider radiation from body "1" at higher temperature (T1 > T2). How much radiation body 1 emits and passes to body 2 determines net thermal energy transfer. According to their arrangement, body "2" will receive energy from body "1". Consider two bodies close. This view factor (F12) shows how much radiation impacts body "2" from body "1". After absorbing stroked radiation, body "2" may reflect it. Body "2"'s radiation may hit body "1" again due to their arrangement. View factor (F21) is the percentage of radiation energy from body "2" that hits body "1". Radiation is also affected by body configuration.

In that case:

– View factor = F_{11} = Fraction of radiation coming from body "1" and interpreted in body "1".

– View factor = F_{22} = Fraction of radiation coming from body "2" and interpreted in body "2".

Then, according to reciprocity relation, $A_1 F_{12} = A_2 F_{21}$; $F_{11} + F_{12} = 1$; $F_{21} + F_{22} = 1$

$$F_{12} = \frac{1}{\frac{1}{\in_1} + \frac{A_1}{A_2} * \left(\frac{1}{\in_2} - 1\right)}$$

Heat transfer by radiation between body "1" and "2" $= Q = q\,A = F_{12}\,\sigma\,A(T_1^4 - T_2^4)$.

7.5. References

Bejan, A. (1996). Models of power plants that generate minimum entropy while operating at maximum power. *Am. J. Phys.*, 64, 1054–1059.

Bergman, T.L., Lavine, A.S., DeWitt, D.P., Incropera, F.P. (2011). *Introduction to Heat Transfer*, 6th edition. John Wiley & Sons, Hoboken.

Callen, H.B. (1985). *Thermodynamics and an Introduction to Thermostatistics*. John Wiley & Sons, Hoboken.

Cengel, Y.A. (2000). *Heat and Mass Transfer: Fundamentals and Applications*, 2nd edition. McGraw-Hill Education, London.

Cengel, Y.A. and Boles, M.A. (2018). *Thermodynamics: An Engineering Approach*. McGraw-Hill Education, London.

Chen, G. (1996). Nonlocal and nonequilibrium heat conduction in the vicinity of nanoparticles. *J. Heat Transfer*, 118(3), 539–543.

Curzon, F.L. and Ahlborn, B. (1975). Efficiency of a Carnot engine at maximum power output. *Am. J. Phys.*, 43, 22–24.

Diller, K.R. and Ryan, T.P. (1998). Special issue on challenges and opportunities in heat transfer applications. *J. Heat Transfer*, 120(4), 809.

Dutta, B.K. (2023). *Heat Transfer: Principles and Applications*, 2nd edition. PHI Learning Pvt. Ltd., New Delhi.

Flik, M.I., Choi, B.-I., Goodson, K.E. (1992). Heat transfer regimes in microstructures. *J. Heat Transfer*, 114(3), 666–674.

Foust, O.J. (ed.) (1972). Sodium chemistry and physical properties. In *Sodium-NaK Engineering Handbook. Volume I. Sodium Chemistry and Physical Properties*. Gordon & Breach, New York.

Gavhane, K.A. (2015). *Unit Operations – II*, 2nd edition. Nirali Prakashan, Mumbai.

Geankoplis, C.J. (1993). *Transport Processes and Unit Operations*, 3rd edition. Prentice Hall, New Jersey.

Holman, J.P. (1986). *Heat Transfer*, 6th edition. McGraw-Hill, New York.

Kern, D.Q. (1950). *Process Heat Transfer*, 2nd edition. McGraw-Hill, New York.

Klemens, P.G. (1969). Theory of the thermal conductivity of solids. In *Thermal Conductivity*, Tye, R.P. (ed.). Academic Press, London.

McCabe, W.L., Smith, J.C., Harriott, P. (1993). *Unit Operations of Chemical Engineering*, 5th edition. McGraw-Hill, New York.

McLaughlin, E. (1969). Theory of the thermal conductivity of fluids. In *Thermal Conductivity*, Tye, R.P. (ed.). Academic Press, London.

Moran, M.J. and Shapiro, H.N. (2004). *Fundamentals of Engineering Thermodynamics*. John Wiley & Sons, Hoboken.

Novikov, I.I. (1958). The efficiency of atomic power stations (a review). *J. Nuclear Energy II*, 7, 125–128.

Welty, J., Rorrer, G.L., Foster, D.G. (2020). *Fundamentals of Momentum, Heat, and Mass Transfer*, 6th edition. John Wiley & Sons, Hoboken.

Zeng, S.Q., Hunt, A., Greif, R. (1995). Geometric structure and thermal conductivity of porous medium silica aerogel. *J. Heat Transfer*, 117(4), 1055–1058.

8

Diffusion and Mass Transfer Coefficients

Diffusion and mass transfer coefficients are essential chemical engineering principles that govern molecules' movement between distinct phases. Diffusion involves the transfer of molecules from regions of high concentration to regions of low concentration. Mass transfer coefficients measure the rate at which mass is transferred between two phases. The initial section of this chapter is dedicated to fundamental principles and factors that affect diffusion processes, as well as analyzing diffusion coefficients and evaluating techniques for measuring these coefficients in gases, liquids and solids. Furthermore, the discussion extends to advancements in understanding diffusion within each phase.

The subsequent section of this chapter provides a more comprehensive analysis of mass transfer coefficients, including their definition, the factors that influence them and the overall mass transfer coefficients. The subject also encompasses interphase mass transfer, theories of mass transfer, techniques for measuring mass transfer coefficients and current advancements in this field.

8.1. Introduction

In chemical process industries, purification of raw materials, separation of desirable from undesirable products and treatment of effluent gases and liquids to remove toxic components are all common challenges engineers face. Although certain mixtures can be effectively segregated using mechanical methods such as cyclones or filtration, complex mixtures often necessitate approaches that go beyond mechanical separation, for example, the removal of pollutants like sulfur dioxide from industrial flue gases, the separation of hydrogen and nitrogen in an ammonia plant, separation of ethanol and acetone-butanol from the fermentation broth, etc. Diffusion-based methods, which exploit the mobility of components at the

---

Chapter written by A.V. RAGHAVENDRA RAO, Rompicherla SRIVIDYA, V. Sravani SAMEERA, K.S.N.V. PRASAD and Thangallapalli SRINIVAS.

molecular level in response to concentration gradients, are of great value in such circumstances. These processes, referred to collectively as mass transfer operations, tackle complex separation issues that surpass the capabilities of traditional mechanical methods.

Diffusion, which is the process of molecules moving spontaneously from higher concentration to lower concentration areas, is crucial in several industrial applications. In addition, mass transfer coefficients are important measures that express the rate at which mass transfers between phases, impacting the effectiveness and performance of many engineering operations.

Molecular diffusion in gases occurs due to the stochastic thermal movement of molecules, causing constituents to move from higher concentration areas to lower concentration areas until a state of equilibrium is reached. As per the kinetic theory of gases, molecules move in straight paths until they collide, and the mean free path measures the distance between these collisions. Fick's law of diffusion is a fundamental equation describing the diffusion process. It states that the molar flow is directly proportional to the concentration gradient. Various factors, such as temperature, concentration gradient, pressure and molecular weight, influence gas diffusion. Applications span from gas separation to exchanging breathing gases and evaporative cooling. Diffusion coefficients quantify the rate at which substances spread, influencing the process of transport phenomena. Different gases exhibit a wide range of diffusion coefficients, which can be determined by several experimental methods such as Stefan Tube, Taylor Dispersion, Gas Chromatography, Gravimetric, Pressure Drop, Infrared Absorption and Electrochemical procedures. Notable progress has been made in various areas, such as multiscale modeling, characterizing non-Fickian diffusion, using diffusion-based microfluidics, controlling diffusion with metamaterials and studying diffusion in quantum gases. These advancements have unveiled distinctive features, such as anomalous diffusion.

The Brownian motion of molecules or ions within a liquid, as opposed to gases, is impacted by increased intermolecular interactions and cohesive forces. Fick's law of diffusion, which applies to liquids, states that the diffusion rate is directly proportional to the concentration gradient and diffusivity. The text discusses several factors that influence the diffusion process in liquids, such as temperature, concentration gradient, viscosity and molecule size. Diffusion in liquids is crucial in water purification, medicine delivery and catalysis. Diffusion coefficients, which are crucial for measuring the movement of molecules, display a broad range of values. The exploration involves many ways of evaluating the diffusivity of liquids, including Crank's equation, Stefan's Law, chromatography, NMR spectroscopy, Taylor Dispersion Technique and electrochemical approaches. Recent advancements in areas such as microfluidics, diffusion in ionic liquids and diffusion in nanofluids are shaping the current state of liquid diffusion research.

Diffusion in solids is a process where atoms gradually travel. This movement is affected by factors such as vacancy, interstitial and dislocation diffusion. The diffusion process is influenced by various factors such as temperature, concentration gradient, crystal structure, flaws and grain size. Applications encompass many uses, including solid-state processes, semiconductor device manufacture and corrosion prevention. Fick's law highlights the significance of the diffusion coefficient. Notable progress has been made in this field, with the introduction of in situ methods, sophisticated microscopy, computational modeling and advancements in nanostructured materials. These breakthroughs enhance our comprehension of diffusion in energy storage, environmental materials and hydrogen embrittlement, influencing the cutting-edge of material science and technology.

Mass transfer coefficients, represented by the symbol "k", measure the efficiency of mass transfer between different phases, which is important in scientific and technical applications. Mass transfer is a process that occurs in biological systems, chemical engineering and environmental engineering. It refers to the movement of constituents, which is impacted by changes in concentration and the area of contact between different phases. The types of mass transfer that occur during interphase include gas–liquid, gas–solid, liquid–liquid and solid–solid. These processes significantly affect various applications such as wastewater treatment, catalysis and extraction. The concentration distribution near interfaces, the overall mass transfer coefficient (K) and mass transfer theories such as film, penetration, surface renewal and boundary layer provide insights into intricate paths. Various factors, including fluid characteristics, flow regime, interfacial area, temperature, pressure, chemical reactions, surfactants, system geometry, impurities and pollutants influence the coefficients.

Various measurement methods, ranging from absorption towers to electrochemical techniques, are used to estimate mass transfer coefficients that are essential for design and optimization. Progress is being made in various fields, such as nanotechnology, biotechnology, data analytics, machine learning and renewable energy applications. This progress is driving the advancement of knowledge and innovation in mass transmission.

8.2. Diffusion mass transfer

Mass transfer involves the transport of matter from a region of higher concentration to a lower concentration, driven by a concentration gradient. This movement can occur in various phases, including gases, liquids and solids. It is pivotal in various chemical engineering processes, including gas absorption, distillation, adsorption and drying. Understanding the principles of mass transfer is essential for designing and optimizing these processes effectively (Dutta 2007).

The primary driving force behind mass transfer is the difference in chemical potential between two regions. Chemical potential represents the energy associated with a molecule in a particular environment. Molecules tend to move from regions of high chemical potential to regions of low chemical potential, seeking a state of equilibrium (Wang et al. 2021).

There are two main mechanisms of mass transfer: molecular diffusion and eddy diffusion (Treybal 1980).

– Molecular diffusion: the spontaneous movement of molecules caused by their thermal energy is referred to as diffusion. Molecules in a high-concentration region are more likely to collide with their neighbors and move into a lower-concentration region. This random movement results in a net flux of molecules from the high-concentration zone to the low-concentration region (McQuarrie and Simon 1997).

– Eddy diffusion: this is the mass transfer caused by bulk fluid motion. When a fluid is in motion, it carries molecules, resulting in a net mass transfer between different regions. Various factors, such as pumps, mixers and temperature gradients, can induce eddy diffusion.

8.2.1. *Diffusion in gases*

Molecular diffusion in gases results from the stochastic thermal movement of molecules. In a fluid mixture of varying composition, the molecules of every component tend to move from regions with higher concentrations to regions of lower concentration. The movement persists until the system reaches a state of equilibrium, characterized by a homogeneous concentration throughout. Figure 8.1 shows the diffusion of gas particles from high to low concentrations. According to the kinetic theory of gases, molecules are postulated to move in a linear trajectory until they encounter another molecule and collide. The distance between two collisions is called the "mean free path" (Loeb 2004). The molecules exhibit random motion, and the overall displacement in a certain direction within a given period is referred to as the diffusion rate (Hartley and Crank 1949).

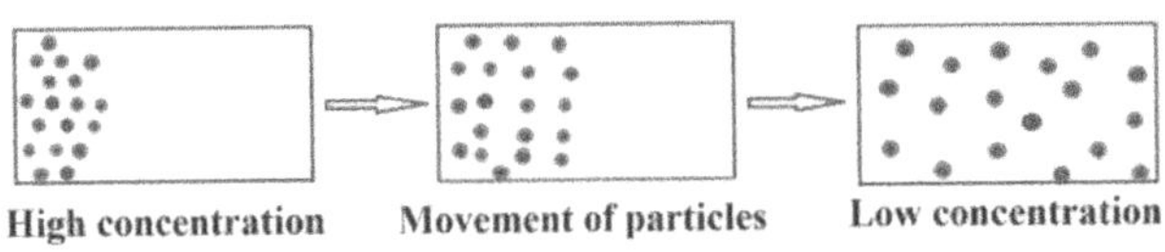

Figure 8.1. *Diffusion of gas particles from high concentration to low concentration*

8.2.1.1. *Fick's law of diffusion*

Thomas Graham (1805–1869), a Scottish chemist, is thought to have conducted the first thorough investigations on diffusion in gases. Graham started the quantitative investigation of gas diffusion when he realized that different types of gases do not align themselves by their densities when they come into contact with the heaviest gas at the bottom. Rather, they permeate one another before coming to an intimate mixing condition. Later, a German physiologist, Adolf Eugen Fick, was inspired to create a mathematical framework for diffusion phenomena by Graham's work on diffusion (Mehrer and Stolwijk 2009). This is named k's law of diffusion and is considered the fundamental equation that describes the diffusion process. Fick's law states that "The concentration gradient of a species in a given direction determines the molar flux of that species concerning an observer moving at the molar average velocity."

For a binary mixture and under steady-state conditions, Fick's law can be mathematically expressed as

$$J_A \propto \frac{dc_A}{dz}$$

$$J_A = -D_{AB}\frac{dc_A}{dz} \qquad [8.1]$$

where:

- J_A is the molar flux of component A in the Z direction, mol/m^2.s;
- C_A is the Concentration of A, mol/m^3;
- Z is the distance of diffusion, m;
- D_{AB} is the diffusion coefficient of the molecule A in mixture, m^2/s.

As diffusion occurs in the direction of decreasing concentration $\frac{dC_A}{dz}$ is negative, but the diffusional flux J_A is a positive quantity so we incorporated "–"ve sign in equation [8.1].

8.2.1.2. *Factors affecting diffusion in gases*

It is critical to investigate the factors governing gas diffusion for scientific understanding and practical applications. It gives important insights into the underlying behavior of gases. The rate of diffusion in gases (Cunningham and

Williams 1980; Chapman and Cowling 1990; Welty et al. 2020) is affected by several factors, including:

– Temperature: the diffusion rate is positively correlated with temperature, as higher temperatures result in increased molecular energy and faster movement.

– Concentration gradient: the diffusion rate is directly related to the concentration gradient. As the concentration gradient becomes more pronounced, the diffusion rate increases.

– Pressure: the diffusion rate is negatively correlated with pressure. The decrease in pressure reduces the mean free path of molecules, representing the average distance traveled by a molecule between collisions.

– Molecular weight: the diffusion rate is inversely proportional to the square root of the molecular weight of the diffusing substance. This is due to the fact that larger molecules possess greater inertia, resulting in a reduced likelihood of rapid movement compared to lighter molecules.

8.2.1.3. *Applications of diffusion in gases*

Diffusion in gases is used in a wide variety of applications (Brown 1918; Wagner 1977; Scholes et al. 2009; Prange 2012; Sanders et al. 2013; Cheng et al. 2016b) including:

– Gas separation: the process of diffusion in gases is employed to separate gases based on their distinct diffusion coefficients. For instance, diffusion is employed to extract helium from natural gas in gases.

– Respiratory gas exchange: the process of diffusion in gases is crucial for the exchange of oxygen and carbon dioxide between the lungs and the blood.

– Evaporative cooling uses the phenomenon of diffusion in gases to chill a surrounding environment by harnessing the evaporation of a liquid. As an illustration, evaporative cooling is employed in personal cooling devices and air conditioners.

Figure 8.2 shows the important applications of diffusion in gases.

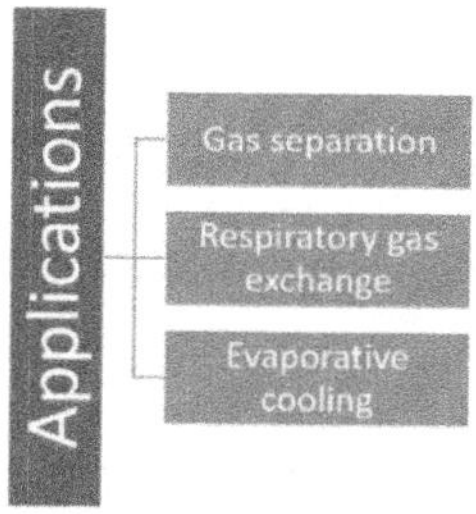

Figure 8.2. *Important applications of diffusion in gases*

8.2.1.4. *Diffusion coefficients in gases*

The diffusion coefficient measures how quickly a substance will diffuse in a particular medium. It is typically measured in units of square meters per second ($m^2\ s^{-1}$) or square centimeters per second ($cm^2\ s^{-1}$). The diffusion coefficient is a fundamental parameter in many transport phenomena, including diffusion, mass transfer and chemical reactions. Here are some examples of diffusion coefficients: the self-diffusion coefficient for hydrogen at 273 K is $1.285 \pm 0.0025\ cm^2\ s^{-1}$, and for nitrogen at 293 K, it is $0.200 \pm 0.008\ cm^2\ s^{-1}$; similarly, interdiffusion coefficients are given as 0.181 for O_2-N_2 at 273 K, as $0.139\ cm^2\ s^{-1}$ for O_2-CO_2 at 273 K (Roberts 1972).

Determination of diffusion coefficients in gases (Marrero and Mason 1972)

1) Experimental diffusion cells:

– Stefan Tube Method: the Stefan Tube Method involves filling a lengthy tube with a motionless gas and sealing one end. The opposite side is unobstructed and leads to a compartment housing the dispersing gas. The diffusion coefficient can be calculated by measuring the diffusion front's movement over time (Heinzelmann et al. 1965).

– Taylor Dispersion Method: this technique entails the examination of the scattering of a dissolved gas put into a stream of carrier gas. The dispersion rate offers insights into the diffusivity (Matsunaga et al. 2002; Cottet et al. 2010).

2) Chromatographic Techniques:

– Gas Chromatography (GC): gas chromatography entails the introduction of a sample into a column packed with a stationary phase. The migration rate observed in the column yields valuable data regarding the diffusion coefficients of the various components (Karaiskakis and Gavril 2004).

3) Gravimetric Methods:

– Wilke–Chang Method: this technique entails quantifying the variations in weight of a diffusion cell over a period as a gas permeates through it. The obtained data can be used to compute the diffusion coefficient (Ferrell and Himmelblau 1967).

4) Pressure Drop Methods:

– Capillary Flow Porometry: the measurement of gas permeability through a porous media is used to determine the gas diffusivity (Gostick et al. 2006).

5) Infrared Absorption Methods:

– Dynamic Light Scattering: the diffusion coefficient can be estimated by studying the Brownian motion of particles in a gas (Fröba et al. 2005).

6) Electrochemical Methods:

– Electrochemical Impedance Spectroscopy (EIS): the diffusion coefficient can be estimated by studying the Brownian motion of particles in a gas (Arico et al. 1992).

8.2.1.5. *Recent advances in diffusion in gases*

1) Multiscale modeling of diffusion in gases:

Various multiscale modeling tools have been devised to investigate the phenomenon of diffusion in gases across a range of lengths and time scales. These techniques integrate atomistic simulations, such as molecular dynamics, with continuum simulations, such as Navier–Stokes equations. This enables researchers to investigate the impact of molecular interactions, pore shapes and flow conditions on diffusion rates (Rama et al. 2010; Mi et al. 2014).

2) Non-Fickian diffusion in gases:

Non-Fickian diffusion is a form of diffusion that deviates from the principles outlined in Fick's laws. This form of diffusion is frequently found in intricate systems, such as porous substances and biological tissues. Recent research has demonstrated that non-Fickian diffusion can be described by fractional diffusion equations, which are more comprehensive than Fick's laws (Petropoulos and Roussis 1968; Frandsen et al. 2007).

3) Diffusion-based microfluidics:

Microfluidic devices that rely on diffusion are now under development for many purposes, such as medicine administration, chemical production and environmental surveillance. These devices employ diffusion to transfer molecules between distinct compartments, enabling accurate regulation of mixing and reaction speeds (Eddings and Gale 2006; Paknahad et al. 2018).

4) Diffusion in metamaterials:

Metamaterials are artificially designed materials that possess qualities that are not naturally found in the environment. Recent research has demonstrated that metamaterials can manipulate the diffusion of gases. Metamaterials can be engineered in a way that allows them to produce waveguides, which can either concentrate or redirect molecules that are spreading out in different directions (Fabre et al. 2006; Restrepo-Flórez and Maldovan 2016).

5) Diffusion in quantum gases:

Quantum gases are gaseous substances with quantum mechanical characteristics, including the simultaneous existence of both wave and particle qualities and the phenomenon of Bose–Einstein condensation. Recent research has demonstrated that quantum gas diffusion processes exhibit notable distinctions from diffusion in classical gases. Quantum gases can demonstrate anomalous diffusion, where the diffusion coefficient varies based on the length scale of the diffusion process (Danani et al. 1997; Yepez 2001).

8.2.2. *Diffusion in liquids*

Diffusion in liquids is similar to diffusion in gases, but it exhibits distinct variations because of the characteristics of liquids. It entails the transportation of molecules or ions from areas of high concentration to areas of low concentration within a liquid medium. Figure 8.3 shows how the purple dye diffuses in the water until its concentration is uniform throughout the liquid. Diffusion in liquids is crucial in multiple disciplines, such as chemistry, biology, food science and environmental studies (Tyrrell and Harris 2013).

Figure 8.3. *Diffusion of purple dye in water*

The stochastic movement of molecules or ions causes diffusion in liquids. Liquids exhibit stronger molecular connections and cohesive forces compared to gases, resulting in an impact on the diffusion process. Liquids often have a slower diffusion rate than gases because of heightened molecular interactions and constrained mobility (McCall et al. 1959).

Fick's law of diffusion, previously described in the context of gas diffusion, also applies to diffusion in liquids. Fick's first law states that the diffusion rate is directly proportional to the concentration gradient and the diffusivity of the material in the liquid (Narasimhan 2004).

8.2.2.1. *Factors affecting diffusion in liquids*

The rate of diffusion in liquids is affected by several factors (Dutta 2007), including:

– Temperature: the diffusion rate increases with temperature as molecules have more energy and move more rapidly.

– Concentration gradient: the diffusion rate is proportional to the concentration gradient. The steeper the concentration gradient, the faster the diffusion rate.

– Viscosity: the diffusion rate is inversely proportional to viscosity. This is because viscosity is a measure of a liquid's resistance to flow, and a more viscous liquid will hinder the movement of molecules.

– Molecular size: the diffusion rate is inversely proportional to the square root of the molecular weight of the diffusing substance. This is because heavier molecules have more inertia and are less likely to move as quickly as lighter molecules.

8.2.2.2. *Applications of diffusion in liquids*

Diffusion in liquids is used in a wide variety of applications (Wesselingh and Krishna 1990), including:

– Water purification: diffusion is used to remove impurities from water. For example, activated carbon filters use diffusion to remove organic contaminants from water.

– Drug delivery: diffusion is used in drug delivery to control the release of drugs into the body. For example, transdermal patches use diffusion to deliver drugs through the skin.

– Catalysis: diffusion is used to bring reactants together at the surface of a catalyst. This can increase the rate of chemical reactions.

Figure 8.4 shows the important applications of diffusion in liquids.

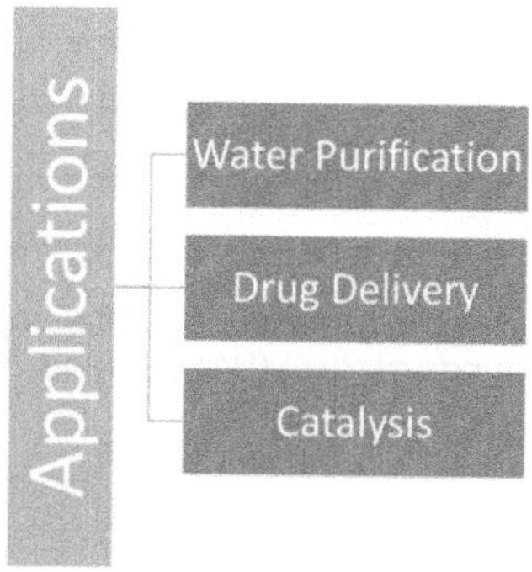

Figure 8.4. *Important applications of diffusion in liquids*

8.2.2.3. *Diffusion coefficients in liquids*

Diffusion coefficients are fundamental parameters that quantify the rate at which molecules move through a liquid medium. Diffusion coefficients in liquids typically range from 10^{-10} to 10^{-9} m²/s at room temperature. For example, the diffusion coefficient of water in water is 2.402 x 10^{-9} m²/s at 25°C, while the diffusion coefficient of sucrose in water is 4.98 x 10^{-10} m²/s at 25°C (Cussler 2009). Table 8.1 shows some liquid phase diffusivities of some compounds at 25°C.

Solvent	**Solute**	**D_{AB}×10^5 cm²/s**
Water	Acetic acid	1.26
Water	Acetone	1.28
Water	Ammonia	1.64
Water	Carbon dioxide	1.92
Water	Chlorine	1.25
Water	Ethanol	1.28
Water	Formic acid	1.50
Water	Methanol	1.60
Water	Nitric acid	2.60
Water	Oxygen	2.41
Water	Sulfuric acid	1.73
Benzene	Acetic acid	2.09
Chloroform	Acetone	2.35

Chloroform	Benzene	2.89
Acetone	Water	4.56
Ethyl alcohol	Water	1.13
***n*-Butyl alcohol**	Water	0.988
Ethyl acetate	Water	3.20

Table 8.1. *Liquid phase diffusivities of some compounds at 25°C (Winkelmann 2018)*

8.2.2.4. *Determination of liquid-phase diffusivity*

1) Crank's equation: Crank's equation is a classic approach for determining liquid-phase diffusivity in a semi-infinite medium. It relates the concentration profile of a solute in the liquid phase to time and distance. The equation is particularly applicable in cases where the liquid-phase concentration is influenced by a sudden change in solute concentration (Asl and Khajenoori 2013).

2) Stefan's Law: Stefan's law applies when a solute is released into a quiescent liquid from a point source. It establishes a relationship between the mass transfer rate, diffusivity and the concentration boundary layer thickness (Pacheco 1998).

3) Chromatography: liquid chromatography techniques can provide information on the diffusivity of solutes in a liquid phase. Chromatographic methods involve separating components based on their differential interactions with stationary and mobile liquid phases. The rate at which a solute moves through the chromatographic system can indicate its diffusivity (Song and Cabooter 2017).

4) NMR (Nuclear Magnetic Resonance) Spectroscopy: NMR spectroscopy can be employed to study the diffusion of solutes in a liquid phase. By monitoring changes in the NMR signal over time, we can derive information about the diffusivity of solutes in the liquid (Söderman et al. 2010).

5) Taylor dispersion technique: this technique involves monitoring the spreading of a solute plug as it moves through a liquid. By analyzing the rate of dispersion, the liquid-phase diffusivity can be determined (Alizadeh et al. 1980).

6) Electrochemical methods: electrochemical techniques like chronoamperometry or chronopotentiometry can study mass transport processes at electrodes. The resulting current or potential-time responses can provide insights into liquid-phase diffusivity (Molina et al. 2012).

8.2.2.5. *Recent advances in diffusion in liquids*

1) Microfluidics based on diffusion in liquids: microfluidic devices are being developed for various applications, including drug delivery, chemical synthesis and environmental monitoring. These devices use diffusion to transport molecules between different compartments, allowing for precise control of mixing and reaction rates (Talebi et al. 2017).

2) Diffusion in inoic liquids: ionic liquids are a type of liquid consisting solely of ions. Ionic liquids possess distinctive characteristics such as low volatility, good thermal stability and potent solvating capability. Recent research has indicated that diffusion in ionic liquids exhibits notable distinctions compared to diffusion in conventional liquids. Diffusion in ionic liquids is frequently described by a power law correlation between the diffusion coefficient and the concentration of molecules undergoing diffusion (Sangoro et al. 2011).

3) Diffusion in nanofluids: nanofluids consist of nanoparticles dispersed in a liquid medium. Nanofluids have several improved characteristics, including enhanced thermal conductivity and heat transfer coefficient. Recent research has indicated that the diffusion process in nanofluids exhibits notable distinctions compared to diffusion in conventional liquids. For instance, the diffusion process in nanofluids can be intensified by the random movement of nanoparticles, known as the Brownian motion (Tripathi et al. 2018).

8.2.3. *Diffusion in solids*

The diffusion process within solids involves the gradual shift of atoms or molecules from areas of high concentration to those of lower concentration (Mehrer 2007). While liquids and gases allow relatively free movement of molecules, diffusion encounters a slower pace within solids because of the robust interatomic bonds anchoring the atoms in position.

8.2.3.1. *Fick's law of diffusion in solids*

Under steady-state conditions, where diffusivity remains unaffected by concentration and bulk flow exists, the diffusive flux of component A within the solid is directly proportional to the concentration gradient in the direction of diffusion.

$$N_A = -D_A \frac{dC_A}{dZ}$$

where D_A is the diffusivity of A through solid.

Understanding and predicting the pace of atomic or molecular movement within a solid heavily rely on the diffusion coefficient. This movement occurs as thermal energy prompts atoms to vibrate, occasionally transitioning into empty lattice sites or interstitial spaces (Gao et al. 2020). While notably slower than diffusion in liquids and gases, diffusion in solids governs crucial phenomena such as solid-state reactions, heat treatment and the fabrication of semiconductor devices.

8.2.3.2. *Mechanisms of diffusion in solids*

1) Vacancy diffusion: this mechanism is the prevalent form of diffusion within solids. It involves atoms moving into vacant lattice sites generated by the absence of an atom (Van Gastel et al. 2001). This sequence continues as atoms effectively "hop" from one vacancy to another.

2) Interstitial diffusion: this process occurs when smaller atoms can fit into the available gaps amidst the lattice's atoms (Borg and Dienes 2012). It is more frequent when smaller atoms move through larger atoms due to size compatibility.

3) Dislocation diffusion: this particular mechanism occurs within dislocations, which are structural imperfections in the crystal lattice. Dislocations offer a swift route for diffusion, enabling atoms to move along these imperfections far more effortlessly than they can within the material's bulk (Hull and Bacon 2011).

8.2.3.3. *Factors affecting diffusion in solids*

1) Temperature: diffusion rates escalate as temperature rises due to atoms possessing greater thermal energy at higher temperatures, making them more prone to vibrating and filling vacancies (Suzuki and Mishin 2005).

2) Concentration gradient: the diffusion rate correlates directly with the concentration gradient. A sharper concentration gradient produces a swifter diffusion rate (Velegol et al. 2016).

3) Crystal structure: the crystal structure of a solid impacts its diffusion rate. Solids characterized by open crystal structures like body-centered cubic and face-centered cubic typically exhibit faster diffusion rates than solids with closed crystal structures such as hexagonal close-packed (Li et al. 2019).

4) Defects: imperfections such as vacancies and dislocations can potentially elevate the diffusion rate (Robson 2020).

5) Grain size: fine-grained materials generally tend to have a swifter diffusion rate than coarse-grained ones. This distinction arises from the higher presence of grain boundaries within fine-grained materials, offering rapid pathways for diffusion (Sen et al. 2023).

8.2.3.4. *Applications of diffusion in solids*

1) Solid-state reactions: many solid-state reactions, including alloy formation and crystal growth, rely fundamentally on diffusion (Jackson 2010).

2) Heat treatment: heat treatment processes such as annealing and tempering leverage diffusion to alter the microstructure and properties of metals (Banerjee 2017).

3) Semiconductor device fabrication: in manufacturing semiconductor devices like transistors, diffusion introduces impurities into silicon, thereby altering its electrical conductivity (Ye et al. 2017).

4) Corrosion: in the corrosion of metals, diffusion plays a role as oxygen and other corrosive elements can permeate into the metal, leading to its degradation (Fuhaid and Niaz 2022).

8.2.3.5. *Recent advances in diffusion in solids*

1) Advanced characterization techniques:

– In situ and operand techniques: new developments in methods such as in situ X-ray diffraction, neutron scattering and operand techniques have enabled scientists to directly observe and analyze diffusion processes within solid materials while they are in use. This real-time understanding offers a more precise grasp of the mechanisms that drive diffusion (Liu et al. 2019).

– Advanced microscopy: using advanced microscopy methods such as electron microscopy and atomic force microscopy, scientists can now visually examine and describe atomic-scale structures, defects and diffusion routes within solid materials (Zhao et al. 2022). This contributes significantly to comprehending diffusion at the nanolevel.

2) Computational approaches:

– Machine learning and computational modeling: the increasing prevalence of combining machine learning algorithms with computational modeling is aiding in predicting diffusion behavior within solids, particularly in intricate systems. These methods consider diverse parameters and structural elements to enhance accuracy (Keith et al. 2021).

– Quantum mechanical simulations: similar to density functional theory (DFT) calculations, these are more frequently applied to comprehend atomic-scale interactions and diffusion mechanisms within solids. These simulations offer insights into how electronic and structural factors influence diffusion (Parfitt et al. 2017).

3) Nanostructured materials:

– Engineered nanostructures: progress in developing and producing nanostructured materials has unlocked fresh opportunities to customize and improve diffusion characteristics (Harish et al. 2023). These materials, such as thin films, nanocomposites and porous structures, present distinctive diffusion routes and expanded surface areas, influencing how materials perform across various applications.

4) Energy storage and conversion:

– Battery technologies: regarding energy storage, recent progress in comprehending and refining diffusion processes within electrode materials is of substantial importance to battery technology. Concentrated endeavors aim to boost diffusion kinetics, thereby elevating the efficiency and longevity of batteries (Meng et al. 2017).

– Fuel cells and catalysis: grasping and controlling diffusion within solid materials are pivotal in creating effective fuel cells and catalysts. Recent studies strive to elevate the mass transport characteristics of solids, aiming to enhance the overall efficiency of energy conversion procedures (Brandon and Brett 2006).

5) Environmental and structural materials:

– Corrosion and degradation studies: progress in the diffusion study is enhancing our comprehension of corrosion and degradation mechanisms within structural materials. This understanding is pivotal in crafting materials that are more robust and long-lasting for diverse applications such as infrastructure and aerospace (Iqbal and Moskal 2023).

– Hydrogen embrittlement: exploring hydrogen diffusion within metals, especially regarding hydrogen embrittlement, has garnered significant attention in research (Laadel et al. 2022). Progress in comprehending these diffusion mechanisms aids in crafting materials that offer enhanced resilience against hydrogen-induced degradation.

8.3. Mass transfer coefficients

The mass transfer coefficient, commonly represented as “k”, quantifies mass transfer efficiency between two phases, including immiscible liquids or gases. It is employed to describe the speed at which a specific constituent in a solution transitions from one phase to another (Geankoplis 2003). Mass transfer is an essential phenomenon that is pivotal in numerous scientific and engineering processes, such as biological systems, chemical engineering and environmental engineering (Faghri et al. 2010).

Rate of mass transfer α concentration driving force (the difference in concentration).

The rate of mass transfers is the area of contact between the phases

– if w_A = rate of mass transfer (kmol/s) of solute;

– Δc_A = concentration driving force between two points;

– a = area of contact between phases, m^2.

$$w_A \; \alpha \; a \, \Delta c_A$$

$$w_A = k_c \, a \, \Delta c_A$$

$$If \; N_A \; is \; the \; flux, we \; can \; write$$

$$N_A = \frac{w_A}{a} = k_c \, \Delta c_A$$

$$k_c = \frac{N_A}{\Delta c_A} \left(\frac{molar \; flux}{concentration \; driving \; force} \right)$$

The inverse of mass transfer coefficient is a measure of mass transfer resistance (Dutta 2007).

8.3.1. *Interphase mass transfer*

The process by which a component transfers from one phase to another across an interface is known as interphase mass transfer. Many chemical engineering activities, such as absorption, distillation, drying and extraction, rely on this process (Burghardt 2010).

There are four main types of interphase mass transfer:

1) Gas–liquid: this kind of interphase mass transfer is the most prevalent and is used in procedures such as wastewater ammonia stripping and CO_2 removal from flue gas. In gas–liquid systems, it is usually believed that the gas phase is well mixed, while the mass transfer resistance of the liquid phase may cause concentration gradients in the liquid phase.

2) Gas–solid: this type of mass transfer is important in processes such as gas adsorption (e.g. drying of solids) and catalytic reactions. The solid phase can be porous in gas–solid systems, allowing the gas to diffuse within its internal structure.

3) Liquid–liquid: this type of mass transfer is used in processes like extraction (e.g. separation of components based on their solubility in different liquids). Both phases are typically well-mixed in liquid–liquid systems, but concentration gradients can still exist at the interface due to mass transfer resistance.

4) Solid–solid: this type of mass transfer is less common but can be important in diffusion bonding and sintering processes. In solid–solid systems, mass transfer occurs through diffusion within the solid lattice.

8.3.2. *Concentration profiles at the interface*

A concentration profile forms close to an interface when a mass transfer occurs across it. As we get farther from the interface, this profile is usually characterized by a reduction in the concentration of the component being transferred. Numerous variables, including the type of mass transfer process, mass transfer coefficients and fluid flow regime, affect the profile's shape (Danckwerts 1970).

8.3.3. *Overall mass transfer coefficient (K)*

The overall mass transfer coefficient (K) is a single quantity that accounts for the whole resistance to mass transfer in both phases and at the interface. It is defined as the mass transfer rate divided by the concentration difference across the interface (Anantharaman and Begum 2017).

$$K = \frac{N}{\Delta c}$$

where:

– N is the mass transfer rate;

– Δc is the concentration difference across the interface.

The overall mass transfer coefficient is a valuable parameter used in designing and analyzing mass transfer equipment.

Relationship between K and individual mass transfer coefficients

The overall mass transfer coefficient can be stated as a function of each phase's individual mass transfer coefficients. These separate coefficients represent the resistance to mass transfer throughout each phase. In gas–liquid systems, for example, the overall mass transfer coefficient can be linked to the individual gas-phase and liquid-phase mass transfer coefficients using the equation below (Basmadjian 2003).

$$\frac{1}{K}=\frac{1}{k_g}+\frac{1}{k_l}$$

where:

– k_g is the gas-phase mass transfer coefficient;

– k_l is the liquid-phase mass transfer coefficient.

This equation highlights the role of individual mass transfer coefficients in determining the overall mass transfer rate across the interface.

Engineers can efficiently design and optimize mass transfer equipment for various chemical engineering applications by grasping the concepts of interphase mass transfer, concentration profiles and overall mass transfer coefficient.

8.3.4. *Mass transfer theories: unveiling the mechanisms of interphase transport*

Mass transfer theories are valuable tools for understanding the complex pathways of mass transfer between interfaces. Comprehending and implementing these theories empower engineers to create and enhance mass transfer procedures for diverse industrial and environmental uses, contributing to progress in energy, medicine and materials science.

1) Film theory:

According to the film theory, a stagnant fluid film exists at the interface between two phases, impeding mass transfer. The concentration gradient drives the component's movement across this film via diffusion. The mass transfer rate is determined by the thickness of the film and the diffusion coefficient. This theory is widely used for gas–liquid and liquid–liquid systems (Lewis and Whitman 1924).

2) Penetration theory:

According to the penetration theory, the interface is a dynamic boundary where fresh fluid elements replace those depleted of the transferred component. The element renewal frequency and the diffusion coefficient within the elements determine the mass transfer rate. This theory is especially useful for understanding turbulent flows and unsteady-state mass transfer (Higbie 1935).

3) Surface renewal theory:

Surface renewal theory incorporates aspects of both film and penetration theories. It proposes that small, turbulent eddies constantly renew the interface.

These eddies carry fresh fluid to the interface while removing depleted fluid, resulting in a dynamic environment for mass transfer. This theory applies especially to high-turbulence situations (Danckwerts 1951).

4) Boundary layer theory:

Boundary layer theory applies principles from fluid dynamics to mass transfer. This study examines the formation of concentration boundary layers and their impact on the mass transfer process. It establishes a relationship between the speed at which mass is transferred and the difference in concentration between the boundary layer. A higher mass transfer coefficient leads to a more effective process (Schlichting and Gersten 2016).

Choosing the correct theory

Several factors influence the selection of the best mass transfer theory:

– System type: different theories are better suited for specific phase combinations (gas–liquid, liquid–liquid, etc.).

– Flow regime: turbulent flows are treated differently than laminar flows.

Chemical engineers can choose the best tool for analyzing and predicting mass transfer phenomena in their specific applications by understanding the strengths and limitations of each theory.

8.3.5. Factors affecting mass transfer coefficients

1) Fluid properties (Song et al. 2017):

– Viscosity: increased viscosity impedes fluid flow and diffusion, resulting in a decrease in k.

– Density: greater densities result in enhanced momentum transfer and mixing, resulting in higher values of k.

– Diffusivity: the intrinsic coefficient of diffusion of the transferred component significantly impacts k. Increased diffusivity promotes accelerated motion and leads to higher k values.

2) Flow regime (Kashid et al. 2011):

– Laminar flow: this is a smooth and organized flow where mass transfer mainly occurs through molecular diffusion, resulting in low k values.

– Turbulent flow: the presence of turbulent eddies increases the mixing and facilitates faster diffusion, leading to considerably higher values of k.

3) Interfacial area: the mass transfer rate increases proportionally with the increase in the available interfacial area. To enhance this, we can use packing materials, baffles and other methods that generate an additional surface area for the phase interaction (Maalej et al. 2003).

4) Temperature: the increase in temperature primarily affects viscosity and diffusivity, leading to their rise. Elevated temperatures reduce viscosity and enhance diffusivity, promoting accelerated mass transfer (Ho et al. 2020).

5) Pressure: pressure generally exerts a relatively minor impact on the value of k compared to other factors. Nevertheless, in gas–liquid systems, elevated pressure can result in greater gas solubility and marginally higher k (Kojima et al. 1997).

6) Chemical reactions: the occurrence of chemical reactions at the interface can substantially impact k. If the reaction depletes the transferred component, the value of k will be greater due to the heightened driving force. On the contrary, if the reaction generates the transferred component, the value of k may be decreased due to the diminished driving force (Aroonwilas et al. 1999).

7) Surface-active agents: the presence of surfactants, which are surface-active agents, can have diverse effects on the value of k, depending on their specific properties and concentration. Certain surfactants can decrease interfacial tension, increasing the rate constant. Conversely, other surfactants can create layers that impede mass transfer and decrease the rate constant. (Burnett and Himmelblau 1970).

8) Geometry of the system: the geometry of the apparatus in which mass transfer takes place can impact the flow patterns, interfacial area and turbulence levels, ultimately influencing the rate constant (k). For instance, baffles in a tank can induce turbulence and enhance the value of k, whereas a lengthy and narrow tube can facilitate laminar flow and decrease k (Cheng et al. 2016b).

9) Impurities and contaminants: impurities and contaminants in the fluids can impact k differently. They can enhance the fluids' thickness, decrease the diffusion rate or adhere to the interface, all resulting in a reduction in k (Vasconcelos et al. 2003).

Understanding how these factors interact and how they affect k allows engineers to optimize mass transfer processes for various applications.

By manipulating these factors, they can control mass transfer rates, improve efficiency and achieve desired product characteristics.

8.3.6. *Measurement of mass transfer coefficients*

The measurement of mass transfer coefficients involves experimental techniques to determine the rate at which a particular component transfers from one phase to another. The mass transfer coefficient is a key parameter in mass transfer processes, and its accurate determination is essential for the design and optimization of various engineering and industrial applications. Here are some common methods used to measure mass transfer coefficients:

1) Experimental methods: absorption and stripping towers: in absorption and stripping processes, towers are used to bring gas and liquid phases into contact. The mass transfer coefficient can be calculated by measuring the concentrations at different heights or stages (Geankoplis 2003).

2) Correlation with empirical equations: Sherwood number correlations: the Sherwood number (Sh) is a dimensionless parameter that relates mass transfer to heat transfer. Empirical correlations exist between the Sherwood and Reynolds numbers, Schmidt numbers, etc. These correlations are used to estimate mass transfer coefficients (Dutta 2007). Table 8.2 shows some mass transfer coefficient correlations for a few simple situations.

Situation	**Condition**	**Correlation**
Flow in circular tube (laminar flow)	$Re \leq 2$	$Sh = 1.62\left[(Re)(Sc)\left(\frac{d}{L}\right)\right]^{\frac{1}{3}}$
Turbulent flow (pipe)	$4000. \leq Re \leq 60{,}000$; $0.6 \leq Sc \leq 3{,}000$	$Sh = 0.023\,(Re)^{0.83}\,(Sc)^{0.33}$
Boundary layer flow (flat plate)	$Re_l < 80{,}000$ $Re_l > 5 \times 10^5$	$j_D = 0.064\,(Re_l)^{-0.5}$ $j_D = 0.037\,(Re_l)^{-0.2}$
Flow wetted wall	$3000. < \boldsymbol{Re'} < 40{,}000$; $0.5 < Sc < 3$	$j_D = 0.0328\,(Re')^{-0.23}$
Gas flow (packed bed)	$10 \leq Re'' \leq 2{,}500$	$j_D = 1.17\,(Re'')^{-0.415}$
Liquid flow (packed bed)	$Re'' < 55$ $3 < Re'' < 10{,}000$	$j_D = 1.09\,(Re'')^{-2/3}$ $Sh = 2 + 1.1(Re)^{0.6}\,(Sc)^{0.33}$

Table 8.2. *Mass transfer coefficient correlations for a few simple situations (Dutta 2007)*

3) Electrochemical Techniques (Cañizares et al. 2006):

– Rotating Disc Electrode (RDE) method: in electrochemical systems, this technique includes monitoring the mass transfer of a material to or from an electrode surface. The gathered data can be used to compute mass transfer coefficients.

– Cyclic Voltammetry: this electrochemical technique can be used to analyze mass transfer mechanisms by analyzing the current response as the applied voltage varies cyclically. The diffusion-limited current can be used to calculate the mass transfer coefficient.

4) Boundary layer analysis: boundary layer thickness measurement: the thickness of the boundary layer can be measured by analyzing concentration patterns near a surface. The mass transfer coefficient is proportional to the thickness of the boundary layer and the concentration gradient (Marciano and Harbeck 1952).

5) Use of tracers: tracer studies: injecting a known tracer chemical into a system allows the movement and concentration of the tracer to be tracked over time. This data can be used to compute mass transfer coefficients (Maraqa 2001).

6) Computational Fluid Dynamics (CFD): CFD simulations can be used to model mass transfer processes in complicated systems. Mass transfer coefficients can be estimated by analyzing the simulated data (Prajongkan et al. 2012).

The method chosen is determined by the specific system, the phase involved (gas–liquid, liquid–liquid, etc.), and the equipment and knowledge available. Multiple methods or a combination of experimental and computational approaches may be applied to obtain complete knowledge of mass transfer in a given system.

8.3.7. *Advances in mass transfer coefficients*

1) Advanced computational modeling:

– The application of CFD and advanced simulation techniques is expanding. Researchers may use high-performance computers to simulate complicated systems, providing precise insights into mass transfer processes in various industrial and environmental applications (Ranganathan and Sivaraman 2011).

2) Nanotechnology and nanomaterials:

– Advances in nanotechnology have resulted in the development of innovative materials with improved mass transfer capabilities. Nanomaterials such as nanoparticles and nanostructured membranes are being investigated for their potential to improve mass transfer coefficients in various processes (Ahmed 2019).

3) Biotechnology and bioengineering:

– Continuous research in biotechnology is being done to optimize mass transfer in bioreactors for fermentation and cell culture operations. Understanding the mass transfer constraints in biological systems improves bioprocess engineering (Garcia-Ochoa and Gomez 2005).

4) Integration of data analytics and machine learning:

– Data-driven techniques like machine learning algorithms analyze enormous data sets created during experiments or simulations. This integration aids in identifying patterns, optimizing processes and the prediction of mass transfer coefficients under varying situations (Hazare et al. 2022).

5) Renewable energy applications:

– In fields such as fuel cells and solar energy, researchers are looking into materials and designs that can improve mass transfer, increasing energy conversion processes' efficiency (Nawayseh et al. 1999).

8.4. Nomenclature

a	Area of contact between phases, m^2
ΔC_A	Concentration driving force of species A between two points, mol/m^3
d	Tube diameter, m
D_{AB}	Diffusion coefficient, m^2/s
h	Heat transfer coefficient $W/m^2.K$
J_A	Molar flux of component A, $kmol/m^2$ s
j_D	A factor defined in Table 8.2
K	Individual mass transfer coefficient
k_g	Gas phase mass transfer coefficient
k_l	Liquid phase mass transfer coefficient
K	Overall mass transfer coefficient
k_c	Mass transfer coefficient (concentration driving force)
L	Characteristic length
L	Length of the tube, m
N_A	Molar flux of species A $kmol/m^2.s$
Re	Reynolds Number, $d v \rho / \mu$

Re_l	Reynolds Number of liquid phase, $dv\rho/\mu$
Re'	Reynolds Number, $dv'\rho/\mu$
Re''	Reynolds Number, $dv''\rho/\mu$
Sc	Schmidt number, $\mu/\rho D$
Sh	Sherwood number, kl/D
v	Velocity of fluid, m/s
v'	Velocity relative to surface of film, m/s
v''	Superficial velocity of the fluid, m/s
w_A	Rate of mass transfer (kmol/s) of solute
Z	Distance of diffusion, m
ρ	Density of the fluid, kg/m^3
μ	Viscosity of a fluid, kg/m.s

8.5. References

Ahmed, M.S. (2019). Nanofluid: New fluids by nanotechnology. In *Thermophysical Properties of Complex Materials*, Shahzad, A. (ed.). IntechOpen, London.

Alizadeh, A., Nieto de Castro, C.A., Wakeham, W.A. (1980). The theory of the Taylor dispersion technique for liquid diffusivity measurements. *International Journal of Thermophysics*, 1, 243–284.

Anantharaman, N. and Begum, K.M.S. (2017). *Mass Transfer: Theory and Practice*. PHI Learning Pvt. Ltd., Patparganj.

Arico, A.S., Alderrucci, V., Antonucci, V., Ferrara, S., Recupero, V., Giordano, N., Kinoshita, K. (1992). *ac* Impedance spectroscopy of porous gas diffusion electrode in sulphuric acid. *Electrochimica Acta*, 37(3), 523–529.

Aroonwilas, A., Veawab, A., Tontiwachwuthikul, P. (1999). Behavior of the mass-transfer coefficient of structured packings in CO2 absorbers with chemical reactions. *Industrial & Engineering Chemistry Research*, 38(5), 2044–2050.

Asl, A.H. and Khajenoori, M. (2013). Subcritical water extraction. In *Mass Transfer – Advances in Sustainable Energy and Environment Oriented Numerical Modeling*, Nakajima, H. (ed.). IntechOpen, London, 459–487.

Banerjee, M.K. (2017). Fundamentals of heat-treating metals and alloys. *Comprehensive Materials Finishing*, 1–49.

Basmadjian, D. (2003). *Mass Transfer: Principles and Applications*. CRC Press, Boca Raton.

Borg, R.J. and Dienes, G.J. (2012). *An Introduction to Solid State Diffusion*. Elsevier, Amsterdam.

Brandon, N.P. and Brett, D.J. (2006). Engineering porous materials for fuel cell applications. *Philosophical Transactions of the Royal Society A: Mathematical, Physical and Engineering Sciences*, 364(1838), 147–159.

Brown, H.T. (1918). The principles of diffusion, their analogies and applications. *Journal of the Chemical Society, Transactions*, 113, 559–585.

Burghardt, A. (2010). Interphase mass transfer. In *Chemical Engineering and Chemical Process Technology - Volume I: Fundamentals of Chemical Engineering*, Pohorecki, R., Bridgwater, J., Molzahn, M., Gani, R., Gallegos, C. (eds). EOLSS Publications, 236.

Burnett Jr, J.C. and Himmelblau, D.M. (1970). The effect of surface-active agents on interphase mass transfer. *AIChE Journal*, 16(2), 185–193.

Cañizares, P., García-Gómez, J., Fernández de Marcos, I., Rodrigo, M.A., Lobato, J. (2006). Measurement of mass-transfer coefficients by an electrochemical technique. *Journal of Chemical Education*, 83(8), 1204.

Chapman, S. and Cowling, T.G. (1990). *The Mathematical Theory of Non-uniform Gases: An Account of the Kinetic Theory of Viscosity, Thermal Conduction and Diffusion in Gases*. Cambridge University Press, Cambridge.

Cheng, W.L., Zhang, W.W., Chen, H., Hu, L. (2016a). Spray cooling and flash evaporation cooling: The current development and application. *Renewable and Sustainable Energy Reviews*, 55, 614–628.

Cheng, X., Xie, Y., Zheng, H., Yang, Q., Zhu, D., Xie, J. (2016b). Effect of the different shapes of air diffuser on oxygen mass transfer coefficients in microporous aeration systems. *Procedia Engineering*, 154, 1079–1086.

Cottet, H., Biron, J.P., Cipelletti, L., Matmour, R., Martin, M. (2010). Determination of individual diffusion coefficients in evolving binary mixtures by taylor dispersion analysis: Application to the monitoring of polymer reaction. *Analytical Chemistry*, 82(5), 1793–1802.

Cunningham, R.E. and Williams, R.J.J. (1980). *Diffusion in Gases and Porous Media*. Plenum Press, New York.

Cussler, E.L. (2009). *Diffusion: Mass Transfer in Fluid Systems*. Cambridge University Press, Cambridge.

Danani, A., Ferrando, R., Scalas, E., Torri, M. (1997). Lattice-gas theory of collective diffusion in adsorbed layers. *International Journal of Modern Physics B*, 11(19), 2217–2279.

Danckwerts, P.V. (1951). Significance of liquid-film coefficients in gas absorption. *Industrial & Engineering Chemistry*, 43(6), 1460–1467.

Danckwerts, P.V. (1970). *Gas-liquid Reactions*. McGraw-Hill, New York.

Dutta, B.K. (2007). *Principles of Mass Transfer and Seperation Processes*. PHI Learning Pvt. Ltd., Patparganj.

Eddings, M.A. and Gale, B.K. (2006). A PDMS-based gas permeation pump for on-chip fluid handling in microfluidic devices. *Journal of Micromechanics and Microengineering*, 16(11), 2396.

Fabre, N., Fasquel, S., Legrand, C., Mélique, X., Muller, M., François, M., Vanbésien, O., Lippens, D. (2006). Towards focusing using photonic crystal flat lens. *Opto-Electronics Review*, 14, 225–232.

Faghri, A., Zhang, Y., Howell, J.R. (2010). *Advanced Heat and Mass Transfer*. Global Digital Press, Kanpur.

Ferrell, R.T. and Himmelblau, D.M. (1967). Diffusion coefficients of hydrogen and helium in water. *AIChE Journal*, 13(4), 702–708.

Frandsen, H.L., Damkilde, L., Svensson, S. (2007). A revised multi-Fickian moisture transport model to describe non-Fickian effects in wood. *Holzforschung*, 61(5), 563–572.

Fröba, A.P., Botero, C., Kremer, H., Leipertz, A. (2005). Mutual diffusion coefficient in fluids by dynamic light scattering. *Diffusion Fundamentals: The Open-Access Journal for the Basic Principles of Diffusion Theory, Experiment and Application*, 2, 70.1–70.2.

Fuhaid, A.F.A. and Niaz, A. (2022). Carbonation and corrosion problems in reinforced concrete structures. *Buildings*, 12(5), 586.

Gao, Y., Nolan, A.M., Du, P., Wu, Y., Yang, C., Chen, Q., Mo, Y., Bo, S.-H. (2020). Classical and emerging characterization techniques for investigation of ion transport mechanisms in crystalline fast ionic conductors. *Chemical Reviews* 120(13), 5954–6008.

Garcia-Ochoa, F. and Gomez, E. (2005). Prediction of gas-liquid mass transfer coefficient in sparged stirred tank bioreactors. *Biotechnology and Bioengineering*, 92(6), 761–772.

Geankoplis, C. (2003). *Transport Processes and Separation Process Principles (Includes Unit Operations)*. Prentice Hall, Hoboken.

Gilliland, E.R. (1934). Diffusion coefficients in gaseous systems. *Industrial & Engineering Chemistry*, 26(6), 681–685.

Gostick, J.T., Fowler, M.W., Ioannidis, M.A., Pritzker, M.D., Volfkovich, Y.M., Sakars, A. (2006). Capillary pressure and hydrophilic porosity in gas diffusion layers for polymer electrolyte fuel cells. *Journal of Power Sources*, 156(2), 375–387.

Harish, V., Ansari, M.M., Tewari, D., Yadav, A.B., Sharma, N., Bawarig, S., Betancourt, M.L.G., Kartalu, A., Bechelany, M., Barhoum, A. (2023). Cutting-edge advances in tailoring size, shape, and functionality of nanoparticles and nanostructures: A review. *Journal of the Taiwan Institute of Chemical Engineers*, 149, 105010.

Hartley, G.S. and Crank, J.F. (1949). Some fundamental definitions and concepts in diffusion processes. *Transactions of the Faraday Society*, 45, 801–818.

Hazare, S.R., Vala, S.V., Patil, C.S., Joshi, A.J., Joshi, J.B., Vitankar, V.S., Patwardhan, A.W. (2022). Correlating interfacial area and volumetric mass transfer coefficient in bubble column with the help of machine learning methods. *Industrial & Engineering Chemistry Research*, 62(5), 2104–2123.

Heinzelmann, F.J., Wasan, D.T., Wilke, C.R. (1965). Concentration profiles in Stefan diffusion tube. *Industrial & Engineering Chemistry Fundamentals*, 4(1), 55–61.

Higbie, R. (1935). The rate of absorption of pure gas into a still liquid during short periods of exposure. *Transactions of the American Institute of Chemical Engineers*, 31, 365–389.

Ho, D., Kim, K., Earmme, T., Kim, C. (2020). Enhancing gas–liquid volumetric mass transfer coefficient. *Journal of Industrial and Engineering Chemistry*, 87, 1–17.

Hull, D. and Bacon, D.J. (2011). *Introduction to Dislocations*. Elsevier, Amsterdam.

Iqbal, A. and Moskal, G. (2023). Recent development in advanced ceramic materials and understanding the mechanisms of degradation of thermal barrier coatings. *Archives of Computational Methods in Engineering*, 30, 4855–4896.

Jackson, K.A. (2010). *Kinetic Processes: Crystal Growth, Diffusion, and Phase Transitions in Materials*. John Wiley & Sons, Hoboken.

Karaiskakis, G. and Gavril, D. (2004). Determination of diffusion coefficients by gas chromatography. *Journal of Chromatography A*, 1037(1–2), 147–189.

Kashid, M.N., Renken, A., Kiwi-Minsker, L. (2011). Influence of flow regime on mass transfer in different types of microchannels. *Industrial & Engineering Chemistry Research*, 50(11), 6906–6914.

Keith, J.A., Vassilev-Galindo, V., Cheng, B., Chmiela, S., Gastegger, M., Müller, K.R., Tkatchenko, A. (2021). Combining machine learning and computational chemistry for predictive insights into chemical systems. *Chemical Reviews*, 121(16), 9816–9872.

Khan, F.A., Fischer, C., Straatman, A.G. (2015). Numerical model for non-equilibrium heat and mass exchange in conjugate fluid/solid/porous domains applied to evaporative cooling and drying. *International Journal of Heat and Mass Transfer*, 80, 513–528.

Kojima, H., Sawai, J., Suzuki, H. (1997). Effect of pressure on volumetric mass transfer coefficient and gas holdup in bubble column. *Chemical Engineering Science*, 52(21–22), 4111–4116.

Laadel, N.E., El Mansori, M., Kang, N., Marlin, S., Boussant-Roux, Y. (2022). Permeation barriers for hydrogen embrittlement prevention in metals – A review on mechanisms, materials suitability and efficiency. *International Journal of Hydrogen Energy*, 47(76), 32707–32731.

Lewis, W.K. and Whitman, W.G. (1924). Principles of gas absorption. *Industrial & Engineering Chemistry*, 16(12), 1215–1220.

Li, Z., Zhao, S., Ritchie, R.O., Meyers, M.A. (2019). Mechanical properties of high-entropy alloys with emphasis on face-centered cubic alloys. *Progress in Materials Science*, 102, 296–345.

Liu, D., Shadike, Z., Lin, R., Qian, K., Li, H., Li, K., Wang, S., Yu, Q., Liu, M., Ganapathy, S. et al. (2019). Review recent development of in situ/operando characterization techniques for lithium battery research. *Advanced Materials*, 31(28), 1806620.

Loeb, L.B. (2004). *The Kinetic Theory of Gases*. Courier Corporation, North Chelmsford.

Maalej, S., Benadda, B., Otterbein, M. (2003). Interfacial area and volumetric mass transfer coefficient in a bubble reactor at elevated pressures. *Chemical Engineering Science*, 58(11), 2365–2376.

Maraqa, M.A. (2001). Prediction of mass-transfer coefficient for solute transport in porous media. *Journal of Contaminant Hydrology*, 50(1–2), 1–19.

Marciano, J.J. and Harbeck, G.E. (1952). Mass-transfer studies. water-loss investigations: Lake Hefner studies. *U.S. Geological Survey, Circular*, 229, 46–70.

Marrero, T.R. and Mason, E.A. (1972). Gaseous diffusion coefficients. *Journal of Physical and Chemical Reference Data*, 1(1), 3–118.

Matsunaga, N., Hori, M., Nagashima, A. (2002). Measurement of mutual diffusion coefficients of gases by the Taylor method: Measurements on H2–Air, H2–N2, and H2–O2 systems. *Heat Transfer – Asian Research: Cosponsored by the Society of Chemical Engineers of Japan and the Heat Transfer Division of ASME*, 31(3), 182–193.

McCall, D.W., Douglass, D.C., Anderson, E.W. (1959). Diffusion in liquids. *The Journal of Chemical Physics*, 31(6), 1555–1557.

McQuarrie, D.A. and Simon, J.D. (1997). *Physical Chemistry: A Molecular Approach*. University Science Books, Sausalito, CA.

Mehrer, H. (2007). *Diffusion in Solids: Fundamentals, Methods, Materials, Diffusion-Controlled Processes*. Springer Science & Business Media, Berlin.

Mehrer, H., and Stolwijk, N.A. (2009). Heroes and highlights in the history of diffusion. *Diffusion Fundamentals: The Open-Access Journal for the Basic Principles of Diffusion Theory, Experiment and Application*, 11(1), 1–32.

Meng, J., Guo, H., Niu, C., Zhao, Y., Xu, L., Li, Q., Mai, L. (2017). Advances in structure and property optimizations of battery electrode materials. *Joule*, 1(3), 522–547.

Mi, L., Jiang, H., Li, J. (2014). The impact of diffusion type on multiscale discrete fracture model numerical simulation for shale gas. *Journal of Natural Gas Science and Engineering*, 20, 74–81.

Molina, Á., Serna, C., Ortuño, J.A., Torralba, E. (2012). Studies of ion transfer across liquid membranes by electrochemical techniques. *Annual Reports Section "C" (Physical Chemistry)*, 108, 126–176.

Narasimhan, T.N. (2004). Fick's insights on liquid diffusion. *Eos, Transactions American Geophysical Union*, 85(47), 499–501.

Nawayseh, N.K., Farid, M.M., Al-Hallaj, S., Al-Timimi, A.R. (1999). Solar desalination based on humidification process – I. Evaluating the heat and mass transfer coefficients. *Energy Conversion and Management*, 40(13), 1423–1439.

Pacheco, M.A. (1998). *Mass Transfer, Kinetics and Rate-Based Modeling of Reactive Absorption*. The University of Texas, Austin.

Paknahad, M., Bachhal, J.S., Hoorfar, M. (2018). Diffusion-based humidity control membrane for microfluidic-based gas detectors. *Analytica Chimica Acta*, 1021, 103–112.

Parfitt, D., Kordatos, A., Filippatos, P.P., Chroneos, A. (2017). Diffusion in energy materials: Governing dynamics from atomistic modeling. *Applied Physics Reviews*, 4(3). doi: 10.1063/1.5001276.

Petropoulos, J.H. and Roussis, P.P. (1968). Study of "Non-Fickian" diffusion anomalies through time lags. III. Simple distance-dependent anomalies in microporous media. *The Journal of Chemical Physics*, 48(10), 4619–4624.

Prajongkan, Y., Piumsomboon, P., Chalermsinsuwan, B. (2012). Computation of mass transfer coefficient and Sherwood number in circulating fluidized bed downer using computational fluid dynamics simulation. *Chemical Engineering and Processing: Process Intensification*, 59, 22–35.

Prange, H. (2012). *Respiratory Physiology: Understanding Gas Exchange*. Springer Science & Business Media, Berlin.

Qureshi, S.M. and Mustafa, R. (2018). Measurement of respiratory function: Gas exchange and its clinical applications. *Anaesthesia & Intensive Care Medicine*, 19(2), 65–71.

Rama, P., Liu, Y., Chen, R., Ostadi, H., Jiang, K., Gao, Y., Zhang, X., Fisher, R., Jeschke, M. (2010). Multiscale modeling of single-phase multicomponent transport in the cathode gas diffusion layer of a polymer electrolyte fuel cell. *Energy & Fuels*, 24(5), 3130–3143.

Ranganathan, P. and Sivaraman, S. (2011). Investigations on hydrodynamics and mass transfer in gas–liquid stirred reactor using computational fluid dynamics. *Chemical Engineering Science*, 66(14), 3108–3124.

Restrepo-Flórez, J.M. and Maldovan, M. (2016). Mass separation by metamaterials. *Scientific Reports*, 6(1), 21971.

Roberts, R.C. (1972). *Molecular Diffusion of Gases*. In *American Institute of Physics Handbook*, 3rd edition, Gray, D.E. (ed.). McGraw-Hill, New York.

Robson, J.D. (2020). Deformation enhanced diffusion in aluminium alloys. *Metallurgical and Materials Transactions A*, 51, 5401–5413.

Sanders, D.F., Smith, Z.P., Guo, R., Robeson, L.M., McGrath, J.E., Paul, D.R., Freeman, B.D. (2013). Energy-efficient polymeric gas separation membranes for a sustainable future: A review. *Polymer*, 54(18), 4729–4761.

Sangoro, J.R., Iacob, C., Naumov, S., Valiullin, R., Rexhausen, H., Hunger, J., Buchner, R., Strehmel, V., Kärgera, J., Kremera, F. (2011). Diffusion in ionic liquids: the interplay between molecular structure and dynamics. *Soft Matter*, 7(5), 1678–1681.

Schlichting, H. and Gersten, K. (2016). *Boundary-Layer Theory*. Springer, Berlin.

Scholes, C.A., Kentish, S.E., Stevens, G.W. (2009). Effects of minor components in carbon dioxide capture using polymeric gas separation membranes. *Separation & Purification Reviews*, 38(1), 1–44.

Sen, S., Glienke, M., Yadav, B., Vaidya, M., Gururaj, K., Pradeep, K.G., Daum, L., Tas, B., Rogal, L., Wilde, G. et al. (2023). Grain boundary self-and Mn impurity diffusion in equiatomic CoCrFeNi multi-principal element alloy. *Acta Materialia*, 119588.

Söderman, O., Price, W.S., Schönhoff, M., Topgaard, D. (2010). NMR diffusometry applied to liquids. *Journal of Molecular Liquids*, 156(1), 38–44.

Song, H. and Cabooter, D. (2017). Relevance and assessment of molecular diffusion coefficients in liquid chromatography. *Chromatographia*, 80(5), 651–663.

Song, D., Seibert, A.F., Rochelle, G.T. (2017). Effect of liquid viscosity on mass transfer area and liquid film mass transfer coefficient for GT-OPTIMPAK 250Y. *Energy Procedia*, 114, 2713–2727.

Suzuki, A. and Mishin, Y. (2005). Atomic mechanisms of grain boundary diffusion: Low versus high temperatures. *Journal of Materials Science*, 40, 3155–3161.

Talebi, S., Abedini, A., Lele, P., Guerrero, A., Sinton, D. (2017). Microfluidics-based measurement of solubility and diffusion coefficient of propane in bitumen. *Fuel*, 210, 23–31.

Treybal, R.E. (1980). Mass transfer operations. *New York*, 466, 493–497.

Tripathi, D., Bhushan, S., Bég, O.A., Akbar, N.S. (2018). Transient peristaltic diffusion of nanofluids: A model of micropumps in medical engineering. *Journal of Hydrodynamics*, 30, 1001–1011.

Tyrrell, H.J.V. and Harris, K.R. (2013). *Diffusion in Liquids: A Theoretical and Experimental Study*. Butterworth-Heinemann, Oxford.

Van Gastel, R., Somfai, E., Van Albada, S.B., Van Saarloos, W., Frenken, J.W.M. (2001). Nothing moves a surface: Vacancy mediated surface diffusion. *Physical Review Letters*, 86(8), 1562.

Vasconcelos, J.M.T., Rodrigues, J.M.L., Orvalho, S.C.P., Alves, S.S., Mendes, R.L., Reis, A. (2003). Effect of contaminants on mass transfer coefficients in bubble column and airlift contactors. *Chemical Engineering Science*, 58(8), 1431–1440.

Velegol, D., Garg, A., Guha, R., Kar, A., Kumar, M. (2016). Origins of concentration gradients for diffusiophoresis. *Soft Matter*, 12(21), 4686–4703.

Wagner, P.D. (1977). Diffusion and chemical reaction in pulmonary gas exchange. *Physiological Reviews*, 57(2), 257–312.

Wang, K., Huang, H., Han, Y. (2021). Quantifying the driving force of silver crystallization by chemical potential difference. *Industrial & Engineering Chemistry Research*, 60(40), 14447–14454.

Welty, J., Rorrer, G.L., Foster, D.G. (2020). *Fundamentals of Momentum, Heat, and Mass Transfer*. John Wiley & Sons, Hoboken.

Wesselingh, J.A. and Krishna, R. (1990). *Mass Transfer*. Ellis Horwood, London.

Winkelmann, J. (2018). Methods for the measurement of the diffusion coefficient of fluids. In *Diffusion in Gases, Liquids and Electrolytes: Nonelectrolyte Liquids and Liquid Mixtures - Part 2: Liquid Mixtures*. Springer, Berlin, Heidelberg, 10–23.

Ye, L., de Jong, M.P., Kudernac, T., van der Wiel, W.G., Huskens, J. (2017). Doping of semiconductors by molecular monolayers: Monolayer formation, dopant diffusion, and applications. *Materials Science in Semiconductor Processing*, 62, 128–134.

Yepez, J. (2001). Quantum lattice-gas model for the diffusion equation. *International Journal of Modern Physics C*, 12(09), 1285–1303.

Zhao, H., Zhu, Y., Ye, H., He, Y., Li, H., Sun, Y., Yang, F., Wang, R. (2022). Atomic-scale structure dynamics of nanocrystals revealed by in situ and environmental transmission electron microscopy. *Advanced Materials*, 2206911.

Part 3

Separation Processes

9

Extraction

In chemical engineering, it is essential to separate different compounds from a mixture of chemicals, especially those required in their desired quality. Extraction is the process of separating parts of a solid or liquid mixture using a fluid that does not mix with other liquids. It can be characterized as either liquid–liquid, where a solvent is used to separate a solute from a liquid phase; liquid–solid extraction, where solutes are taken out of solid phases with the help of solvents; or supercritical extraction, where supercritical conditions are used as the solvent for a fluid. Products such as pharmaceuticals and cosmetics are more acceptable in their natural form than their chemical and synthetic derivatives. Compounds that are used in the manufacturing of drugs and supplements of food are desired at their highest form of purity and in chemically stable form, to be well adapted in the biological system. In this case, extraction is a crucial step for separating different chemical mixtures based on their physical, chemical properties, including their affinity towards different molecules. In analytical chemistry, a number of approaches are employed to isolate specific chemical compounds, which includes processes such as maceration, cold and hot extraction, extraction by microwave and sonication, as well as solvent extraction. These extraction techniques vary in different aspects depending on the quality and other physio-chemical properties, and hence different parameters such as temperature, pH, polarity of solvent, volatility, molecular size and chemical reactivity come into consideration while extraction is being carried out. This chapter intends to give an in-depth examination of qualities and parameters, besides their principles and act on various strategies of extraction.

9.1. Introduction

Extraction plays a crucial role in the realm of analytical chemistry, serving as an essential method for the isolation and concentration of specific compounds from intricate mixtures. Its importance lies in its capacity to extract particular elements from varied matrices, facilitating in-depth analyses across fields such as

---

Chapter written by Juli DAS, Diptanu DE, Aanchal MITTAL, G.C. JEEVITHA and Sangeeta GARG.

pharmaceuticals, environmental science, and food and forensic chemistry. The necessity for extraction arises from the complex composition of samples, where the target substances may be present in low concentrations or amid a diverse array of other compounds. This process acts as a vital stage in sample preparation, allowing for heightened sensitivity and selectivity in subsequent analytical procedures. Smaller sizes enhance both penetration and diffusion as particle size improves efficiency of extraction. Temperature interferes with solubility and diffusion dynamics; however, excessive temperatures can create solvent loss and initiate unwanted contaminants. Diverse extraction methods cater to various applications and sample characteristics, with liquid–liquid extraction (LLE), solid-phase extraction (SPE), supercritical fluid extraction (SFE) and microwave-assisted extraction (MAE) being a few examples. Each technique offers unique advantages based on factors such as solubility, speed and environmental considerations. Traditional extraction approaches, including maceration, reflux and percolation, rely on organic solvents and are characterized by prolonged extraction periods.

In contrast, contemporary methods such as Supercritical Fluid Chromatography (SFC), Pressurized Liquid Extraction (PLE) and MAE offer efficiency gains with reduced solvent usage and shorter extraction times. Although maceration is suitable for thermolabile components, its drawback lies in the extended duration required for extraction. Recent research highlights the effectiveness of maceration for specific substances. Emerging techniques such as SFC, PLE and MAE are becoming increasingly popular due to their efficiency, abbreviated extraction durations and environmentally friendly attributes.

9.2. Conventional extraction methods

Conventional extraction techniques are time-tested methods that are applied in different sectors to eliminate required components from raw materials. These techniques include maceration, steam distillation, cold press extraction, solvent extraction, decoction, soxhlet extraction, percolation and acid-base extraction. While steam distillation vaporizes volatile molecules for essential oils, solvent extraction mixes target components using solvents like hexane ethanol. Soxhlet extraction is an efficient method for extracting lipids and perfumes because it recycles solvents in cycles. Herbal extracts are macerated, and edible oils are extracted using cold press. Whereas decoction boils plant material for herbal teas and infusions, percolation allows solvents to flow through porous materials. Organic chemistry and pharmaceuticals employ acid–base extraction, which is based on variations in acidity.

9.2.1. *LLE*

LLE is a separation technique that exploits the differential solubility of compounds in two immiscible liquids, typically a polar solvent (often water) and a nonpolar organic solvent. When a mixture is introduced to this two-phase system, components distribute themselves between the phases based on their relative solubilities, with polar substances favoring the aqueous phase and nonpolar ones preferring the organic phase (Figure 9.1). The efficiency of this separation is quantified by the partition coefficient, which is the ratio of a component's concentration in the organic phase to its concentration in the aqueous phase (Pena-Pereira et al. 2009). This coefficient is influenced by various factors including the chemical nature of the solvents and solutes, pH, temperature and the presence of other dissolved substances.

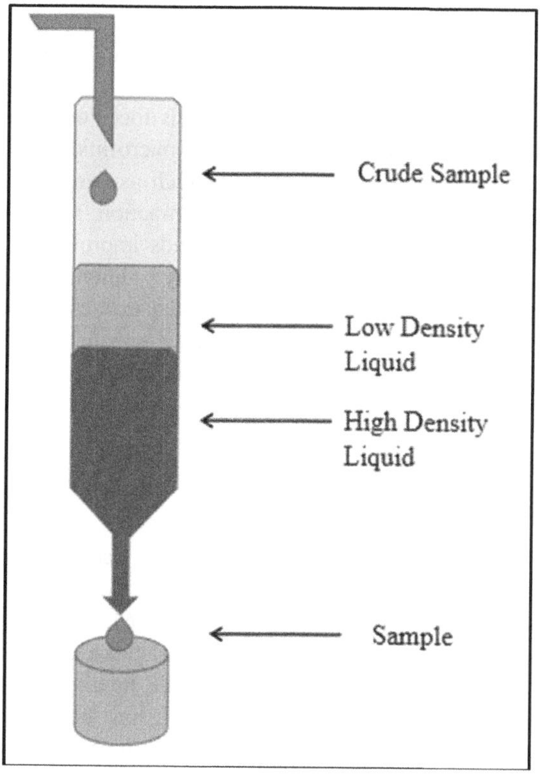

Figure 9.1. *Liquid–liquid extraction technique (prepared for this work)*

In pharmaceutical, chemical and analytical applications, LLE is a versatile and frequently used technology that may be tailored to specific separation needs by adjusting these parameters. In the pharmaceutical business, LLE is a basic and incredibly versatile process that is vital at different phases of drug research and manufacture. It is used to remove contaminants from medication formulations to isolate and purify active pharmaceutical ingredients (APIs). Antibiotics may be extracted from fermentation combinations, valuable molecules can be recovered from complex solutions, and LLE is particularly good at extracting natural compounds from plant sources. It also has a significant impact on biotechnology, helping to extract proteins and enzymes. LLE is essential for sample preparation in chromatographic analysis, solvent recovery and purification, breaking azeotropes and wastewater treatment, in addition to direct pharmaceutical synthesis (Cardoso and Dasgupta 1995). Because LLE may separate chemicals according to how soluble they are in various phases, it is a crucial instrument for guaranteeing the effectiveness, safety and quality of pharmaceutical procedures. Numerous uses of LLE in the pharmaceutical industry, ranging from environmental management and large-scale production to research and development, demonstrate its significance (Bailes et al. 1981).

Recent advancements in LLE for pharmaceuticals focus on improving efficiency, selectivity and sustainability. Innovations include microfluidic devices that reduce solvent use and enhance control, green solvents such as deep eutectic solvents and ionic liquids, and integrated systems combining extraction, separation and analysis. Novel extractants such as functionalized ionic liquids improve API targeting, while continuous LLE processes offer better scalability. Other developments include advanced centrifugal contactors, electric field-assisted extraction and bionic systems mimicking natural processes. These advancements enhance LLE's role in drug discovery, manufacturing and environmental sustainability (Cardoso and Dasgupta 1995).

9.2.2. *SPE*

SPE is an exceedingly promising technique used for separating and concentrating analytes from complex matrices. The selection of the extraction sorbent is crucial, requiring a rigorous understanding of the interaction mechanisms between the sorbent and the target analyte. This process involves several phases, including sorbent and column selection based on analyte properties, sample matrix characteristics and the specific analytical protocol (Figure 9.2). Depending on the contaminants present, either normal-phase or reverse-phase separation methods can be employed (Żwir-Ferenc and Biziuk 2006). In SPE, three primary methods – selective extraction, washing and elution – are used to separate analytes from interferences. These methods enable the retention of desired components, rinse away impurities and elute adsorbed molecules of interest. Solvent selection, guided by the principle of "like dissolves like", is fundamental to the success of SPE applications (Ötles and Kartal 2016).

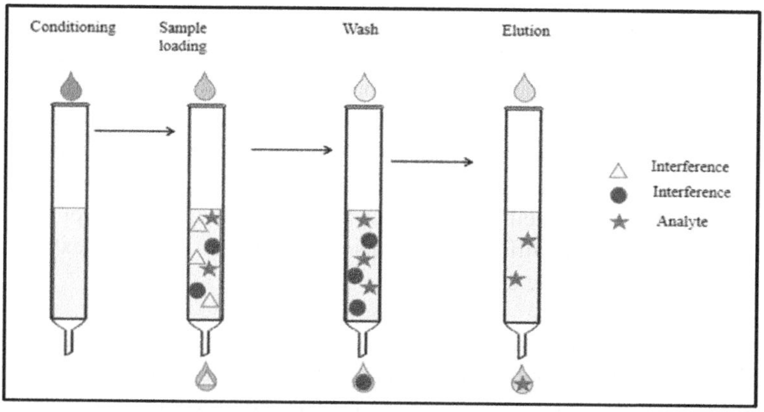

Figure 9.2. *Solid-phase extraction technique (prepared for this work). For a color version of this figure, see www.iste.co.uk/kumararya/chemicalengineering1.zip*

Sorbents are chosen based on specific criteria such as the nature of the food matrix, the characteristics of the target analytes and potential interferent chemicals. Various interaction mechanisms, including van der Waals forces, hydrogen bonding, dipole interactions and ionic exchange, play critical roles in the extraction of analytes. Silica, due to its modifiability and stability when coupled with nonpolar or polar chains, is a popular sorbent type.

SPE not only aids in the isolation and separation of analytes but also facilitates concentration by allowing analytes to elute from the column in a minimal solvent volume. This sample preparation method is adaptable and widely used across many domains of analytical chemistry, including environmental analysis, pharmaceutical and bioanalysis, and food and beverage analysis. Compared to conventional LLE methods, SPE offers advantages such as reduced organic solvent usage, capability for onsite sample processing, and integration into automated systems for high-throughput analysis (Badawy et al. 2022). In contemporary analytical chemistry, SPE is considered to be vital due to its versatility in accommodating various sorbent chemistries and formats. This versatility allows SPE to be tailored for a broad range of analyte classes and sample matrices, making it a cornerstone technique in modern analytical methodologies.

9.2.3. *Solvent extraction*

Researchers have used a broad range of solvents such as methanol, ethyl alcohol and hexane used for extracting antioxidants from diverse plant parts, such as seeds and leaves. Researchers must use different polarity solvents in order to extract phenolic compounds with accuracy. Interestingly, methanol and other highly polar solvents have been shown to be efficient antioxidants. Many studies demonstrate how well solvents such as acetone, methanol and N,N-dimethylformamide (DMF) remove antioxidants from plant sources. To remove water interference, researchers frequently choose dried plant powder. The choice of solvent is determined by the polarity of the solute. Common solvents have the following polarity order: hexane, water, methanol, acetone, ethyl acetate and chloroform (Altemimi et al. 2017). Table 9.1 represents the different solvents that can be used for extraction of compounds with their characteristics.

In many different sectors, solvent extraction is an adaptable method that is essential for isolating and purifying substances according to how soluble they are in various solvents. By concentrating trace analytes from complicated matrices, it increases detection sensitivity in analytical chemistry. By making it easier to recover and recycle metals from ores and electronic trash, hydrometallurgy promotes sustainable resource management. Solvent extraction is critical in the nuclear industry for removing uranium and plutonium from spent nuclear fuel, which is important for waste management as well as energy production. This is especially true for procedures like PUREX. This technique is used in wastewater treatment to get rid of organic pollutants, which helps with environmental cleanup. It makes it possible to extract bioactive substances and purify biomolecules for use in medicine and research in the fields of biochemistry and pharmaceuticals. Solvent extraction is a technique used in the food, beverage and cosmetic industries to separate tastes, scents and essential oils from natural sources in order to improve the quality and marketability of products.

Solvent	Polarity index	Preferred samples	Characteristics	Functional groups	References
Hexane	0	Nonpolar compounds, lipids, hydrocarbons	Nonpolar compounds, aliphatic hydrocarbons	C-H bonds, alkanes	Bhuiya et al. (2020)
Dichloromethane	3.1	Lipids, alkaloids	Moderately polar, suitable for a wide range of compounds	C-Cl bonds	Cequier-Sánchez et al. (2008)
Ethyl Acetate	4.4	Organic acids, alkaloids	Moderate polarity, versatile solvent	C = O (carbonyl), C-O bonds	Boonsongsawat et al. (2010)

Acetone	5.1	Polar and nonpolar compounds	Polar, suitable for a variety of compounds	C = O (carbonyl)	Remler, (1922)
Methanol	5.1	Polar compounds	Polar, miscible with water, suitable for polar compounds	C-OH (hydroxyl)	Alkahtani et al. (2022)
Ethanol	5.1	Polar compounds	Polar, miscible with water, commonly used in biological extractions	C-OH (hydroxyl)	Yeasmen and Islam (2015)
Isopropanol	5.1	Proteins, nucleic acids	Polar, miscible with water, used for nucleic acid extraction	C-OH (hydroxyl)	Nuchdang et al. (2022)
Water	10	Polar compounds	Highly polar, essential for extraction of hydrophilic compounds	Hydrogen bonding	Castro-Puyana et al. (2017)

Table 9.1. *Table showing different properties of solvents used in extraction techniques*

9.2.4. *Maceration*

Maceration, a simple extraction process known for its user-friendly nature, has the drawback of prolonged extraction time. Tailored for heat-sensitive chemicals, it involves immersion of plant components within a solvent at room temperature for a longer time interval, maintaining a steady state. The process encompasses placing cleaned and air-dried plant parts in a solvent containing a closed container for a duration of four to six days, with intermittent agitation. The soluble components dissolve, and the damp solid material is filtered and pressed for solution recovery.

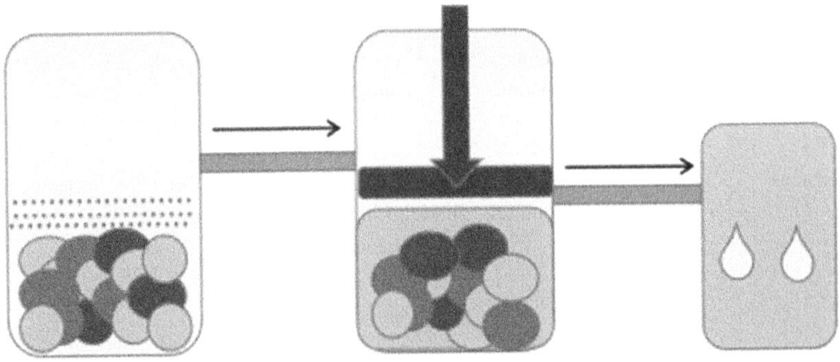

Figure 9.3. *Extraction technique with maceration process (prepared for this work). For a color version of this figure, see www.iste.co.uk/kumararya/chemicalengineering1.zip*

Coagulation and settling are facilitated, and the obtained extract is concentrated at a lower pressure to result in the crude plant extract (Figure 9.3). Due to its stationary nature relying on molecular diffusion, maceration is time-consuming and uses a closed system to prevent solvent evaporation. Common solvents include methanol and methanol–water combinations. Various maceration processes exist, such as simple maceration for crude pharmaceuticals like orange tinctures, double maceration for concentrated infusions and triple maceration for drugs such as Quassia and Senna. While maceration has benefits such as small sample size requirements, potent swelling qualities and energy efficiency, drawbacks include incomplete drug extraction, a lengthy process and a substantial solvent requirement (Tambun et al. 2021).

Maceration is a flexible extraction technique that is applied in many different sectors. It takes substances from medicinal plants, such as alkaloids and phenolic compounds, to make herbal cures and medical products. It creates oil macerates for skincare in cosmetics. It improves flavors and colors in food and drink, which is important for alcoholic beverages and culinary goods. Mucilage extraction is used in pharmaceuticals for skin and intestinal health. For delicate scents, cold maceration of essential oils is beneficial. It is also essential for the manufacturing of liqueurs and nutraceuticals, but depending on the compound, it may vary in efficiency and take a long time.

9.2.5. *Infusion and percolation*

The process of creating new infusions involves briefly submerging the source material in hot or cold water. During this steeping phase, the easily soluble components in the raw materials dissolve into the water, creating diluted solutions that capture the essence of the raw materials' advantageous qualities. This process is frequently used to produce a liquid infusion with a subtly concentrated concentration of its soluble components, often employed in making tinctures and fluid extracts. A percolator, a thin, cone-shaped device open at either end, is typically used for this technique. The solid parts are humidified in a designated menstruum and left in a sealed container for approximately four hours. The mixture is then packed, and the percolator's top is shut. To form a shallow layer above the bulk, more menstruum is added, and the closed percolator is engaged to macerate the whole day. Next, the percolator is opened, allowing the fluid to flow gradually. When the filtrate reaches about three-fourths of the intended amount for the final product, more menstruum is added as needed. After pressing the residual solid residue, or marc, the released liquid is mixed with the percolate. To get the desired volume, enough menstruum is added, and the mixed liquid is clarified by filtering or letting it settle and then decanting it, as shown in Figure 9.4, ensuring the extraction of active ingredients and the creation of a purified liquid product (Singh et al. 2008).

Recent advancements in infusion and percolation techniques have significantly improved the quality, control and efficiency of extraction operations across various applications. Modern coffee and tea brewing tools offer precise control over steeping conditions, uniform contact and preservation of volatile ingredients, with the flexibility to operate in microgravity environments. Enhanced herbal extraction uses advanced percolation equipment with temperature control and inert components to maintain high-quality outputs (Hidayat and Wulandari 2021).

Environmentally friendly methods, such as green solvents and microwave-assisted techniques, are gaining traction for bioactive compound extraction. These techniques are improving the extraction of bioactive chemicals and antimicrobial agents from natural sources for pharmaceutical drug development. In addition, the food and beverage industry is integrating infusion and percolation with other processes for efficient extraction of beneficial compounds from plant sources and customized beverage preparation (Abubakar and Haque 2020).

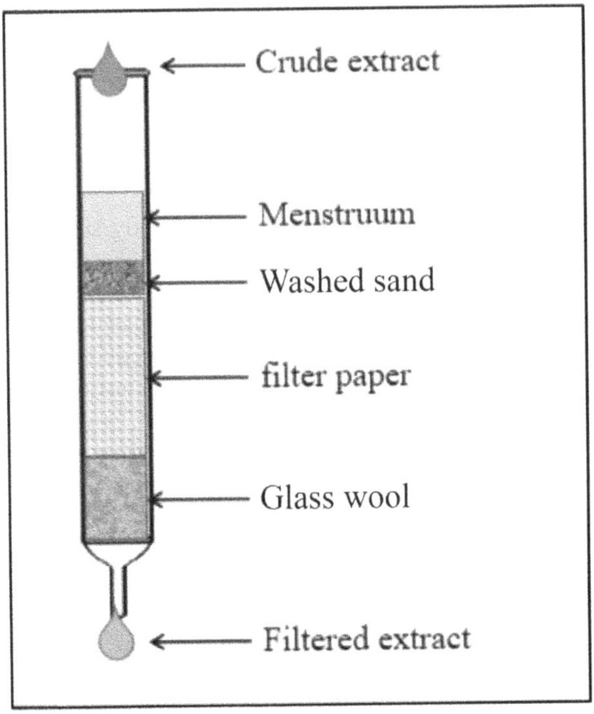

Figure 9.4. *Extraction technique using percolation (prepared for this work). For a color version of this figure, see www.iste.co.uk/kumararya/chemicalengineering1.zip*

9.2.6. *Hydrodistillation*

The extraction of essential phytol from plant sources by hydrodistillation is a conventional approach where water and heat are used to release and collect vaporous chemicals, such as essential oils, from plant matter. In this process, water is added to the plant material and the mixture is heated, producing steam that carries the volatile components away from the plant material. The steam is then condensed back into liquid form, resulting in a blend of water and essential oil.

Due to density variations, the essential oils float on the surface and can be separated from the water (Figure 9.5). Hydrodistillation is widely used for producing essential oils for aromatherapy, perfumery and flavoring (Okoh et al. 2010).

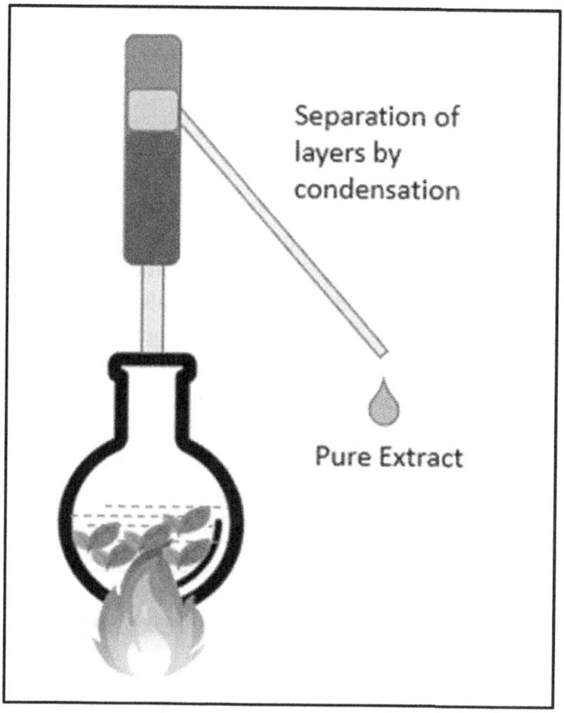

Figure 9.5. *Extraction technique using hydrodistillation process (prepared for this work). For a color version of this figure, see www.iste.co.uk/kumararya/chemicalengineering1.zip*

Beyond its traditional use in essential oil extraction, hydrodistillation has seen several innovative advancements. Researchers have introduced a process control system using digital twins for autonomous, climate-neutral manufacturing of phytochemicals, optimizing extraction parameters such as temperature, pressure and flow rates. Another notable development is Ohmic-Assisted Hydro Distillation (OAHD) (Gavahian and Farahnaky 2018), which integrates ohmic heating with hydrodistillation to shorten extraction times, reduce energy consumption and enhance distillate quality. In addition, hydrodistillation is being used to extract bioactive compounds from plants and marine organisms for pharmaceutical purposes, including wound healing and drug development. The technique is also combined with other separation methods, such as SFE, to create more efficient and selective systems. Furthermore, essential oils obtained via hydrodistillation are showing promise as natural preservatives in the food industry, capable of inhibiting the growth of harmful microorganisms and extending product shelf life (Irshad et al. 2020). These advancements highlight the versatility and potential of hydrodistillation in various sectors, from autonomous manufacturing to pharmaceutical applications and food preservation.

9.2.7. *Hydrodiffusion*

Hydrodiffusion extraction, which is often known as water distillation, is an acknowledged process that involves extracting essential oils from plant materials using steam. In this method, steam is pumped from the upper surface of the plant material, allowing it to pass through the botanicals and absorb most volatile substances such as essential oils. The fragrant compounds, which are delivered by steam, proceed through the plant substances that allow for the gentle release of essential oils. The steam condenses following that, resulting in a combination of water and essential oil (Figure 9.6). Because of differences in density, essential oils frequently float in water and can be extracted. The capacity of hydrodiffusion to extract essential oils at lower temperatures maintains the subtle aroma of the botanicals while decreasing the risk of thermal damage (Vian et al. 2008).

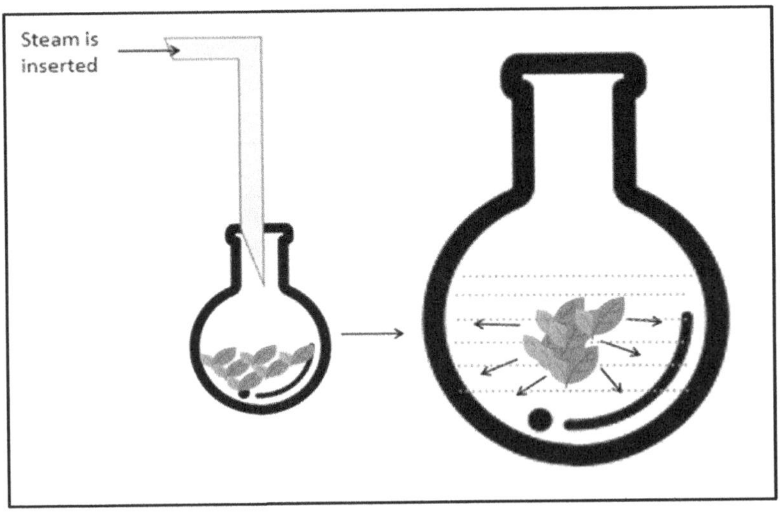

Figure 9.6. *Extraction technique using hydrodiffusion process (prepared for this work). For a color version of this figure, see www.iste.co.uk/kumararya/chemicalengineering1.zip*

Confortin et al. (2024) have used pressurized liquid along with a microwave-hydrodiffusion technique and gravity to extract bioactive compounds such as senecionine, integerrimine and quinic acid from *Solanum viarum*, which are also composed of pyrrolizidine alkaloids. Moraes et al. (2022) use gravity and microwave hydrodiffusion combinedly to extract essential oils from fruit juices rich in phenolic compounds, in which microwaves cause dipolar rotation and ionic conduction. This produces heat and results in the diffusive release of bioactive compounds into the solvent environment while gravity pulls them down, followed by extraction and isolation. Similarly, López et al. (2020) compared hydrodiffusion methods with steam distillation to extract bioactive compounds from nettle leaves and discussed that the addition of extracts of hydrodifusion assisted with microwave technology increased the foamability of the shampoo.

9.2.8. *Decoction*

This familiar method for extracting therapeutic compounds from plants is decoction extraction, which involves boiling the sample in water. The process of decoction starts with choosing and chopping the necessary plant material, then simmering it in water for an extended amount of time. The boiling procedure allows the active chemicals to be liberated from harder plant portions such as roots or bark. After simmering, the mixture is strained to separate the liquid form extract within the solid plant material. Decoction is often consumed orally and is used in herbal therapy and traditional medicine. It is necessary to examine the specific plant, its intended application and seek assistance from healthcare specialists or herbalists before using decoctions for therapeutic purposes (Zhang et al. 2018).

Martins et al. (2018) used decoction to isolate phenolic extracts to identify salvianolic A isomer, as well as other phytochemical compounds. This was observed that engaging decoction technique, 32.4% of the phenolic compounds were isolated, from hydroethanolic extracts, and these compounds reduced the tumor cells of different human cancer cell lines, and also exhibited relevant antioxidant properties, anti-inflammatory as well as antiproliferative properties. Martins et al. (2014) have compared the antioxidant and antibacterial properties in phenolic compounds obtained from extractions of oregano and discussed that this method provided the maximum concentration of flavonoids as well as phenolic compounds, as compared to infusion or hydroethanolic extractions. The samples also had antimicrobial properties against gram positive bacteria and *Escherichia coli.*

9.2.9. *Steam distillation*

Steam distillation is commonly used for isolation of volatile and heat-sensitive compounds, especially essential oils from plants. It works on the principle that the targeted compounds can be vaporized at temperatures below their boiling points when steam is introduced, which will reduce the risk of thermal degradation (Cassel et al. 2009).

Typical components of a steam distillation system are a boiling flask, a condenser and a receiving flask. Water is boiled to produce steam in the boiling flask, which is then passed through the plant material or sample that will cause the volatile chemicals to evaporate. The vaporized mixture of steam and essential oils is condensed back into liquid form in the condenser, and the resulting liquid that contains water and the essential oils is collected in the receiving flask (Figure 9.7).

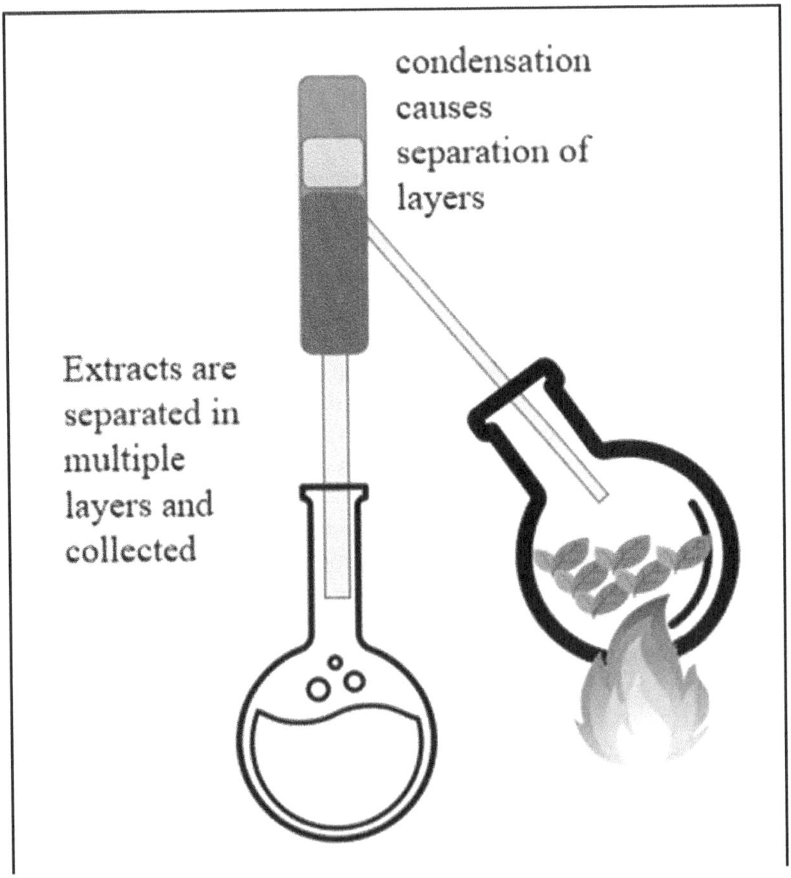

Figure 9.7. *Extraction using distillation technique (prepared for this work). For a color version of this figure, see www.iste.co.uk/kumararya/chemicalengineering1.zip*

One of the significant benefits of steam distillation is the potential to extract essential oils without introducing them to high temperatures and keeping their aromatic qualities. Furthermore, steam distillation is useful for extracting compounds with higher boiling points that would degrade using traditional distillation methods.

While steam distillation is a helpful technique, its efficiency for specific chemical types may be limited, and wastes may contain nonvolatile or high-boiling point substances. To improve extraction efficiency for specific applications, modifications such as the inclusion of a solvent or the use of vacuum steam distillation may be employed. Overall, steam distillation is important in the extraction of aromatic compounds and the synthesis of vital oils in the sectors of perfumery, pharmaceuticals and food (Chemat et al. 2006).

De Santis et al. (2022), used steam distillation to extract bioactive compounds from leaves of *Rubus idaeus* L. which is treated as waste or a by-product from different agro- and food industrial sources. They discovered that these extracts were rich in polyphenols and exhibited antioxidant activity and also responded as anticarcinogenic agent for tumor cell lines. Due to their antioxidant and antibacterial properties, they proposed their application in cosmetics and food industries as nutraceutical supplements.

Tavares et al. (2020) carried out extractions of bioproducts like essential oils, high value products like phenolic compounds and hydrolates from biomass of *Cistus ladanifer* L. waste residues using steam distillation and characterized their antioxidant activity with high phenolic contents for the inhibition of super oxides. They further purified catechins, gallocatechin, gallic acid, hydroxycinnamic acid and salicylic acid from those extracts with relatively higher concentrations.

9.2.10. *Extractive distillation*

Distillation extraction is the separation of components based on the variable evaporation rates of volatile substances within a still under precise temperature, pressure and subsequent condensation conditions. Extractive distillation is a type of distillation that uses a separation solvent which has non-volatility, an elevated boiling point and miscibility with the mixture without generating an azeotropic combination (Figure 9.8). The solvent's interactions with the fusion components change their relative volatility which allows the separation via conventional distillation. The highest product is the most volatile original component, while the bottom result is a mixture of the solvent and the remaining component that separates rapidly due to the lack of an azeotrope. Extraction is frequently used in pharmaceuticals, fragrances, essential oils and cosmetics, whereas distillation is largely used in the petroleum and solvent industries (Lei et al. 2003).

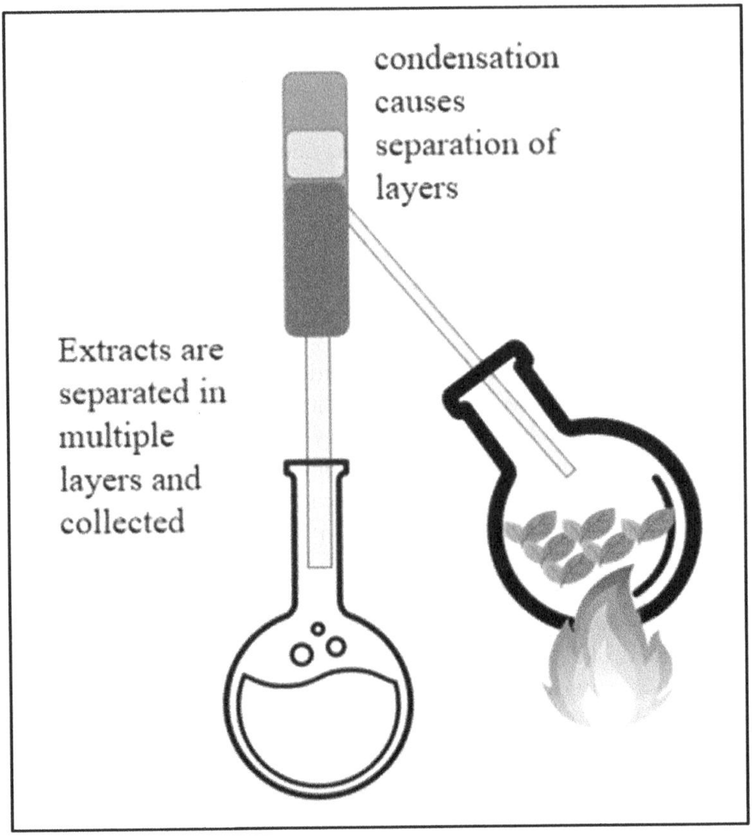

Figure 9.8. *Extraction using distillation technique (prepared for this work). For a color version of this figure, see www.iste.co.uk/kumararya/chemicalengineering1.zip*

The goal of recent developments in extractive distillation technology is to increase sustainability and energy efficiency. The creation of energy-intensified configurations, like the extractive dividing wall column (EDWC), which simplify the separation of several components in a single unit, is one of the major breakthroughs (Shen et al. 2015). Using ionic liquids as entrainers is one way to make operations more environmentally friendly. Hybrid technology integration, such as combining adsorption or distillation with membranes, also shows potential in gas separation. The goal of process intensification techniques, such as heat-integrated and thermally connected distillation columns, is to use less energy. The design and optimization of these processes have been enhanced by developments in modeling and simulation, and novel operation tactics have optimized both batch and continuous extractive distillation, increasing overall efficiency and practicality across a range of sectors (Aquilon et al. 2023).

9.2.11. *Sublimation*

Sublimation is the process by which a substance transitions directly from its solid phase to the vapor phase, bypassing the liquid phase altogether. This unique property makes sublimation a promising purification technique, particularly in situations where traditional methods may not be as effective. Continuous and multistage sublimation processes have the potential to be crucial in large-scale industrial applications, offering an efficient method for the separation and purification of materials. The process involves heating the material to induce sublimation, collecting the resulting vapor and then condensing it back into a solid state (Figure 9.9). Widely employed in industries such as pharmaceuticals, fragrance and chemical production, sublimation extraction is a common approach for isolating and purifying volatile chemicals such as organic and aromatic molecules. It enables the acquisition of pure and concentrated compounds without the necessity for intermediate liquid phases (Abubakar and Haque 2020).

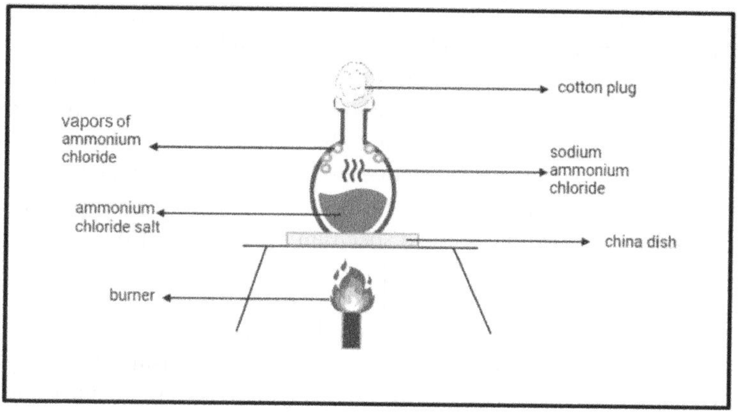

Figure 9.9. *Extraction technique using sublimation process (prepared for this work). For a color version of this figure, see www.iste.co.uk/kumararya/chemicalengineering1.zip*

The rising applications and advances of sublimation technology are highlighted by recent advancements in the field. Eco-friendly substrates, such bamboo, organic cotton and recycled materials, are becoming more and more popular because they appeal to customers who care about the environment. The fashion industry is adopting all-over printing, which uses specialized technology to cover an item of clothing's whole surface. In addition, sublimation is widely used to create personalized home decor pieces such as pillows and blankets. Niche markets are catered to by the technology's capacity to produce small batches and limited editions. Furthermore, soft signage in advertising is increasingly using sublimated fabric banners and flags. Industrial uses include long-lasting prints on unique materials for safety devices and signage, while dye-sublimation printing is becoming more popular than conventional inkjet printing in some applications because of its quick drying times.

9.2.12. *Chromatography*

Chromatography is depicted to be versatile for the differentiation, characterization, as well as refining of different compounds mixed in mixtures. In the circumstances of extraction, chromatography is a technique used to extract specific substances from a given sample. The basic aim of chromatography comprises the passage of a mobile phase (liquid or gaseous) through a stationary phase, which results in the separation of different compounds according to their response along with the stationary phase. Various chromatographic techniques exist, including thin-layer chromatography (TLC) (Figure 9.10), gas chromatography, paper chromatography, column chromatography and high-performance liquid chromatography (HPLC) as well as others (Coskun et al. 2016). Every type of these techniques possesses its specific applications and is preferred according to the quality of the sample and the components to be separated. For instance, in the case of extraction, SPE is a common technique that falls under chromatography and is used for the separation of analytes from a mixture of samples. These numerous ways provide versatility in extracting a broad spectrum of chemicals from various sample matrices.

Chromatography is now more widely used and has better capabilities in more domains because of recent advancements in the field. To ensure the safety of the product, contaminants such as pesticides and mycotoxins are found using liquid chromatography (LC), gas chromatography (GC) and mass spectrometry (MS) in cannabis analysis. Using methods like pyrolysis-gas chromatography-mass spectrometry (Pyr-GC/MS), chromatography is also being used to detect microplastics and associated chemicals in food and environmental samples. While multicolumn chromatography systems provide more effective biomolecule analysis, new chromatographic columns in the pharmaceutical industry are improving the analysis of complicated substances. HPLC technology is becoming increasingly accessible for educational and small lab settings thanks to the trend towards portable and user-friendly equipment. For thorough analysis, chromatography and high-resolution mass spectrometry (HRMS) are being combined more and more (López-Ruiz et al. 2022).

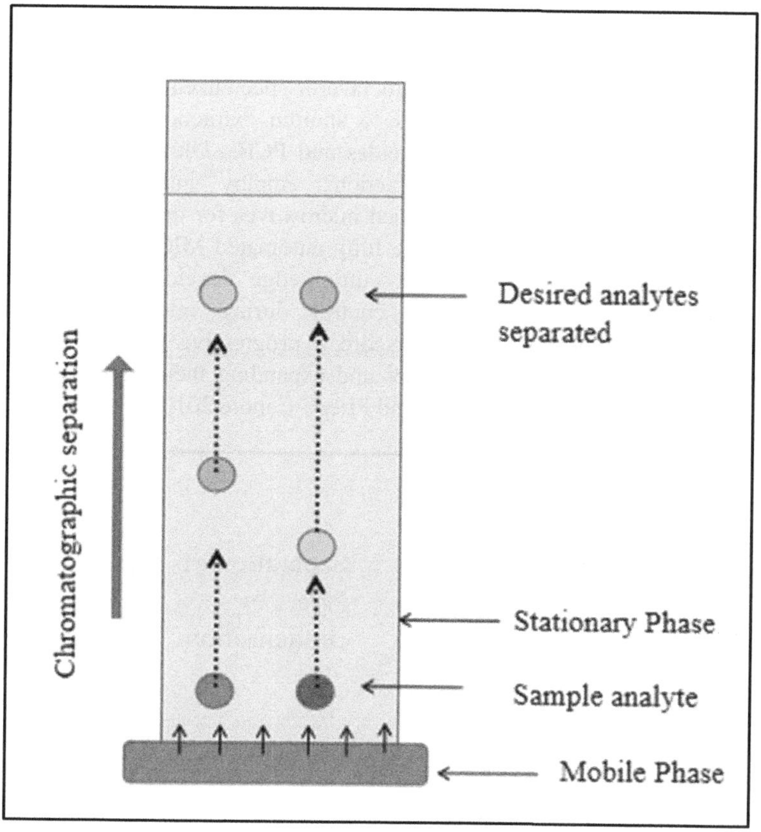

Figure 9.10. *Chromatography technique (TLC) (prepared for this work). For a color version of this figure, see www.iste.co.uk/kumararya/chemicalengineering1.zip*

9.2.13. *Extraction by soxhlet*

Soxhlet extraction methods have evolved greatly to solve numerous problems and improve sample preparation efficiency, specifically for determination of fat content. It is a traditional method broadly used for extracting chemicals, particularly in evaluating fat content in diverse samples, and requires delicate preparation of a solid sample. Within the soxhlet equipment, the sample is held in a porous thimble, which consists of a condenser and a flask. The flask is filled with an organic solvent like hexane, which starts a cyclical extraction in which the solvent evaporates, condenses, ascends and returns again to the sample. This continuous cycle ensures thorough extraction by maintaining contact with a fresh solvent. The solvent is now enriched with extracted chemicals and is concentrated by evaporation (Figure 9.11). This method faces drawbacks such as prolonged extraction times, solvent wastage and difficulties in automation while offering advantages like repeated contact with fresh extractant (López-Bascón and De Castro 2020). High-pressure soxhlet extraction introduces a stainless-steel autoclave or specialized extractors which use solvents or gases under high pressures to shorten extraction time period, with applications extending to isolating pesticides and PCBs. Ultrasound (US)-assisted and microwave-assisted soxhlet extractions employ innovative techniques, minimizing cycles and introducing focused microwaves for open vessel extraction, irradiation and improved efficiency. The fully automated MIC V prototype, along with the compact Accesox, represents cutting-edge developments, overcoming limitations and providing advanced control during microwave irradiation. Collectively, these modifications underscore a progressive trajectory in soxhlet extraction methods, optimizing processes and expanding their applicability across diverse analytical scenarios (De Castro and Priego-Capote 2010).

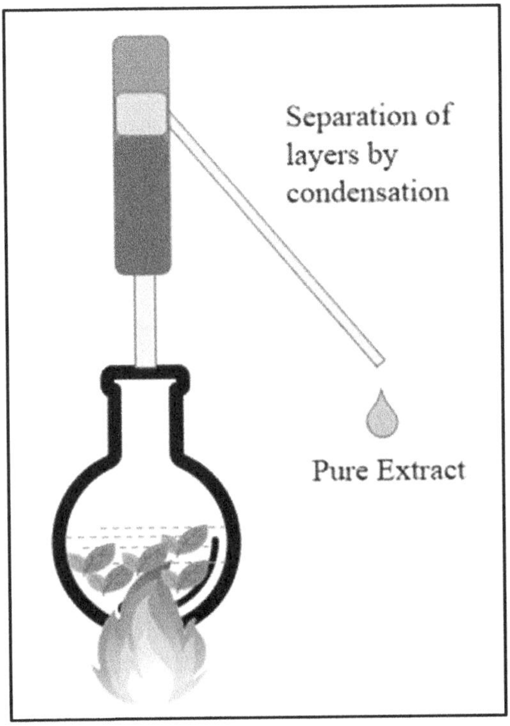

Figure 9.11. *Soxhlet extraction technique (prepared for this work). For a color version of this figure, see www.iste.co.uk/kumararya/chemicalengineering1.zip*

Abbas et al. (2021) extracted bioactive compounds like flavonoids and phenolics from *Lagenaria siceraria* using soxhlet extraction technique and compared it with processes by combining microwave and ultrasonic waves, where they observed that the highest yield was observed for ultrasonic-assisted extraction (UAE), followed by MAE and solvent extraction, and they concluded that soxhlet extraction technique would be more productive when combinedly used with ultrasonic waves and microwaves for extraction of bioactives. Similarly, Fuente et al. (2022) reported that the efficiency of soxhlet extraction can be enhanced when optimized with the assistance of microwaves to extract bioactive essential oils from salmon, which allowed as high as more than 90% of oil recovery from the salmon.

9.2.14. *Solvent-free microwave extraction*

Solvent-free microwave extraction (SFME), a more environmentally conscious adaptation of MAE, which uses microwave technology for the efficient process of extraction of bioactive compounds from natural products. The consequences result in reduced pollution, lower handling costs and minimized labor. SFME works by using microwave irradiation to cause the rotation of in situ water within plant cells (Li et al. 2013). This process increases pressure, causing the breakdown of cell walls and the subsequent release of target molecules.

SFME significantly deviates from conventional separation by facilitating mass and heat transfer from the inside to the outside of plant cells. The efficiency of SFME is influenced by factors such as microwave power, water content and extraction time, which can be adjusted using response-surface methodology (RSM). Various SFME-related techniques, which include improved SFME such as microwave hydro-diffusion and gravity (MHG), vacuum microwave hydro-distillation (VMHD), microwave steam diffusion (MSDf), microwave steam distillation (MSD), pressurized SFME (PSFM) and microwave dry-diffusion and gravity (MDG), offer various strategies for solvent-free extraction. These procedures provide benefits such as energy efficiency, enhanced extraction effectiveness and speed (Armenta et al. 2019).

Idris et al. (2020) optimized extraction of bioactives from *Centella asiatica* using SFME method, and isolated asiaticoside and discussed that vacuum followed by microwaves were the most important factors for enhanced yields which signified higher accuracy levels in SFME process. Essential oil extracts from *Thymus mastichina* were obtained by Araujo et al. (2021) using the SFME process and were observed to be similar quantitatively as well as qualitatively to those obtained by using hydrodistillation process in terms of aromatic profile, as well as possess similar antimicrobial activities in contrast to hydrodistillation and steam distillation. However, this method contributes to higher yield while producing less waste, consuming less solvent and energy, which is reported to be an advantage over steam distillation.

9.2.15. *Magnetic extraction*

Magnetic extraction, often known as magnetic separation, is a technique that separates materials based on their magnetic properties. Magnets and specialized equipment are used in this method to attract, catch and separate magnetic particles from nonmagnetic elements in a combination. Magnetic separators, for example, generate magnetic fields to aid in the sorting process. This approach has a wide range of uses, including mineral processing for ore concentration, recycling to separate ferrous and nonferrous metals, and even in the food industry to ensure product integrity by removing magnetic contaminants. Magnetic extraction effectiveness is determined by factors such as particle size, magnetic susceptibility and magnetic field intensity. Constant advancements in material technologies and separator designs aims to improve the efficacy and adaptability of this separation process, making it a useful tool for resource recovery and environmental goals (Cridland et al. 1968).

Recently, magnetic solid-phase extraction (MSPE) has found more uses in a wider range of industries, demonstrating its effectiveness and adaptability in analytical chemistry. In phytochemical analysis, MSPE uses an external magnetic field to streamline the extraction process, making it easier to isolate phytochemicals from plant sources. To improve monitoring efforts, it is used in food and environmental safety to identify and measure pollutants, such as pesticides in fruits and organochlorine pesticides in water. For environmental and public health monitoring, MSPE's function in heavy metal detection – often in conjunction with atomic spectroscopy – has shown to be essential. Moreover, MSPE is used in clinical and biochemical analysis to enhance research and diagnostics by extracting and analyzing pharmaceuticals and biologically significant chemicals (Yin et al. 2021).

9.3. Modern extraction methods

Traditional extraction methods, which are commonly used in many different sectors, employ solvent extraction, steam distillation, soxhlet extraction, maceration, cold-press extraction, percolation, decoction and acid–base extraction. Aromatic oils, lipids and medicinal substances are among the raw materials from which these procedures are crucial in isolating the appropriate components. These techniques have limitations, including poor selectivity, extended extraction periods and large solvent usage, even with their historical relevance. Modern extraction methods, such as US-assisted extraction (UAE) and SFE, have arisen in response to these difficulties, which emphasize maximizing effectiveness and reducing environmental impact. The use of these cutting-edge techniques marks a change towards extraction procedures that are more economically and environmentally sound yet meeting modern efficiency and environmental requirements.

9.3.1. *Accelerated solvent extraction*

Accelerated solvent extraction (ASE) was developed by Dionex Corporation in 1995 which is an automated process used for thoroughly extracting organic molecules from solid matrices. This patented method enhances extraction capability by using conventional solvents at high temperatures and pressures which exceed the boiling points of these solvents. The mixture of increased temperature and pressure make sure quick, safe and effective extractions complete the operation in 15–25 minutes with minimal solvent volumes (Figure 9.12). ASE is effective in removing both environmental pollutants and various natural goods. Its advantages include fast extractions of sample sizes ranging from 1 to 100 g that significantly cause reduction in solvent usage and versatility in managing matrices with diverse acidity and alkalinity (Mottaleb and Sarker 2012).

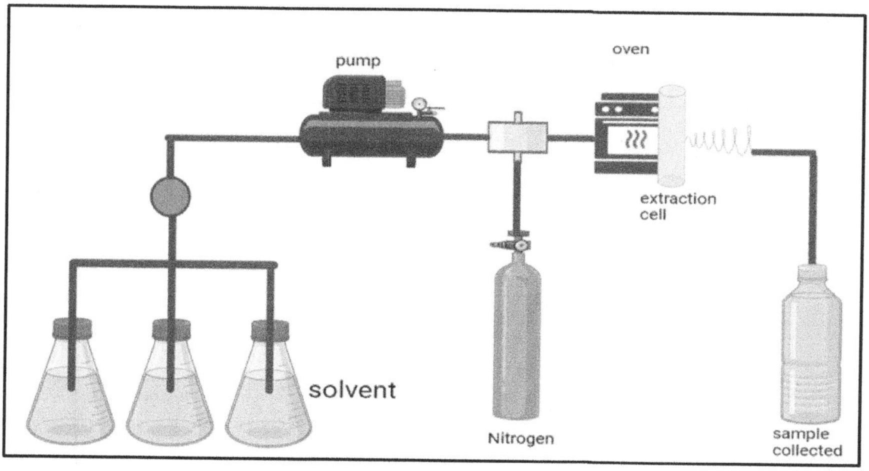

Figure 9.12. *Accelerated solvent extraction technique (prepared for this work). For a color version of this figure, see www.iste.co.uk/kumararya/chemicalengineering1.zip*

Extraction parameters include the solvent used, whose polarity needs to be quite similar to the target compounds, and the use of combining solvents with different polarities for different compound classes. Temperature is an important influence on solvent viscosity and extraction efficiency. The recommended temperature range for ASE applications is 75–125°C, with environmental applications often requiring 100°C. Pressure, typically about 1,500 psi, is critical for keeping solvents liquid. The static cycle introduces fresh solvent during extraction, assisting in the maintenance of a good extraction equilibrium, and is useful for samples with high analyte concentrations or difficult matrices. The total extraction time may be shortened by modifying the static time (Li et al. 2010).

9.3.2. *MAE*

Extraction using microwave exposes analytes to solvents via cell rupture, allowing for desorption from the matrix of the plant. Heat produced by free water molecules in plant structures causes localized heating and cell wall expansion, allowing plant constituents to exit from cells. The solvent and matrix's dielectric susceptibility interact to influence microwave energy usage. Additional extraction solvents improve efficiency, and to retain thermolabile components from being destroyed, the sample matrix is immersed in a solvent that is transparent to microwave radiation, such *n*-hexane (Figure 9.13). Extraction, microwave power and matrix properties such as particle size are all factors that influence MAE extraction yield. MAE optimization includes selecting a suitable extractor, determining the process type (closed or open) and selecting a solvent. The orthogonal array technique is then used to control variables such as temperature, microwave power and extraction duration. It is critical to select an appropriate collection and filtering (or centrifugation) procedure, and the optimum extract can be properly kept for further analysis (Delazar et al. 2012).

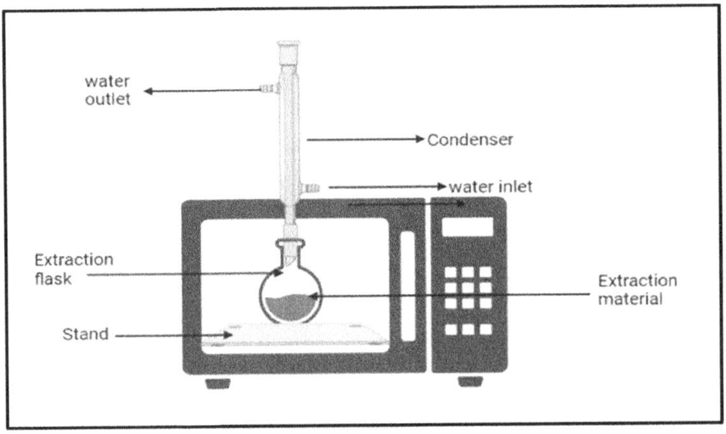

Figure 9.13. *Microwave-assisted extraction technique (prepared for this work). For a color version of this figure, see www.iste.co.uk/kumararya/chemicalengineering1.zip*

Due to its immense advantage in terms of speed, efficiency and lower solvent utilization, MAE has been accepted widely as a green technology with some modifications for the extraction of different bioactive compounds. In certain cases where, due to influence of temperature, there is a chance of degradation of phytocompounds, MAE is carefully applied to extract the thermoliable compounds as well as essential oils (Talebi et al. 2004). MAE also uses nitrogen gas (Yu et al. 2009) in order to prevent the oxidative degradation of bioactive compounds, whereas ultrasonication is also combinedly used in case of Ultrasonic Microwave Assisted

Extraction (UMAE) which increases momentum and energy and influences the penetration of solvents into the cells of plant tissues, resulting in lowers solvent usage and quick extraction of bioactive eluents into the solvent (Chen et al. 2010).

9.3.3. *UAE*

Extraction using US depends on the mechanical impact caused by cavitation caused by US application to enhance mass transfer between the solvent and the sample. This method allows chemicals to be extracted at lower temperatures and for shorter periods of time. UAE is a simple procedure that requires no complicated maintenance or skill, and it requires a lower amount of solvent than conventional techniques, making it cost-efficient. The methodology consistently yields more chemicals than traditional extraction methods, perhaps tripling the extracted component levels (González et al. 2022).

UAE extracts particular chemicals from a wide range of plant matrices employing solvents and ultrasonic energy. The fundamental process is acoustic cavitation, in which the development and collapse of bubbles cause localized effects in the plant's cellular matrix such as erosion, pore formation, shear stress, enhanced absorption and swelling (Figure 9.14). As a result, bioactive compounds dissolve in the solvent. UAE can be performed directly using a sonotrode immersed in the sample-solvent mixture or indirectly with an ultrasonic bath capable of simultaneously processing several samples. UAE is a highly efficient extraction technology that allows for the quick extraction of bioactive components at low temperatures with minimal energy and solvent requirements (Kumar et al. 2021).

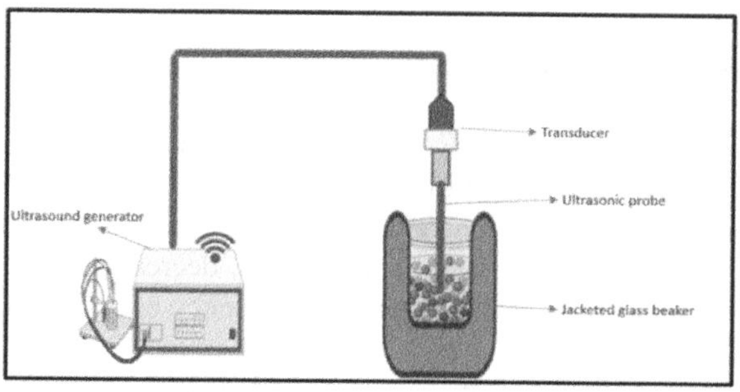

Figure 9.14. *Ultrasound-assisted extraction technique (prepared for this work). For a color version of this figure, see www.iste.co.uk/kumararya/chemicalengineering1.zip*

UAE happens to be a diverse technique which is influenced by several critical parameters. The ideal ultrasonic power range, as defined by amplitude percentage and power density, improves extraction yields by enhancing cavitation effects. The frequency of ultrasonic waves, which commonly ranges between 20 and 120 kHz, has an effect on efficiency, mechanical force balance and radical production. UAE efficiency is influenced by the duty cycle, which is the ratio of pulse duration to cycle time, with pulsed US offering energy savings. Extraction affects the solvent and its concentration, whereas solute release is affected by the liquid-to-solid ratio (LSR) and temperature. The extraction of substances such as pectin and phenolic compounds is heavily impacted by pH, with lower pH boosting yield through improved fragmentation and solubility. Increasing the UAE duration provides for more cavitation effects and solute release, but increasing the sonication time may cause structural damage and reduced yields. Understanding and enhancing these variables are critical for the efficient and long-term use of UAE in the extraction of bioactive compounds from diverse plant matrices.

The UAE method of extraction has been widely explored for the extraction of proteins in different plant-based raw materials such as soybean, rapeseeds, ricebran and legumes and is also known for modifications of proteins as well as formulations of concentrates and isolates (Moulton and Wang 1982). Apart from protein extraction ultrasonication is applied to change the size and shape of the cells of plants, resulting in cavitational breakdown and release of necessary phytocompounds into the extraction medium. UAE have also been known to find its interests in isolation and extraction of several bioactive compounds from microalgae (Carreira-Casais et al. 2021). Some of the well practiced have been reported to have isolated proteins, carbohydrates like polysaccharides, like carrageenans and alginates (Cesário et al. 2018), as well as lipids such as cylglycerols, sterols and waxes (Chen et al. 2018).

9.3.4. *Enzyme-assisted extraction*

Enzyme-assisted extraction (EAE) is a technology that serves an environmentally acceptable substitute that replaces traditional solvent-based techniques by extracting useful chemicals from plant materials using enzymes. EAE hypes the economic value of extracted goods, decreases waste and manages renewable plant resources. It provides energy efficiency and simplifies processing procedures and also protects the inherent qualities of extracted goods. EAE has been acknowledged by the French Environment and Energy Management Agency (ADEME) as a workable substitute for volatile organic compounds in industrial operations. The work of EAE is by breaking down substances that are soluble in water and trapping oil which allows it to run away naturally. In EAE, enzymes are essential because they function as biological catalysts, accelerating chemical processes under more benign circumstances and with more selectivity. The selection of enzymes involves a thorough comprehension of the composition of raw materials, cell wall structure, enzymatic activities, and pretreatments and working conditions (Muniglia et al. 2014).

Extraction of different plan materials and bioactive compounds has been widely practiced using different industrial enzymes such as cellulase, pectinase and hemicellulase. This practice has also been used on dairy products to increase the yields of total solids and protein contents (Costa et al. 2021). Industries also use enzymes such as alpha amylase and glucoamylase to release the reducing sugars to form the complex oligosaccharides (Rempel et al. 2018).

9.3.5. *SFE*

SC-CO_2 technology makes use of carbon dioxide and high pressure to destroy germs without compromising the nutritional value or palatability of bioactive ingredients in food and pharmaceuticals. It is a promising substitute for traditional pasteurization techniques that may remove these kinds of chemicals (Sairam et al. 2012).

The solubility of the analyte to be extracted and other compound in the desired solvent determines how effective an extraction process is. SFE has been preferred as an industry standard due to its shorter extraction times, low use of organic solvents, ability to handle thermosensitive materials and environmentally friendly production of cleaner extracts (Ahmad et al. 2019).

Like CO_2, nitrous oxide is also used for its solvating and separation qualities, although alkanes work better as solvents for nonpolar solutes. Even though SC-CO_2 technology is quite versatile, it is crucial to carefully choose co-solvents to combine efficacy with safety and environmental concerns (Mendiola et al. 2013).

Recent applications involving supercritical CO_2 extraction have demonstrated enriched plant extracts with different range of phytochemicals, showing properties of antioxidant, anticancer, antimicrobial, antidiabetic and anti-inflammatory properties. For example, Ribeiro Grijo et al. (2019) found that cannabis flower extracts possess high neutral cannabinoids and had potential as antitumor properties on cervical cells. In addition, extracts of polyphenolic compounds from olive oil residues bore significant antioxidant and therapeutic properties. The selection of SFE depends on conditions of operations such as temperature, pressure and type of solvent, which influences the functional properties of the extracts as well as poses defiance for obtaining high-yield extractions.

Innovative strategies to mitigate CO_2 emission, such as using CO_2 in the form of supercritical fluid, have been discussed widely to tackle both climate change and industrial pollution. This approach could be helpful in the reduction of emissions while effectively remediating the heavy metals and organic pollutants from the environment, presenting a sustainable solution for industrial processes (Costa-Boddeker et al. 2020).

9.3.6. *Pressurized hot water extraction*

The process referred to as pressurized hot water extraction (PHWE) is defined by a sequence of successive operations carried out inside an extraction cell that holds sample materials and a sizable amount of sand. First, under pressure and at a high temperature, solutes are removed from the sample matrix's active spots. In the following step, the extraction fluid diffuses into the matrix, causing solutes from the sample matrix to partition into the extraction fluid. The solutes are chromatographically eluted from the cell and placed into a collecting container as the last step (Figure 9.15).

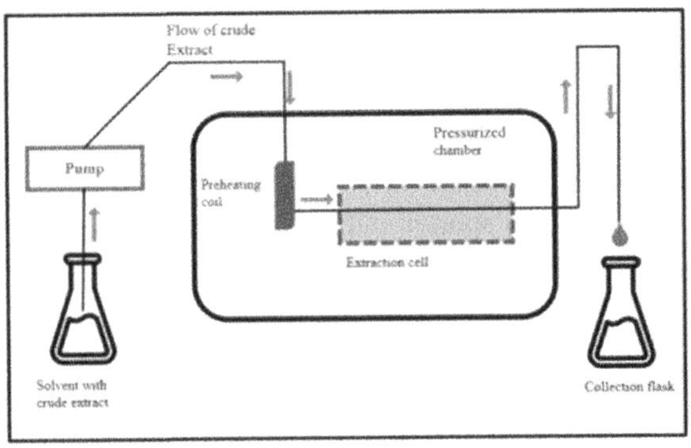

Figure 9.15. *Pressurized hot water extraction technique (prepared for this work). For a color version of this figure, see www.iste.co.uk/kumararya/chemicalengineering1.zip*

The process of extracting a chemical from PHWE follows a thermodynamic model and consists of two steps: first, the components are required to be freed using the sample matrix original binding sites and then it must be eluted out of the sample. KD, or partitioning coefficient of thermodynamics, controls this elution process. The enhanced effects of mass transfer and solubility and a greater disruption of surface equilibria are credited with the increased efficiency of PHWE extraction (Teo et al. 2010).

A number of studies have revealed that more than half of the extractions using PHWE techniques have been used to extract phenolic compounds followed by polysaccharides and proteins (Plaza and Marina 2023). Their immense interest in pharmaceutical and medical uses has grown over the years, and therefore, the extraction using techniques like PHWE has also increased with time.

9.3.7. *Pulsed electric field-assisted extraction*

Pulsed electric field (PEF) applies a moderate-to-high electric field potential, typically between 100 and 300 V/cm in batch mode and 20 and 80 kV/cm in continuous mode extraction. As far as PEF mechanisms are concerned, there are two schools of thought.

One perspective focuses on increasing the solubility of liquids by speeding up chemical processes occurring within the biological cell membrane. According to this viewpoint, PEF speeds up the extraction process by encouraging interactions between different molecules inside the cell membrane (Vorobiev and Lebovka 2011).

The technique of electroporation, which involves applying an external electrical force to promote cell membrane permeability, is the focus of the second viewpoint. This method involves placing the targeted materials between electrodes in an electric field with a high voltage. Electroporation opens protein channels and produces hydrophilic holes in the cell membrane. Consequently, the membrane's structural integrity is weakened, which increases permeability. High-voltage potential of electric pulses is applied, and this results in the electric field which destroys the membrane's structural integrity (Figure 9.16). As a result, this technique extracts the plant material (Ranjha et al. 2021).

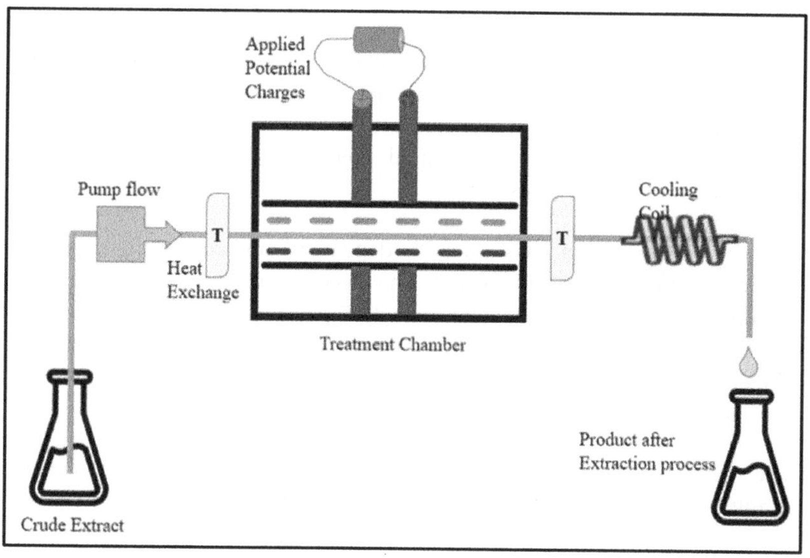

Figure 9.16. *Pulsed electric field-assisted extraction technique (prepared for this work). For a color version of this figure, see www.iste.co.uk/kumararya/chemicalengineering1.zip*

Pulsed electric field-assisted extraction has been emergingly used in the extraction of various bioactive compounds from food industry and waste materials, while reducing time, energy and solvent utilization. By using this technique, several researchers have extracted and purified bioactive compounds such as polyphenols and anthocyanins from grape peels and stem (Corrales et al. 2008), flavonoids and polyphenols from peels of orange (Luengo et al. 2013), compounds like Hesperidin and Narirutin from peels of discarded cirus unshiu peels, ascorbic acid and carotenoids from oranges (Athanasiadis et al. 2022) and polyphenols from olive leaves (Pappas et al. 2021).

In a study, Bobinaitė et al. (2015) applied pulsed electric field (PEF) pretreatment to uplift cell membrane permeability in blueberries, enhancing the yield of juice and other bioactive compound extracted from press cake, and discussed that PEF treatment of 5 kV/cm and 10 kJ/kg, increases the extraction yield of the phenolic content and anthocyanins by ~ 79% and ~ 106%, respectively, whereas Zhou et al. (2015) applied PEF for the extraction of anthocyanins from blueberries, achieving extraction with 60% ethanol, 0.1% HCl at 20 kV/cm. This method yielded 223.13 mg/L of anthocyanins, showing a ~ 1% higher yield than US extraction at lower temperatures and shorter times.

9.4. Conclusion

In biology, chemistry and pharmaceuticals, there is an effective interaction between traditional and contemporary approaches to extraction. Modern procedures have introduced remarkable efficiency and accuracy, pushing the frontiers of yield with purity and sustainability, while conventional methods, founded in established principles, continue to play an essential role due to their simplicity, broad application and cost-effectiveness.

Scientific development has mostly depended on conventional extraction procedures, like maceration and soxhlet. These techniques have advantages but also have disadvantages, such as extended extraction periods, high solvent consumption and the possibility of sensitive component degradation. Newer technologies, such as SFE, MAE and UAE, give advantages such as shorter processing times, more selectivity and lower environmental impact.

The choice between conventional and modern extraction procedures in chemistry is determined by the specific needs of substances and the objectives of the investigation. Traditional procedures are appropriate for routine analysis, whereas modern methods excel in difficult extractions. Bioactive substance extraction is critical in domains such as biology and medicine. Traditional methods are more suitable for large-scale production, although they may be better suited for sensitive biomolecule extraction. Excellent extraction processes should balance environmental sustainability, cost-effectiveness and efficiency in the future. Hybrid solutions that blend traditional and modern ways may become popular. Extraction strategies will evolve in response to changing commercial and scientific needs.

9.5. References

Abbas, M., Ahmed, D., Qamar, M.T., Ihsan, S., Noor, Z.I. (2021). Optimization of ultrasound-assisted, microwave-assisted and Soxhlet extraction of bioactive compounds from *Lagenaria siceraria*: A comparative analysis. *Bioresource Technology Reports*, 15, 100746.

Abubakar, A.R. and Haque, M. (2020). Preparation of medicinal plants: Basic extraction and fractionation procedures for experimental purposes. *Journal of Pharmacy and Bioallied Sciences*, 12(1), 1–10.

Ahmad, T., Masoodi, F.A., Rather, S.A., Wani, S.M., Gull, A. (2019). Supercritical fluid extraction: A review. *Journal of Biological and Chemical Chronicles*, 5(1), 114–122.

Alkahtani, J., Elshikh, M.S., Dwiningsih, Y., Rathi, M.A., Sathya, R., Vijayaraghavan, P. (2022). In-vitro antidepressant property of methanol extract of *Bacopa monnieri*. *Journal of King Saud University – Science*, 34(8), 102299.

Altemimi, A., Lakhssassi, N., Baharlouei, A., Watson, D.G., Lightfoot, D.A. (2017). Phytochemicals: Extraction, isolation, and identification of bioactive compounds from plant extracts. *Plants*, 6(4), 42.

Aquilon, A.F., Cargullo, D.M., Onayan, J., Sarno, J., Molino, V.M., Lopez, E.C.R. (2023). Recent advances in extractive distillation. *Engineering Proceedings*, 56(1), 11.

Araujo, A.R., Périno, S., Fernandez, X., Cunha, C., Rodrigues, M., Ribeiro, M.P., Jordao, L., Silva, L.A., Rodilla, J., Coutinho, P., Chemat, F. (2021). SFME of *Thymus mastichina* essential oil: Influence on their chemical composition and on the antioxidant and antimicrobial activities. *Pharmaceuticals*, 14(8), 709.

Armenta, S., Garrigues, S., Esteve-Turrillas, F.A., de la Guardia, M. (2019). Green extraction techniques in green analytical chemistry. *TrAC Trends in Analytical Chemistry*, 116, 248–253.

Athanasiadis, V., Chatzimitakos, T., Kotsou, K., Palaiogiannis, D., Bozinou, E., Lalas, S.I. (2022). Optimization of the extraction parameters for the isolation of bioactive compounds from orange peel waste. *Sustainability*, 14(21), 13926.

Badawy, M.E.I., El-Nouby, M.A.M., Kimani, P.K., Lim, L.W., Rabea, E.I. (2022). A review of the modern principles and applications of solid-phase extraction techniques in chromatographic analysis. *Analytical Sciences*, 38, 1457–1487. doi: 10.1007/s44211-022-00190-8.

Bailes, P.J. (1981). Solvent extraction in an electrostatic field. *Industrial & Engineering Chemistry Process Design and Development*, 20(3), 564–570.

Bhuiya, M.M.K., Rasul, M., Khan, M., Ashwath, N., Mofijur, M. (2020). Comparison of oil extraction between screw press and solvent (n-hexane) extraction technique from beauty leaf (*Calophyllum inophyllum* L.) feedstock. *Industrial Crops and Products*, 144, 112024.

Bobinaitė, R., Pataro, G., Lamanauskas, N., Šatkauskas, S., Viškelis, P., Ferrari, G. (2015). Application of pulsed electric field in the production of juice and extraction of bioactive compounds from blueberry fruits and their by-products. *Journal of Food Science and Technology*, 52(9), 5898–5905.

Boonsongsawat, T., Shotipruk, A., Tantayakom, V., Prasitchoke, P., Chandavasu, C., Boonnoun, P., Muangnapoh, C. (2010). Solvent extraction of biologically derived 1, 3-propanediol with ethyl acetate and ethanol cosolvent. *Separation Science and Technology*, 45(4), 541–547.

Cardoso, A.A. and Dasgupta, P.K. (1995). Analytical chemistry in a liquid film/droplet. *Analytical Chemistry*, 67(15), 2562–2566.

Carreira-Casais, A., Otero, P., Garcia-Perez, P., Garcia-Oliveira, P., Pereira, A.G., Carpena, M., Soria-Lopez, A., Simal-Gandara, J., Prieto, M.A. (2021). Benefits and drawbacks of ultrasound-assisted extraction for the recovery of bioactive compounds from marine algae. *International Journal of Environmental Research and Public Health*, 18(17), 9153.

Cassel, E., Vargas, R.M.F., Martinez, N., Lorenzo, D., Dellacassa, E. (2009). Steam distillation modeling for essential oil extraction process. *Industrial Crops and Products*, 29(1), 171–176.

Castro-Puyana, M., Marina, M.L., Plaza, M. (2017). Water as green extraction solvent: Principles and reasons for its use. *Current Opinion in Green and Sustainable Chemistry*, 5, 31–36.

Cequier-Sánchez, E., Rodriguez, C., Ravelo, Á.G., Zarate, R. (2008). Dichloromethane as a solvent for lipid extraction and assessment of lipid classes and fatty acids from samples of different natures. *Journal of Agricultural and Food Chemistry*, 56(12), 4297–4303.

Cesário, M.T., da Fonseca, M.M.R., Marques, M.M., de Almeida, M.C.M. (2018). Marine algal carbohydrates as carbon sources for the production of biochemicals and biomaterials. *Biotechnology Advances*, 36(3), 798–817.

Chemat, F., Lucchesi, M.E., Smadja, J., Favretto, L., Colnaghi, G., Visinoni, F. (2006). Microwave accelerated steam distillation of essential oil from lavender: A rapid, clean and environmentally friendly approach. *Analytica Chimica Acta*, 555(1), 157–160.

Chen, Y., Gu, X., Huang, S., Li, J., Wang, X., Tang, J. (2010). Optimization of ultrasonic/microwave assisted extraction (UMAE) of polysaccharides from inonotus obliquus and evaluation of its anti-tumor activities. *International Journal of Biological Macromolecules*, 46(4), 429–435. doi: 10.1016/j.ijbiomac.2010.02.003.

Chen, Z., Wang, L., Qiu, S., Ge, S. (2018). Determination of microalgal lipid content and fatty acid for biofuel production. *BioMed Research International*, 2018(1), 1503126.

Confortin, T.C., Todero, I., Luft, L., Schmaltz, S., de Freitas Ferreira, D., Barin, J.S., dos Santos, M.S.N., dos Mazutti, M.A., Zabot, G.L., Tres, M.V. (2024). Microwave hydrodiffusion and gravity and pressurized-liquid extraction for obtaining bioactive compounds from *Solanum viarum*. *Revista Brasileira de Ciências Ambientais (RBCIAMB)*, 59, e2070–e2070.

Corrales, M., Toepfl, S., Butz, P., Knorr, D., Tauscher, B. (2008). Extraction of anthocyanins from grape by-products assisted by ultrasonics, high hydrostatic pressure or pulsed electric fields: A comparison. *Innovative Food Science and Emerging Technologies*, 9(1), 85–91.

Coskun, O. (2016). Separation techniques: Chromatography. *Northern Clinics of Istanbul*, 3(2), 156.

Costa, C., Azoia, N.G., Coelho, L., Freixo, R., Batista, P., Pintado, M. (2021). Proteins derived from the dairy losses and by-products as raw materials for non-food applications. *Foods*, 10(1), 135.

Costa, R.D.S., de Almeida, S.S., Cavalcanti, E.D.C., Freire, D.M.G., Moura-Nunes, N., Monteiro, M., Perrone, D. (2022). Enzymes produced by solid state fermentation of agro-industrial by-products release ferulic acid in bioprocessed whole-wheat breads. *Food Research International*, 140, 109843.

Costa-Boddeker, S., Thuyên, L.X., Hoelzmann, P., de Stigter, H.C., van Gaever, P., Huy, H.Đ., Smol, J.P., Schwalb, A. (2020). Heavy metal pollution in a reforested mangrove ecosystem (Can Gio Biosphere Reserve, Southern Vietnam): Effects of natural and anthropogenic stressors over a thirty-year history. *Science of the Total Environment*, 716, 137035. doi: 10.1016/j.scitotenv.2020.137035.

Criland, N. (1968). Section III: Management: Magnet extraction. *International Ophthalmology Clinics*, 8(1), 213–229.

De Castro, M.L. and Priego-Capote, F. (2010). Soxhlet extraction: Past and present panacea. *Journal of Chromatography A*, 1217(16), 2383–2389.

De Santis, D., Carbone, K., Garzoli, S., Laghezza Masci, V., Turchetti, G. (2022). Bioactivity and chemical profile of *Rubus idaeus* L. leaves steam-distillation extract. *Foods*, 11(10), 1455.

Delazar, A., Nahar, L., Hamedeyazdan, S., Sarker, S.D. (2012). Microwave-assisted extraction in natural products isolation. *Natural Products Isolation*, 864, 89–115.

de la Fuente, B., Pinela, J., Mandim, F., Heleno, S.A., Ferreira, I.C., Barba, F.J., Berrada, H., Caleja, C., Barros, L. (2022). Nutritional and bioactive oils from salmon (*Salmo salar*) side streams obtained by Soxhlet and optimized microwave-assisted extraction. *Food Chemistry*, 386, 132778.

Gavahian, M. and Farahnaky, A. (2018). Ohmic-assisted hydrodistillation technology: A review. *Trends in Food Science & Technology*, 72, 153–161

González, M.J.A., Carrera, C., Barbero, G.F., Palma, M. (2022). A comparison study between ultrasound-assisted and enzyme-assisted extraction of anthocyanins from blackcurrant (*Ribes nigrum* L.). *Food Chemistry X*, 13, 100192.

Hidayat, R. and Wulandari, P. (2021). Methods of extraction: Maceration, percolation and decoction. *Eureka Herba Indonesia*, 2(1), 68–74.

Hwang, H.-J., Kim, H.-J., Ko, M.-J., Chung, M.-S. (2021). Recovery of hesperidin and narirutin from waste citrus unshiu peel using subcritical water extraction aided by pulsed electric field treatment. *Food Science and Biotechnology*, 30(2), 217–226.

Idris, F.N., Nadzir, M.M., Abd Shukor, S.R. (2020). Optimization of solvent-free microwave extraction of *Centella asiatica* using Taguchi method. *Journal of Environmental Chemical Engineering*, 8(3), 103766.

Irshad, M., Subhani, M.A., Ali, S., Hussain, A. (2020). Biological importance of essential oils. *Essential Oils – Oils of Nature*, 1, 37–40.

Kumar, K., Srivastav, S., Sharanagat, V.S. (2021). Ultrasound-assisted extraction (UAE) of bioactive compounds from fruit and vegetable processing by-products: A review. *Ultrasonics Sonochemistry*, 70, 105325.

Lei, Z., Li, C., Chen, B. (2003). Extractive distillation: A review. *Separation & Purification Reviews*, 32(2), 121–213.

Li, Y., Pang, T., Guo, Z., Li, Y., Wang, X., Deng, J., Zhong, K., Lu, X., Xu, G. (2010). Accelerated solvent extraction for GC-based tobacco fingerprinting and its comparison with simultaneous distillation and extraction. *Talanta*, 81(1–2), 650–656.

Li, Y., Fabiano-Tixier, A.S., Vian, M.A., Chemat, F. (2013). Solvent-free microwave extraction of bioactive compounds provides a tool for green analytical chemistry. *TrAC Trends in Analytical Chemistry*, 47, 1–11.

López-Bascón, M.A. and De Castro, M.L. (2020). Soxhlet extraction. In *Liquid-Phase Extraction*, Poole, C.F. (ed.). Elsevier, Amsterdam.

López-Hortas, L., Le Juge, C., Falqué, E., Domínguez, H., Torres, M.D. (2020). Bioactive extracts from edible nettle leaves using microwave hydrodiffusion and gravity and distillation extraction techniques. *Process Biochemistry*, 94, 66–78.

López-Ruiz, R., Marín-Sáez, J., Garrido Frenich, A., Romero-González, R. (2022). Recent applications of chromatography for analysis of contaminants in cannabis products: A review. *Pest Management Science*, 78(1), 19–29.

Luengo, E., Álvarez, I., Raso, J. (2013). Improving the pressing extraction of polyphenols of orange peel by pulsed electric fields. *Innovative Food Science and Emerging Technologies*, 17, 79–84.

Martins, N., Barros, L., Santos-Buelga, C., Henriques, M., Silva, S., Ferreira, I.C. (2014). Decoction, infusion and hydroalcoholic extract of *Origanum vulgare* L.: Different performances regarding bioactivity and phenolic compounds. *Food Chemistry*, 158, 73–80.

Martins-Gomes, C., Taghouti, M., Schäfer, J., Bunzel, M., Silva, A.M., Nunes, F.M. (2018). Chemical characterization and bioactive properties of decoctions and hydroethanolic extracts of *Thymus carnosus* Boiss. *Journal of Functional Foods*, 43, 154–164.

Mendiola, J.A., Herrero, M., Castro-Puyana, M., Ibáñez, E. (2013). Supercritical fluid extraction. In *Natural Product Extraction Principles and Applications*, Prado, J. and Rostagno, M. (eds). RSC Publishing, Cambridge.

Moraes, D.P., Farias, C.A., Barin, J.S., Ballus, C.A., Barcia, M.T. (2022). Application of microwave hydrodiffusion and gravity for phenolic compounds extraction from fruits. *Food and Bioprocess Technology*, 15(9), 1936–1947.

Mottaleb, M.A. and Sarker, S.D. (2012). Accelerated solvent extraction for natural product isolation. *Natural Products Isolation*, 864, 75–87. doi: 10.1007/978-1-61779-624-1_4.

Moulton, K.J. and Wang, L.C. (1982). A pilot-plant study of continuous ultrasonic extraction of soybean protein. *Journal of Food Science*, 47(4), 1127–1129. doi: 10.1111/j.1365-2621.1982.tb07632.x.

Muniglia, L., Claisse, N., Baudelet, P.H., Ricochon, G. (2014). Enzymatic aqueous extraction (EAE). In *Alternative Solvents for Natural Products Extraction*, Chemat, F.and Vian, M.A. (eds). Springer, Berlin, Heidelberg.

Nuchdang, S., Phruetthinan, N., Paleeleam, P., Domrongpokkaphan, V., Chuetor, S., Chirathivat, P., Phalakornkule, C. (2022). Soxhlet, microwave-assisted, and room temperature liquid extraction of oil and bioactive compounds from palm kernel cake using isopropanol as solvent. *Industrial Crops and Products*, 176, 114379.

Okoh, O.O., Sadimenko, A.P., Afolayan, A.J. (2010). Comparative evaluation of the antibacterial activities of the essential oils of *Rosmarinus officinalis* L. obtained by hydrodistillation and solvent-free microwave extraction methods. *Food Chemistry*, 120(1), 308–312.

Ötles, S. and Kartal, C. (2016). Solid-Phase Extraction (SPE): Principles and applications in food samples. *Acta Scientiarum Polonorum Technologia Alimentaria*, 15(1), 5–15.

Pappas, V.M., Lakka, A., Palaiogiannis, D., Athanasiadis, V., Bozinou, E., Ntourtoglou, G., Makris, D.P., Dourtoglou, V.G., Lalas, S.I. (2021). Optimization of pulsed electric field as standalone "green" extraction procedure for the recovery of high value-added compounds from fresh olive leaves. *Antioxidants*, 10(10), 1554.

Pena-Pereira, F., Lavilla, I., Bendicho, C. (2009). Miniaturized preconcentration methods based on liquid–liquid extraction and their application in inorganic ultratrace analysis and speciation: A review. *Spectrochimica Acta Part B: Atomic Spectroscopy*, 64(1), 1–15.

Plaza, M. and Marina, M.L. (2023). Pressurized hot water extraction of bioactives. *TrAC Trends in Analytical Chemistry*, 17, 39–54. doi: 10.1016/j.trac.2015.02.022.

Ranjha, M.M.A., Kanwal, R., Shafique, B., Arshad, R.N., Irfan, S., Kieliszek, M., Kowalczewski, P.Ł., Irfan, M., Khalid, M.Z., Roobab, U., Aadil, R.M. (2021). A critical review on pulsed electric field: A novel technology for the extraction of phytoconstituents. *Molecules*, 26(16), 4893.

Remler, R.F. (1922). The solvent properties of acetone. *Journal of the Franklin Institute*, 194(4), 520.

Remler, R.F. (1923). The solvent properties of acetone. *Industrial & Engineering Chemistry*, 15(7), 717–720.

Rempel, A., Machado, T., Treichel, H., Colla, E., Margarites, A.C., Colla, L. (2018). Saccharification of *Spirulina platensis* biomass using free and immobilized amylolytic enzymes. *Bioresource Technology*, 263, 163–171.

Ribeiro Grijo, D., Lazarin Bidoia, D., Vataru Nakamura, C., Vieitez Osorio, I., Cardozo-Filho, L. (2019). Analysis of the antitumor activity of bioactive compounds of *Cannabis* flowers extracted by green solvents. *Journal of Supercritical Fluids*, 149, 20–25. doi: 10.1016/j.supflu.2019.03.012.

Richter, B.E., Jones, B.A., Ezzell, J.L., Porter, N.L., Avdalovic, N., Pohl, C. (1996). Accelerated solvent extraction: A technique for sample preparation. *Analytical Chemistry*, 68(6), 1033–1039.

Sairam, P., Ghosh, S., Jena, S., Rao, K.N.V., Banji, D. (2012). Supercritical fluid extraction (SFE) – An overview. *Asian Journal of Research in Pharmaceutical Science*, 2(3), 112–120.

Shen, W., Benyounes, H., Gerbaud, V. (2015). Extractive distillation: Recent advances in operation strategies. *Reviews in Chemical Engineering*, 31(1), 13–26.

Singh, J. (2008). Maceration, percolation and infusion techniques for the extraction of medicinal and aromatic plants. *Extraction Technologies for Medicinal and Aromatic Plants*, 67, 32–35.

Talebi, M., Ghassempour, A., Talebpour, Z., Rassouli, A., Dolatyari, L. (2024). Optimization of the extraction of paclitaxel from *Taxus baccata* L. by the use of microwave energy. *Journal of Separation Science*, 27(13), 1130–1136. doi: 10.1002/jssc.200401754.

Tambun, R., Alexander, V., Ginting, Y. (2021). Performance comparison of maceration method, soxhletation method, and microwave-assisted extraction in extracting active compounds from soursop leaves (*Annona muricata*): A review. In *IOP Conference Series: Materials Science and Engineering*, 1122(1), 012095. IOP Publishing, Bristol.

Tavares, C.S., Martins, A., Miguel, M.G., Carvalheiro, F., Duarte, L.C., Gameiro, J.A., Roseiro, L.B. (2020). Bioproducts from forest biomass II. Bioactive compounds from the steam-distillation by-products of *Cupressus lusitanica* Mill. and *Cistus ladanifer* L. wastes. *Industrial Crops and Products*, 158, 112991.

Teo, C.C., Tan, S.N., Yong, J.W.H., Hew, C.S., Ong, E.S. (2010). Pressurized hot water extraction (PHWE). *Journal of Chromatography A*, 1217(16), 2484–2494.

Vian, M.A., Fernandez, X., Visinoni, F., Chemat, F. (2008). Microwave hydrodiffusion and gravity, a new technique for extraction of essential oils. *Journal of Chromatography A*, 1190(1–2), 14–17.

Vorobiev, E. and Lebovka, N. (2011). *Pulse Electric Field-Assisted Extraction*. CRC Press, Boca Raton.

Yeasmen, N. and Islam, M.N. (2015). Ethanol as a solvent and hot extraction technique preserved the antioxidant properties of tamarind (*Tamarindus indica*) seed. *Journal of Advanced Veterinary and Animal Research*, 2(3), 332–337.

Yin, S.J., Zhao, J., Yang, F.Q. (2021). Recent applications of magnetic solid phase extraction in sample preparation for phytochemical analysis. *Journal of Pharmaceutical and Biomedical Analysis*, 192, 113675.

Yu, Y., Chen, B., Chen, Y., Xie, M., Duan, H., Li, Y., Duan, G. (2009). Nitrogen-protected microwave-assisted extraction of ascorbic acid from fruit and vegetables. *Journal of Separation Science*, 32(23–24), 4227–4233. doi: 10.1002/jssc.200900487.

Zhang, Q.W., Lin, L.G., Ye, W.C. (2018). Techniques for extraction and isolation of natural products: A comprehensive review. *Chinese Medicine*, 13(20), 1–26.

Zhou, Y., Zhao, X., Huang, H. (2015). Effects of pulsed electric fields on anthocyanin extraction yield of blueberry processing by-products. *Journal of Food Processing and Preservation*, 39(6), 1898–1904.

Żwir-Ferenc, A. and Biziuk, M. (2006). Solid phase extraction technique – Trends, opportunities, and applications. *Polish Journal of Environmental Studies*, 15(5), 677–697.

10

Distillation

Distillation is a technique that separates two or more components of a liquid mixture by using the difference in their respective volatilities or boiling points. It is considered to be one of the most important separation processes in the field of chemical engineering. For example, distillation columns can be classified according to their operation, which may be batch, semi-batch or continuous. This is one of the ways that distillation columns can be classified. Because of its versatility, batch distillation is frequently used for the separation and purification of chemicals that are present in small amounts or ratios. It also allows product separation with varied specifications inside the same column. The batch distillation column can function with a constant reflux ratio, which results in a continuous change in the composition of the distillate. Therefore, this chapter covers distillation principles, formulae and procedures for binary and multiple-component systems, as well as calculations for the number of trays required for the distillation column.

10.1. Introduction

Distillation is a separation and purification technique used to separate two or more components of a liquid mixture based on the difference of their respective boiling point using vaporization and condensation.

The steps involved in distillation are:

– entering liquid feed into a distillation tower called distilland;

– heating of liquid mixture or feed in the stripping section;

– converting the liquid phase into the vapor phase;

– condensing vapors in the rectifying section;

Chapter written by Neha DHIMAN and Girish GUPTA.

– recovering condensate or distillate;

– recovering waste from the bottom.

Distillation is one of the most widely used mass-transfer operations on an industrial scale. It can be used for the separation of volatile as well as nonvolatile components from a mixture. A liquid mixture contains two components, i.e. component A (more volatile) and component B (less volatile) and is subjected to heat to separate both components. A more volatile component (i.e. component A) more readily evaporates compared to a less volatile component (i.e. component B). As a result, at any time from the start of the distillation operation, the stripping section is richer in the less volatile component (B) than in the more volatile component (A). At the same time, the rectifying section is richer in the amount of component A (more volatile) compared to component B (less volatile). The reason behind the different proportions of less and more volatile components, in respective sections, is a difference in the extent of vaporization of each component (Treybal 1981; Gavhane 2011).

While considering the processes involved in the mass transfer unit operations, other than distillation, there may be some similarities between them, for example, in the case of evaporation and drying operation, the component under study is subjected to heat. Similarly, in the distillation of liquid mixtures, heat is supplied to convert liquid into a vapor phase. These processes can be differentiated based on the required or desired product of that unit operation.

– In the case of *distillation*, heat is supplied to the liquid component mixture in order to convert it into the vapor phase and the desired product is the collected condensate or distillate.

– In the case of *evaporation*, the liquid compound is heated below its boiling point to reduce the amount of diluents from it and the condensate is not required. The desired product is the concentrate of liquid.

– *The drying* operation is initiated with the evaporation of moisture content on heating and summed up with the collection of the desired product, i.e. the only dried solid residue.

10.2. Applications of distillation process

– *Recovery of solvent after the extraction process*: the distillation process involves the separation of valuable solvent from effluents, waste materials and other impurities. Distillation is a frequently used technique for recovery, involving the heating of waste fluids to vaporize the solvent. The vapor is then collected and cooled, causing it to condense back into a liquid state. The waste product is separated and up to 90% of the initial solvent is retrieved and can be reused without any decrease in effectiveness.

– *Purification of solvents*: distillation is the combined procedure of transforming a substance into vapor and then returning it to its liquid state through condensation. This approach is employed to purify liquids that undergo boiling without breakdown and include contaminants that are not volatile. This method can also be used to separate liquids with significant disparities in their boiling points.

– *Refining of petroleum products*: the refinement is accomplished using a process known as fractional distillation, which takes place in a distillation tower or fractionating column. In the context of industries other than oil, fractional distillation is a technique that involves heating a combination of two or more fluids of varying densities to the point where it boils. This technique is used to separate the mixture into its constituent fluids. What we refer to as the "distillates" are these separated fluids. When discussing the process of refining oil, the distillates refer to the numerous products used, such as gasoline, diesel fuel or kerosene. During the various stages of the fractionating column, each of these products exhibits a distinct density. The gases with the lowest density are located at the top of the column, while the heavier fuels and bitumen find their density at the bottom.

– *Separation of two or more liquids from a liquid mixture*: it is possible to separate mixtures that are composed of two or more pure liquids by the use of distillation, which is an efficient process. There is a process of purification known as distillation, in which the components of a liquid mixture are first vaporized and then condensed and separated. A combination is heated in the process of simple distillation and the component that is the most volatile is the one that vaporizes at the lowest temperature. After traveling through a condenser, that is cooled, the vapor enters a condition of condensation and returns to its liquid state. Distillate is the term used to describe the condensate that is collected.

10.3. Relative volatility

Relative volatility is defined as the ratio of the vapor pressure of one component in its pure state to that of another (Treybal 1981; Gavhane 2011). For a liquid mixture comprising two components, A and B, the relative volatility of component A with respect to component B is denoted α_{AB} and the relative volatility of component B with respect to component A is expressed as α_{BA}.

$$the\ relative\ volatility\ of\ A\ concerning\ B, \alpha_{AB} = \frac{v_A}{v_B} \qquad [10.1]$$

$$\textit{the relative volatility of B concerning A, } \alpha_{BA} = \frac{v_B}{v_A} \quad [10.2]$$

where v_A represents the vapor pressure of pure component A; v_B signifies the vapor pressure of pure component B.

Conforming to Raoult's law of partial pressure, the vapor pressure of a component is equal to the ratio of the partial pressure to its mole fraction in the liquid state.

$$\text{For component A, } v_A = \frac{p_A}{x_A} \quad [10.3]$$

$$\text{For component B, } v_B = \frac{p_B}{x_B} \quad [10.4]$$

where p_A is the partial pressure of component A; p_B is the partial pressure of component B; x_A is the mole fraction of component A in the liquid phase; x_B is the mole fraction of component B in the liquid phase.

For the distillation of two components (designated as A and B) within a liquid mixture, the relevance of the mole fraction extends from the liquid phase to encompass the vapor phase. Accordingly, under a total pressure denoted as P, the expression for the mole fractions of components A and B in their vapor phase is articulated as follows:

$$p_A = Py_A \quad [10.5]$$

$$p_B = Py_B \quad [10.6]$$

where P is the total pressure; y_A is the mole fraction of component A in the vapor phase and y_B is the mole fraction of component B in the vapor phase.

Relative volatility should not be equal to 1, otherwise no separation is possible. Therefore, $\alpha_{AB} = 1$, when either $y_A = y_B$ and $x_A = x_B$. When considering both components at the same time, the mole fraction of both the components in the liquid, as well as vapor phases, should not be equal to each other. Therefore, no separation will occur.

Relative volatility can also be expressed in terms of mole fractions of single components in liquid and vapor phases, say for component A,

$$\alpha_{AB} = \frac{y_A(1-x_A)}{(1-y_A)x_A} \quad [10.7]$$

Then, $\alpha_{AB} = 1$, if $y_A = x_A$. At the same time, the mole fraction of component A in the liquid phase should not be equal to its mole fraction in the vapor phase. Therefore, no separation is possible.

10.4. Classification of distillation

The distillation process can be classified into different types based on laboratory and industrial-scale applications.

10.4.1. *Simple distillation*

Simple distillation is used to separate the components from liquid mixtures featuring one volatile component, designated as the desired product, while the remaining components are nonvolatile (Subrahmanyam et al. 2019). This method, also known as differential distillation, segregates the various constituents of a liquid mixture based on disparities in their individual volatility and vapor pressure.

10.4.1.1. *Principle*

A liquid mixture is introduced into a distillation flask and is subjected to heat. The initiation of liquid boiling transpires when the vapor pressure of each constituent attains equilibrium with the atmospheric pressure. Throughout the distillation process at any given point, a specific mole fraction of each component in the liquid phase undergoes conversion to the vapor phase of that particular. Subsequently, vapor condensation is facilitated by a condenser, which provides a refrigerated surface for condensation. The separation is contingent upon the relative volatility of each component concerning the other. A higher relative volatility of the desired component facilitates more effective separation. Consequently, nonvolatile impurities are neglected to the bottom or waste, while the purified component is accumulated in the form of condensate or distillate (Gavhane 2011).

10.4.1.2. *Construction*

It consists of a simple heated distillation flask attached with a condenser on the side arm having a circulated water jacket. This assembly is attached to a condensate collecting flask (Figure 10.1).

10.4.1.3. *Working*

A simple heated distillation flask is filled with a liquid mixture of up to two thirds of its volume. Small porcelain pieces are added to it. The thermometer is also attached to the neck of the flask to measure the temperature and is not dipped into the liquid mixture. Water is circulated through the outer shell of the condenser. The flask is heated, the liquid starts boiling and begins to convert into a vapor phase. Temperature is noted using a thermometer, which is equal to the boiling point of that component. These vapors are condensed using the condenser attached to the flask (Subrahmanyam et al. 2019).

Let us assume that the distillation flask initially contained L_1 moles of a liquid mixture of components A and B. At any time t, L moles of liquid are left. For component A, x moles are present in the liquid phase, and y moles in the vapor phase. With time, the composition of the mixture changes, let us contemplate a change in the quantity of liquid, denoted as dL, and a marginal alteration of dx in the mole fraction of component A in the liquid phase (Treybal 1981).

The comprehensive equilibrium for component A can be expressed as follows:

$$xL = (x - dx)(L - dL) + yL \tag{10.8}$$

$$xL = xL - xdL - Ldx + dx\, dL + ydL \tag{10.9}$$

dx and dL are the small changes, so the product of dxdL becomes very small and can be neglected, equation [10.9] can be written as,

$$(y - x)\, dL = Ldx \tag{10.10}$$

$$\frac{dL}{L} = \frac{dx}{(y-x)} \tag{10.11}$$

On integrating equation [10.11], we get

$$\int_{L2}^{L1} \frac{dL}{L} = \int_{x2}^{x1} \frac{dx}{(y-x)} \tag{10.12}$$

where:

– L1 is the initial moles of liquid;

– L2 is the moles of liquid left;

– x1 is the initial moles of component A in the liquid phase;

– x2 is the moles left of component A in the liquid phase.

Equation [10.12] is known as the Rayleigh equation.

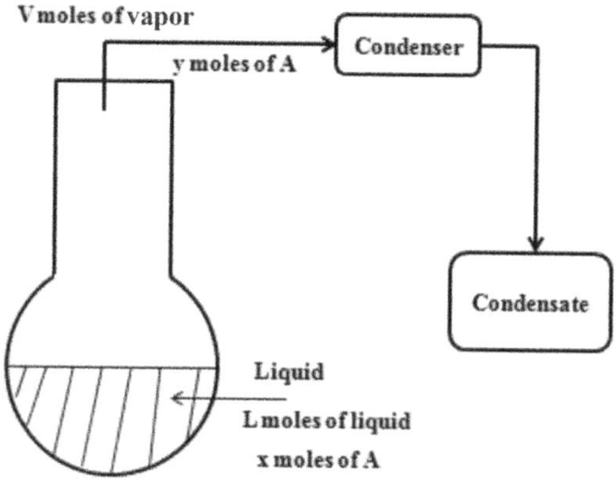

Figure 10.1. *Simple distillation (prepared for this work)*

10.4.2. *Equilibrium or flash distillation*

This form of distillation may be conducted in a batch-wise or continuous manner. Flash or equilibrium distillation is applicable to separate multiple components within a liquid mixture. The boiling point of these components lies close to each other, for example, petroleum products can be separated using flash or equilibrium distillation (Gavhane 2011; Subrahmanyam et al. 2019).

10.4.2.1. *Principle*

In this distillation process, a multicomponent liquid mixture is partially vaporized. Then, equilibrium is maintained between the vapor and liquid phase of the liquid mixture, and finally, the components can be separated easily.

10.4.2.2. *Construction*

The flash distillation chamber consists of a heating chamber fitted with the heating element. The chamber is attached to a feeding pump. An outlet is attached with a vapor–liquid separator at the other top end of the chamber with a pressure-reducing valve (Figure 10.2).

10.4.2.3. *Working*

In flash distillation, a liquid mixture feed is pumped into the heating chamber through a heater at a particular pressure. Then it flows through a pressure-reducing valve and the liquid mixture becomes partially vaporized. The pressure drops when the equilibrium is reached. A flash occurs in the hot liquid when passed from high-pressure to low-pressure volume. Then, the vapors are separated from the top through the condenser and liquid from the bottom of the separator.

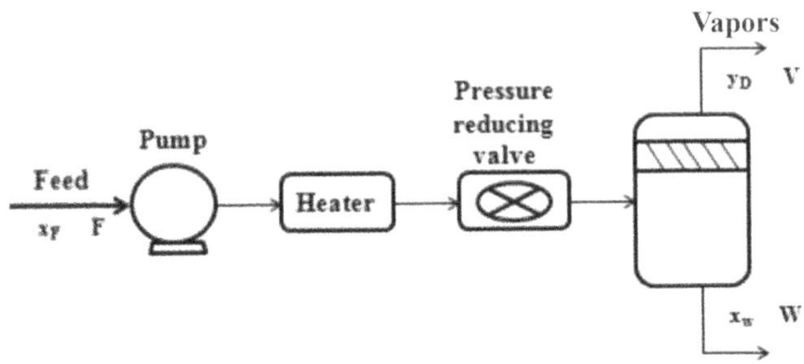

Figure 10.2. *Equilibrium or flash distillation (prepared for this work)*

The overall balance can be written as,

$$F = V + W \quad [10.13]$$

where:

– V is the moles of vapor;

– W is the moles of liquid.

$$F\,x_F = V\,y_D + W\,x_w \quad [10.14]$$

where:

– x_F is the mole fraction of more volatile component in feed;

– y_D is the mole fraction of more volatile component in the vapor phase at equilibrium;

– x_w is the mole fraction of more volatile component in liquid phase at equilibrium.

$$x_F = \frac{V}{F} y_D + \frac{W}{F} x_W \quad [10.15]$$

$$x_F = \frac{V}{F} y_D + \frac{(F-V)}{F} x_W \quad [10.16]$$

Let us assume V/F = f, a fraction of feed that is converted into the vapor phase, equation [10.16] becomes

$$x_F = f\,y_D + (1-f)x_W \qquad [10.17]$$

$$y_D = -\frac{(1-f)}{f}x_W + \frac{x_F}{f} \qquad [10.18]$$

If $y_D = y$ and $x_w = x$, then

$$y = -\frac{(1-f)}{f}x + \frac{x_F}{f} \qquad [10.19]$$

For flash distillation, y = x at equilibrium

$$x = -\frac{(1-f)}{f}x + \frac{x_F}{f} \qquad [10.20]$$

From equation [10.20], we get $x = x_F$

Therefore, the line started from (x_F, x_F) coordinates on a 45° diagonal line having slope –(1-f)/f and the point of intersection with the equilibrium curve is (y_D, x_w) when y_D and x_w are in equilibrium.

10.4.3. *Steam distillation*

Steam distillation is particularly applicable for separating substances or components from a liquid mixture that exhibits sensitivity to high temperatures or undergoes decomposition under ordinary pressure during distillation. Therefore, to separate these temperature-sensitive materials using a distillation process, partial pressure of the less volatile components is reduced by introducing the vapors of an immiscible liquid, and the desired component gets distilled at a lower temperature before its decomposition. This method is used for the separation of organic compounds and an immiscible liquid vapor used is steam (Treybal 1981).

10.4.3.1. *Principle*

Its principle is based on the separation of heat-sensitive substances or components by reducing the partial pressure of that component in addition to an immiscible liquid, i.e. water or steam in the form of a vapor phase. Steam is the most widely used inert vapor for organic compounds. Due to the comparatively elevated boiling points of organic compounds in contrast to water, typically exceeding 100°C, facile separation is achieved through condensation collection. For instance, bromobenzene, with a boiling point of 156°C, and water boiling at 100°C, form a mixture that boils at 95°C without undergoing decomposition. Consequently, bromobenzene can be effectively distilled at a temperature below its boiling point.

10.4.3.2. *Construction and working*

The setup consists of a round bottom flask containing the liquid mixture being separated. A leg of a glass tube from the steam generator is dipped in the liquid mixture in a round bottom flask and the flask is also attached with a condenser of circular water flowing jacket followed by a receiving flask to collect the distillate of the desired product (Figure 10.3).

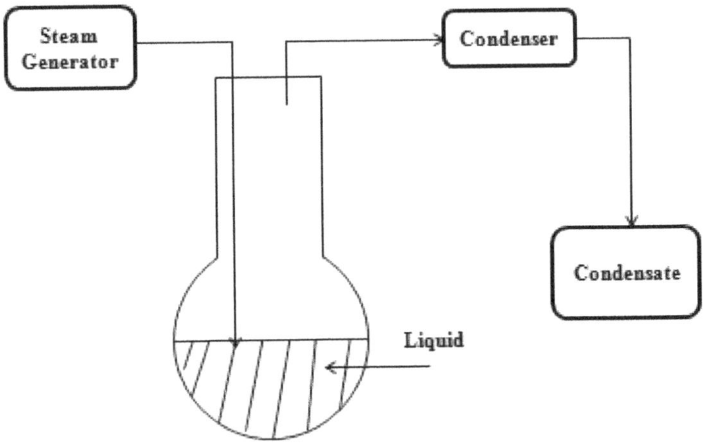

Figure 10.3. *Steam distillation (prepared for this work)*

10.4.4. *Azeotropic and extractive distillation*

Azeotropic mixtures are liquid mixtures of the same composition in the vapor and liquid phases. They are also known as constant boiling mixtures. The azeotropic mixture separation poses challenges with simple distillation due to the resistance to separation arising from identical compositions in both the liquid and vapor phases. This difficulty arises despite simple distillation methods based on the principle of the high relative volatility of liquid mixtures (Carter 2000; Sinha and De 2012).

Azeotropic mixtures or constant boiling mixtures are the mixtures of two or more different liquids which can be classified into two types based on the comparison between the boiling point of the mixture with each component.

10.4.4.1. *Minimum boiling point azeotropes*

It is the azeotropic mixture of liquids, boiled at low temperature compared to either of the component boiling points. It exhibits a substantial positive deviation from Raoult's law, as exemplified by the minimum boiling azeotrope observed in the ethanol and water mixture. Ethanol boils at 251.5 K and water at 373 K, while its minimum boiling point azeotrope boils at 351.1 K, which is lesser than that of both the components. It is also known as positive azeotropes.

10.4.4.2. *Maximum boiling point azeotropes*

It refers to an azeotropic liquid mixture characterized by a boiling point considerably higher than that of either of its components. Such mixtures exhibit substantial negative deviations from Raoult's law. For instance, the hydrogen chloride and water combination represents a maximum boiling azeotrope. Hydrogen chloride with a boiling point of 188 K, and water, boiling at 373 K, form an azeotrope with a maximum boiling point of 381.6 K, surpassing the boiling points of both individual components.

The azeotropic mixtures, commonly known as negative azeotrope, can be separated through homogenous azeotropic distillation, heterogeneous azeotropic distillation and extractive distillation.

10.4.4.3. *Azeotropic distillation*

Within this methodology, a supplementary component, termed an entrainer, is introduced to an azeotropic mixture, leading to the formation of a novel boiling point azeotropic solution. For instance, the addition of benzene into an ethanol and water mixture, or the inclusion of an ester in the combination of acetic acid and water, exemplifies this approach.

In azeotropic distillation, an entrainer is incorporated into the azeotropic mixture to improve the separation of azeotropes by altering the relative volatility of the liquid components. This leads to the formation of a new azeotrope with any one of the components. The resulting azeotrope can exhibit a minimum or maximum boiling point (Carter 2000; Sinha and De 2012).

The azeotropic distillation arrangement comprises two columns and a decanter (Figure 10.4). For example, in the case of acetic acid and water separation, feed is entered into the first distillation column and benzene as an entrainer is also added to this column along with the feed. In addition to the entrainer, a minimum boiling azeotrope is formed with water. High-purity ethanol is attainable from the lower segment of the primary distillation column, and the vapors emanating from the upper portion of the column comprise the benzene and water azeotrope. These vapors undergo condensation in the decanter, resulting in the formation of two distinct liquid phases: an organic layer and an aqueous layer. The top layer is a benzene-rich layer containing water and the bottom layer is a water-rich layer containing benzene. The top layer of benzene is again returned to the first column as an entertainer and the bottom layer is forwarded to the second distillation column for the recovery of benzene from the top of the column in the form of distillate sent back to the first column for recycling and pure water is collected from the bottom. If an entrained or third component is richer in the distillate of the first column, then it is known as azeotropic distillation.

Ethanol + Water $\xrightarrow{\textbf{Benzene (Entrainer)}}$ **Ethanol + (Water + Benzene)**
(Distillate)

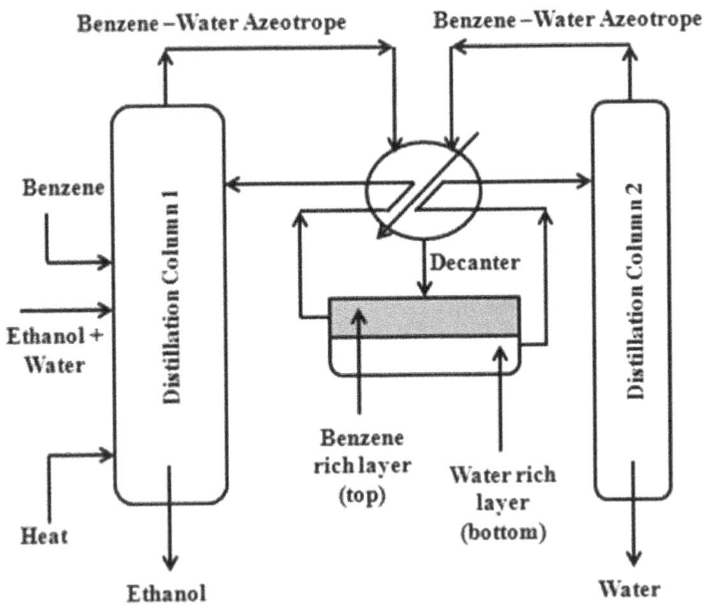

Figure 10.4. *Azeotropic distillation (prepared for this work)*

10.4.4.3.1. Homogeneous azeotrope

A homogeneous azeotropic mixture is a liquid mixture in which the components of the mixture are completely miscible in each other, for example, alcohol and water are completely miscible.

10.4.4.3.2. Heterogeneous azeotrope

A heterogeneous azeotropic mixture is a liquid mixture in which the components are not completely miscible in each other, for example, chloroform and water are not completely miscible.

10.4.4.4. *Extractive distillation*

Extractive distillation shares similarities with the azeotropic distillation process, specifically involving a third component, commonly referred to as an entrainer. One thing that differs from the azeotropic distillation process is that the trainer or third component is richer in the bottom product of the first column called extractive distillation (Figure 10.5). Upon introduction of an entrainer, a maximum boiling point azeotrope is generated at the base of the column (Sinha and De 2012; Subrahmanyam et al. 2019). This azeotrope is subsequently directed to the solvent recovery column, where the trainer is reclaimed from the bottom and reintroduced to the initial column. Concurrently, the second component is retrieved from the upper section of the solvent recovery column.

$$A + B \xrightarrow{\text{(Entrainer)}} A + \underset{\text{(Bottom)}}{(B + \text{Entrainer})}$$

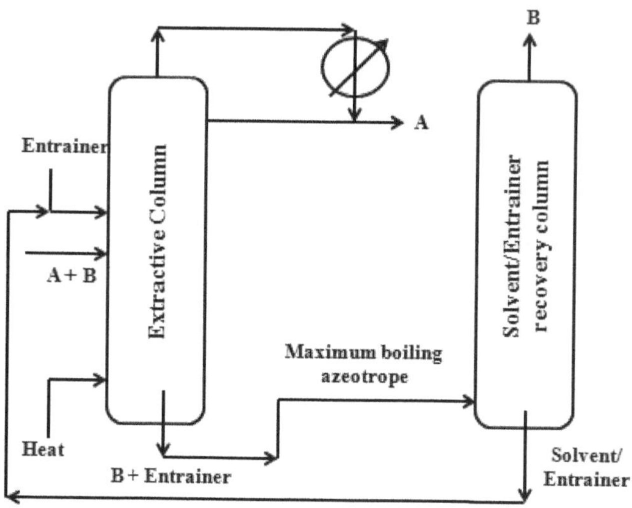

Figure 10.5. *Extractive distillation (prepared for this work)*

10.4.5. *Fractional distillation*

10.4.5.1. *Principle*

Fractional distillation is used for a liquid mixture separation comprising miscible liquids, relying on the distinction in their boiling points. This method is applied to, for instance, mixtures separation such as acetone and water, or chloroform and benzene. It is particularly well suited for mixtures where the components exhibit a difference in boiling point less than 25°C. In cases where the boiling point difference exceeds 25°C, simple distillation is employed for component separation (Treybal 1981).

10.4.5.2. *Construction*

In the process of simple distillation, the flask is directly affixed to the condenser to facilitate the condensation of vapors. In contrast, fractional distillation incorporates a fractionating column within the distillation flask and condenser setup. The fractionating column serves the purpose of partially condensing vapors before they reach the condenser. The vapors generated through partial condensation are subsequently directed back to the still, thereby augmenting the vapor composition with an increased concentration of the more volatile component (Subrahmanyam et al. 2019).

10.4.5.3. *Working*

The liquid mixture of miscible components is taken in a heated distillation flask. Generally, steam is used for heating purposes. Then condensation of vapors occurs in a sequence in the attached fractionating column (Figure 10.6).

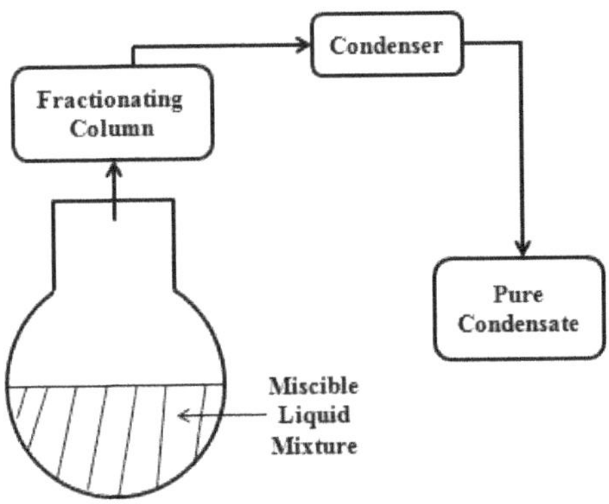

Figure 10.6. *Fractional distillation (prepared for this work)*

Let us consider the example of the separation of two miscible liquids say, a mixture of benzene and toluene containing 20% benzene and 80% of toluene. Benzene is a more volatile component with a low boiling point and toluene is less volatile having a high boiling point. The boiling point diagram for the separation of two miscible liquids followed by a sequence of fractional distillation is shown in Figure 10.7.

The first curve in the figure shows the initial composition of a mixture of benzene and toluene containing 20% benzene and 80% toluene. When the above composition mixture reaches its boiling point, the vapors collected from this liquid mixture have the composition of 54% benzene and 46% toluene, as shown by the first condensation curve. These vapors are condensed and further heated until boiling. These vapors contain 76% benzene and 24% toluene and are condensed, as shown by the second condensation curve. The residue collected after the second condensation curve is again heated to its boiling point and the collected condensate contains 97% benzene and 3% toluene shown by the third condensation curve. These vapors contain almost pure benzene (i.e. 97% benzene), and the amount of toluene in this composition is approximately negligible. The final vapors, following the third condensation curve, are condensed via the column affixed at the apex of the fractionating column.

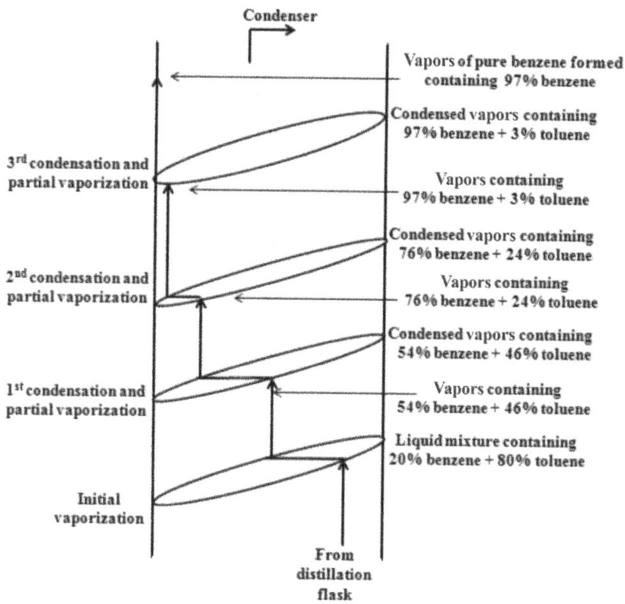

Figure 10.7. *Boiling point diagram of two miscible liquids for fractional distillation (prepared for this work)*

10.4.6. *McCabe–Thiele method: calculation of the number of theoretical plates*

The calculation of several theoretical plates for the separation of two components through the distillation column is an important aspect of this unit operation. The McCabe–Thiele method is a mathematical–graphical method that provides an insight into the simulation program and the selection of designing parameters of the distillation column for binary distillation (Gavhane 2011). Feed is entered from the center of the distillation column. The distillation column is divided into two equal parts, i.e. upper part above the feeding section is a rectifying section in which the vapor phase exists and the below part of the liquid phase is the stripping section (Figure 10.8). Condensed vapors or distillate can be collected from the top of the rectifying section and residue/waste from the bottom of the stripping section.

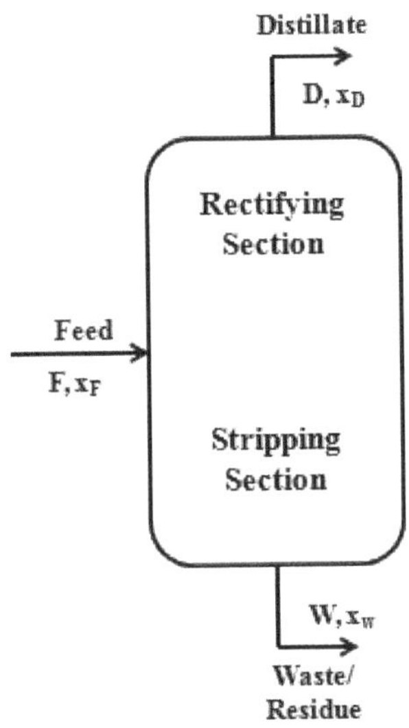

Figure 10.8. *Distillation column (prepared for this work)*

10.4.6.1. *Steps involved in the McCabe–Thiele method for the calculation of the number of theoretical plates*

– Overall material balance to determine x_F, F, x_D, D, x_w, W.

– Plot the equilibrium curve and 45° diagonal line with x = y data.

– Draw the operating line for rectifying the section from (x_D, x_D).

– Find the slope and intercept of the operating line of the rectifying section.

– Draw the operating line for the stripping section (x_w, x_w).

– Find the slope and intercept of the operation line of the stripping section.

– Point of intersection of operating lines of rectifying and stripping section.

– Draw feed line, i.e. q-line from (x_F, x_F) passes through the point of intersection of rectifying and stripping section.

– Draw rectangular steps starting from (x_D, x_D) between the operating line of the rectifying section and the equilibrium curve pass the q-line and follow between the operating line of the rectifying section and the equilibrium curve till reaches or passes the point (x_w, x_w).

– Let "n" number of rectangular steps be drawn.

– A number of theoretical plates are n including reboiler, otherwise (n-1) theoretical plates are considered.

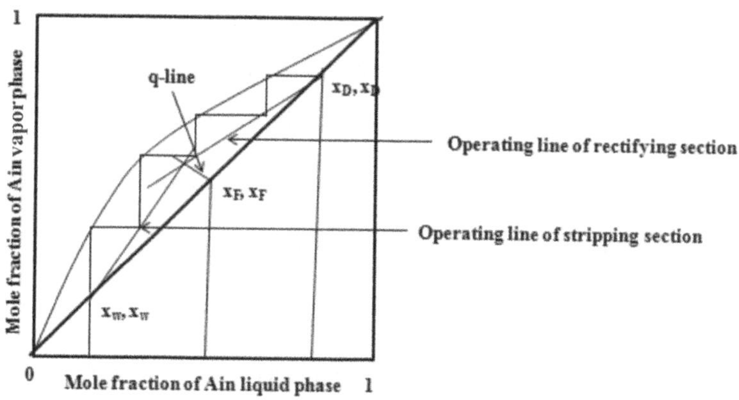

Figure 10.9. *Plot for the determination of the number of theoretical plates (prepared for this work)*

10.4.6.2. *Assumptions*

The main assumptions of the McCabe–Thiele method are as follows:

– the constant molal flow rate of vapor between the feed plate and top plate;

– the constant molal flow rate of liquid between the feed and bottom plate.

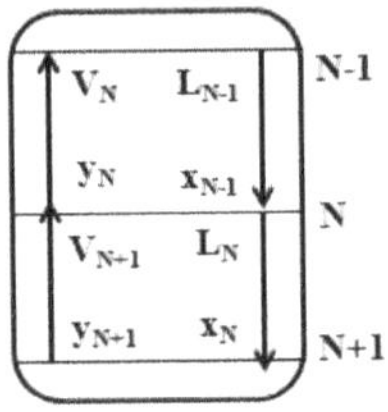

Figure 10.10. *Vapor or liquid entering and leaving the plates of the distillation column (prepared for this work)*

Applying material balance for the Nth plate, we get

$$V_{N+1} + L_{N-1} = V_N + L_N \quad [10.21]$$

where:

– V is the flow rate of the number of moles of vapors per hour;

– L is the flow rate of the number of moles of liquid per hour.

Material balance of more volatile component mole fraction,

$$V_{N+1}y_{N+1} + L_{N-1}\,x_{N-1} = V_N\,y_N + L_N x_N \quad [10.22]$$

where:

– y is the mole fraction of more volatile component in the vapor phase;

– x is the mole fraction of more volatile component in liquid phase.

The more volatile component is readily converted from the liquid to the vapor phase. Therefore, the rectifying section is richer in more volatile components.

10.4.6.3. *Overall material balance*

At steady-state condition, the overall material balance is given by

$$F = D + W \quad [10.23]$$

Material balance of more volatile component mole fraction,

$$F\,x_F = D\,x_D + W\,x_w \quad [10.24]$$

10.4.6.4. *Rectifying section*

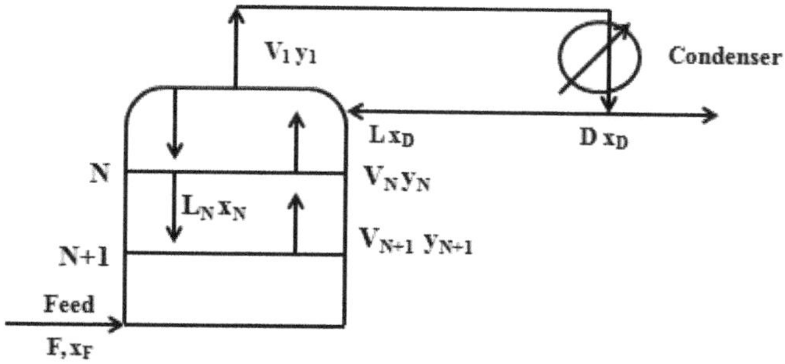

Figure 10.11. *Rectifying section (prepared for this work)*

For a constant molal flow rate of vapor and constant molal flow rate of liquid between each plate for the rectifying section,

$$L_1 = L_2 = L_N = L \quad [10.25]$$

$$V_1 = V_2 = V_N = V_{N+1} = V \quad [10.26]$$

Overall material balance for the rectifying section, we get

$$V_{N+1} = L_N + D \quad [10.27]$$

From equation [10.25] and [10.26], equation [10.27] becomes

$$V = L + D \quad [10.28]$$

Material balance of more volatile component mole fraction,

$$V\, y_{N+1} = L\, x_N + D\, x_D \quad [10.29]$$

By combining equations [10.28] and [10.29], we get

$$y_{N+1} = \frac{L}{L+D} x_N + \frac{D}{L+D} x_D \quad [10.30]$$

Equation [10.30] is the operating line of the *rectifying section.*

The equation of the 45° diagonal line is y = x, and the intersection point of the 45° diagonal line and the operating line of the rectifying section is determined by putting y = x in equation [10.30].

$$x = \frac{L}{L+D}\,x + \frac{D}{L+D}\,x_D \qquad [10.31]$$

By solving equation [10.31], we get $x = x_D$.

$y = x$, therefore, $y = x = x_D$

The operating line of rectifying section intersects the 45° diagonal line at (x_D, x_D).

10.4.6.5. *Stripping section*

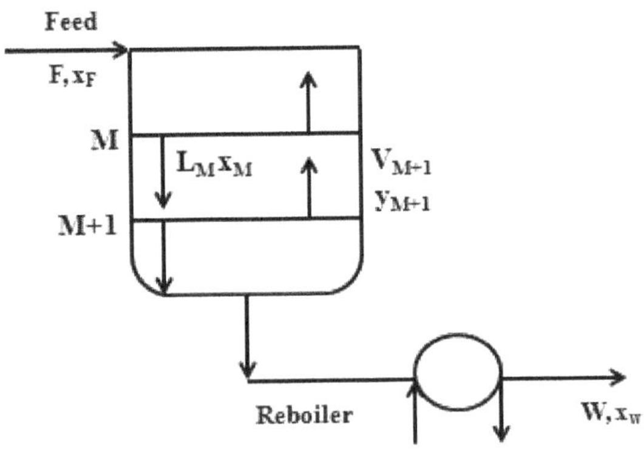

Figure 10.12. *Stripping section (prepared for this work)*

For a constant molal flow rate of vapor and constant molal flow rate of liquid between each plate for the stripping section,

$$L_M = L_{M+1} = \bar{L} \qquad [10.32]$$

$$V_M = V_{M+1} = \bar{V} \qquad [10.33]$$

Overall material balance for the stripping section, we get

$$V_{M+1} = L_M - W \qquad [10.34]$$

From equations [10.32] and [10.33], equation [10.34] becomes

$$\bar{V} = \bar{L} - W \quad [10.35]$$

Material balance of more volatile component mole fraction,

$$\bar{V}y_{M+1} = \bar{L}x_M - Wx_w \quad [10.36]$$

By combining equations [10.35] and [10.36], we get

$$y_{M+1} = \frac{\bar{L}}{\bar{L}-W}x_N - \frac{W}{\bar{L}-W}x_w \quad [10.37]$$

Equation [10.37] is the operating line of the *stripping section.*

The equation of the 45° diagonal line is y = x, and the intersection point of the 45° diagonal line and the operating line of the stripping section is determined by putting y = x in equation [10.37].

$$x = \frac{\bar{L}}{\bar{L}-W}\,x - \frac{W}{\bar{L}-W}x_w \quad [10.38]$$

By solving equation [10.38], we get $x = x_w$.

$y = x$, therefore, $y = x = x_w$

The operating line of the stripping section intersects the 45° diagonal line at (x_w, x_w).

10.4.6.6. *q-line*

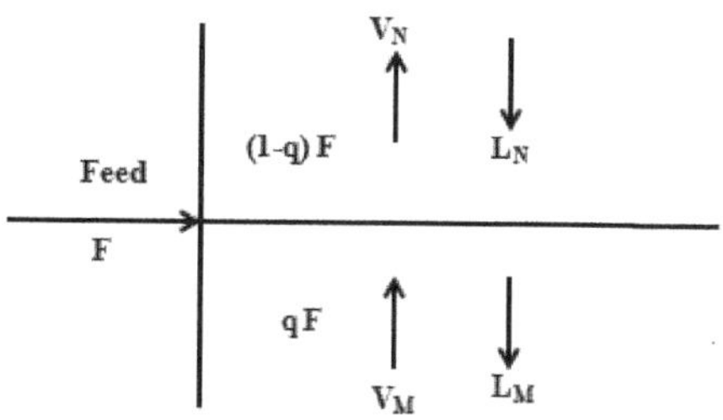

Figure 10.13. *Liquid and vapor above and below the feed plate (prepared for this work)*

Vapor will add to V_M to give V_N and the liquid will be added to L_N to give L_M.

q is the number of moles of saturated liquid produced on the feed plate by each mole of feed added to the tower (Treybal 1981). It is also defined as the ratio of heat required to vaporize one mole of feed at entering conditions to the molar latent heat of feed vaporization.

$$L_M = L_N + qF \quad [10.39]$$

$$V_N = V_M + (1-q)\,f \quad [10.40]$$

For the constant molal rate of vapor and liquid, we get

$$\bar{L} = L + qF \quad [10.41]$$

$$V = \bar{V} + (1-q)\,f \quad [10.42]$$

The operating lines for the rectifying and stripping section are,

$$y = \frac{L}{L+D}x + \frac{D}{L+D}x_D \quad [10.43]$$

$$y = \frac{\bar{L}}{\bar{L}-W}x - \frac{W}{\bar{L}-W}x_W \quad [10.44]$$

From equation [10.30], we get

$$Ly + Dy = Lx + Dx_D$$

$$L\,(y-x) = D\,(x_D - y) \quad [10.45]$$

From equations [10.41] and [10.44], we get

$$y = \frac{L+qF}{L+qF-W}\,x - \frac{W}{L+qF-W}x_W$$

$$Ly + qFy - Wy = Lx + qFx - Wx_W$$

$$L\,(y-x) = qF\,(x-y) + W\,(y - x_W) \quad [10.46]$$

From equations [10.45] and [10.46],

$$D\,(x_D - y) = qF\,(x-y) + W\,(y - x_W)$$

$$Dx_D - Dy = (qx - qy)F + Wy - Wx_W$$

$$Dx_D + Wx_w = (qx - qy)F + (D + W)y \qquad [10.47]$$

Substituting equations [10.23] and [10.24] in equation [10.47] and rearranging,

$$Fx_F = (qx - qy)F + Fy$$

$$x_F = qx - (q - 1)y$$

$$(q - 1)y = qx - x_F$$

$$y = \frac{q}{(q-1)}x - \frac{x_F}{(q-1)} \qquad [10.48]$$

This is termed the q-line equation and intersects the 45° diagonal line at (x_F, x_F).

10.4.6.7. *Reflux ratio*

Reflux is the portion of the vapors returned to the previous still. Forward flow is the portion of condensate collected as a product.

Reflux ratio is the ratio of kg of reflux to the kg of forward flow.

$$R = L/D = \text{kg of reflux/kg of forward flow} \qquad [10.49]$$

Using equations [10.30] and [10.49],

$$y_{N+1} = \frac{R}{R+1}x_N + \frac{1}{R+1}x_D \qquad [10.49a]$$

If R = ∞, then slope R/(R+1) = 1 – [1/(R+1)] = 1, which states that the operating line of rectifying a stripping section coincides with the 45° diagonal line, which leads to the possibility of a minimum number of theoretical plates.

10.4.7. *McCabe–Thiele method (binary mixtures)*

The McCabe–Thiele method is a graphical approach to determining the number of theoretical stages required for a binary distillation process.

– *Construct the equilibrium curve*: plot the equilibrium curve (y vs. x) for the binary mixture using the vapor-liquid equilibrium data.

– *Plot the operating lines*: draw the operating lines for the rectifying and stripping sections.

The rectifying section operating line equation is,

$$y = \frac{R}{R+1}x + \frac{x_D}{R+1} \qquad [10.50]$$

where R is the reflux ratio and x_D is the distillate composition.

For the stripping section, the equation is,

$$y = \frac{R}{R+1}x - \frac{x_B}{R+1} \qquad [10.51]$$

where x_B is the bottom composition.

10.4.7.1. *Step-off stages*

Starting from the distillate composition on the y-axis, draw horizontal and vertical steps between the equilibrium curve and the operating lines until you reach the desired bottom composition. Each step represents one theoretical stage.

10.4.8. *Fenske–Underwood–Gilliland (FUG) method (multicomponent mixtures)*

The FUG method is a combination of three methods used for multicomponent distillation.

10.4.8.1. *Fenske equation*

This equation determines the minimum number of theoretical stages required at total reflux (no product withdrawal, infinite reflux ratio)

$$N_{min} = \frac{Log\left(\frac{x_{D,i}/x_{D,j}}{x_{B,i}/x_{B,j}}\right)}{Log(\alpha_{ij})} \qquad [10.52]$$

where $x_{D,i}$ and $x_{D,j}$ are the mole fractions of components i and j in the distillate, $x_{B,i}$ and $x_{B,j}$ are the mole fractions in the bottoms, and α_{ij} is the relative volatility of component i to component j.

10.4.8.2. *Underwood equations*

Calculate the minimum reflux ratio R_{min},

$$\sum_{i=1}^{n} \frac{qz_i}{\alpha_i - \theta} = 1 \qquad [10.53]$$

and

$$\sum_{i=1}^{n} \frac{x_{D,i}}{\alpha_i - \theta} = R_{min} + 1 \qquad [10.54]$$

where θ is the root of the first equation, z_i is the mole fraction of component i, in the feed, α_i is the relative volatility of component I, and $x_{D,i}$ is the mole fraction of component i, in the distillate.

10.4.8.3. *Gilliland correlation*

Relates the actual number of stages N, minimum number of stages N_{min}, actual reflux ratio R and minimum reflux ratio R_{min}.

$$\frac{N - N_{min}}{N+1} = 0.75\left(\frac{R - R_{min}}{R+1}\right)^{0.5608} \qquad [10.55]$$

10.4.9. ***Practical application and software tools***

Modern process simulation software like ASPEN PLUS, HYSYS, PRO/II and CHEMCAD are used to automate these calculations, providing accurate and efficient design solutions for binary and multicomponent distillation systems.

10.4.9.1. *ASPEN PLUS*

– *Define the components and thermodynamic model*: specify the components and select an appropriate thermodynamic model for the mixture.

– *Set up the distillation column*: input feed stream conditions, column specifications and desired separation criteria.

– *Run the simulation*: use the built-in distillation tools to calculate the number of stages, reflux ratio and other designs.

– *Analyze results*: review the simulation output, including stage-by-stage compositions, temperature profiles and efficiency calculations.

– *Widely used process simulation software for modeling distillation, including reactive distillation.*

10.4.9.2. *HYSYS*

Another popular simulation software for process design and optimization.

10.4.9.3. *PRO/II*

Process simulation software by AVEVA for distillation and other chemical processes.

10.4.9.4. *CHEMCAD*

Comprehensive chemical process simulation software for designing distillation columns.

10.4.10. *Flooding velocity and calculation*

Flooding velocity is a critical parameter in distillation column design, as it determines the maximum vapor velocity at which a column can operate without experiencing excessive liquid holdup and loss of efficiency. Flooding occurs when the upward vapor flow rate is so high that it prevents the downward flow of liquid, causing the liquid to accumulate in the column and potentially leading to operational failure.

10.4.10.1. *Empirical correlations*

Various empirical correlations are used to estimate the flooding velocity in distillation columns. These correlations often consider factors, i.e. column diameter, tray spacing, liquid and vapor densities, and surface tension.

10.4.10.2. *Fair and Fenske correlation*

$$\boldsymbol{u_f = C_s\left(\frac{\rho_L-\rho_V}{\rho_V}\right)^{0.5}} \qquad [10.56]$$

where u_f is the flooding velocity, ρ_L is the liquid density, ρ_V is the vapor density, and C_s is a constant that depends on the tray type and column geometry.

10.4.10.3. Leva's correlation

$$\boldsymbol{u_f = C_s\left(\frac{\rho_L-\rho_V}{\rho_V}\right)^{0.5}\left(\frac{\sigma}{20}\right)^{0.2}} \qquad [10.57]$$

where σ is the surface tension of the liquid.

10.4.11. *Tray types*

Trays are critical components in distillation columns, providing the surface area for vapor–liquid contact. Different tray types are used depending on the specific requirements of the distillation process, such as efficiency, capacity and pressure drop (Lockett 1986).

10.4.11.1. *Sieve trays*

– *Description*: sieve trays consist of perforated plates with uniform holes through which vapor rises. The liquid flows across the tray, providing a large contact area for mass transfer.

– *Advantages*: simple construction, low cost, high capacity.

– *Disadvantages*: susceptible to weeping (liquid falling through the holes at low vapor flow rates), moderate efficiency.

– *Applications*: suitable for a wide range of applications, especially where a high capacity is required.

10.4.11.2. *Valve trays*

– *Description*: valve trays feature movable valves or discs adjusted with the vapor flow rate, optimizing vapor and liquid distribution.

– *Advantages*: high efficiency, flexible operation over a wide range of flow rates, reduced weeping and entrainment.

– *Disadvantages*: more complex construction and higher cost compared to sieve trays.

– *Applications*: preferred in applications requiring high efficiency and flexible operation.

10.4.11.3. *Bubble cap trays*

– *Description*: bubble cap trays have riser caps with slots or holes that direct vapor into the liquid, creating bubbles for enhanced contact.

– *Advantages*: high efficiency, effective at low vapor rates, minimal weeping.

– *Disadvantages*: higher pressure drop, complex construction and higher cost.

– *Applications*: used in processes requiring high separation efficiency and stable operation at varying flow rates.

10.4.11.4. *Dual flow trays*

– *Description*: dual flow trays have both vapor and liquid passing through the same openings without any moving parts.

– *Advantages*: simple design, high capacity, minimal maintenance.

– *Disadvantages*: less efficient than other tray types, prone to maldistribution at low vapor rates.

– *Applications*: suitable for fouling systems where maintenance and cleaning are difficult.

10.4.11.5. *Practical considerations*

– *Selection criteria*: tray-type selection depends on process requirements, such as separation efficiency, operating range and cost constraints. Engineers often use pilot plant data, vendor recommendations and simulation software to make informed decisions.

– *Maintenance*: tray maintenance involves regular inspection for damage, corrosion and fouling. Proper maintenance ensures consistent performance and extends the life of the distillation column.

10.5. References

Carter, S.J. (2000). *Cooper and Gunn's Tutorial Pharmacy*, 6th edition. CBS Publishers & Distributors, New Delhi.

Gavhane, K.A. (2011). *Mass Trasfer-II*, 9th edition. Nirali Prakashan, Pune.

Lockett, M.J. (1986). *Distillation Tray Fundamentals*. Cambridge University Press, Cambridge.

Sinha, A.P. and De, P. (2012). *Mass Transfer Principles and Operations*, 1st edition. PHI Learning Private Limited, New Delhi.

Subrahmanyam, C.V.S., Thimmasetty. J., Suresh, S., Devi, V.K. (2019). *Pharmaceutical Engineering Principles and Practice Unit Operations*, 3rd edition. Vallabh Prakashan, Delhi.

Treybal, R.E. (1981). *Mass-Trasfer Operations*, 3rd edition. McGraw Hill, Singapore.

11

Mathematical Modeling of Binary and Multicomponent Distillation

Distillation is a standard chemical process used in industries for various purposes ranging from component separation to attaining desired purity. Numerous types of distillation processes are employed for multiple purposes. To comprehend and assimilate all of these, it is vital to have a comprehensive understanding of this procedure. Any process can be better understood and analyzed by its mathematical modeling. This chapter presents the distillation process and mathematical modeling methods. Different feed compositions are used to assess the process, including binary distillation (simple distillation), multicomponent distillation and batch and continuous operation. Making the necessary assumptions, defining the variables, formulating the mass and energy balance and then using preexisting mathematical formulas and equations to solve them yield the mathematical modeling of these processes. This aids in providing a clear-cut perception of the distillation process.

11.1. Introduction

Distillation is the method of separating specific liquid components by selective boiling and then condensing them. This process occurs due to different driving forces, like the difference in vapor pressure and the difference in boiling point between the components in the mixture. Distillation has a lot of significance in the chemical process industry due to its wide variety of applications. Its importance mainly lies in its ability to separate the individual components from the mixture with maximum possible purity. This ability of the distillation process has found applications in various industries like pharmaceutical, petrochemical, food and beverages, essential oil productions, etc., where a product's purity is crucial for its effectiveness, safety and quality. As a chemical engineer, it is necessary to have a

---

Chapter written by Ananya MUNNANGI, MUTHAMIZHI K., Karthikeyan C., K. TAMILSELVI and A.B. MOHAMMED BASITH.

complete understanding of this process. The best way to understand any system or process is through mathematical modeling.

Mathematical modeling involves representing a real-world problem or plausible situation using mathematical terms, typically equations. These formulas will later be used to get a complete understanding of the problem and to discover further possible new aspects of the problem. Mathematical modeling can also be used to understand and design the chemical processes, which will not only help in understanding the process but also develop the processes according to their demand and application. The main components of mathematical modeling are decision variables, constraints, objective functions and parameters, which are used appropriately to define a process, situation, or real-life problem. This chapter gives a brief insight into the mathematical modeling of the distillation process. In this chapter, we focus on the modeling of various distillation processes.

11.2. Ideal simple distillation

Mostly, the distillation process handles multicomponent feed, which can be approximated into binary or pseudo-binary systems to simplify the complexity of the process for better understanding. This chapter starts by providing insight into a much simpler distillation process – ideal simple distillation – to reduce the complexity of the problem to its elementary form. Simple distillation is a separation technique often used to purify liquids by exploiting differences in their boiling point. This process is particularly effective for binary mixtures, where two components with distinct boiling points are present. Some assumptions are made to reduce the complexity of the process and define the ideality.

Assumptions start with assuming a two-component feed mixture (binary system), with constant relative volatility throughout the whole length of the distillation column and in each tray. This means the liquid on the tray and the vapor leaving the tray are in equilibrium, and a basic vapor–liquid equilibrium can be used.

$$y_n = \frac{\alpha x_n}{(\alpha-1)x_n+1} \quad [11.1]$$

where

- y_n is the mole fraction of the more volatile component in the vapor phase;
- x_n is the mole fraction of the more volatile component in the liquid phase;
- α is the relative volatility of the more volatile component.

The feed mixture flows into the column with a flow rate of F (mol/hr) at its bubble point and has an Xf composition of x_f. The feed enters the column at tray N_f as a single stream. The vapor from the top is entirely condensed within a condenser and is directed into the reflux drum, which can hold the liquid of M moles. The drum's contents are assumed to be well mixed, and the composition is labeled x_d. The liquid in the drum is at its bubble point. Reflux is returned to the top tray (N_t) of the column at a rate R, while the overhead distillate product is drawn off at a rate D. Any delay (or dead time) in the vapor line from the column top to the reflux drum, and in the reflux line back to the top tray is ignored. This assumption is generally valid in industrial-scale columns and does not hold in small-scale laboratory columns. It is important to note that dynamically, the vapor leaving the column and the condensed reflux entering the column are not equal. Their equality occurs only at steady-state conditions.

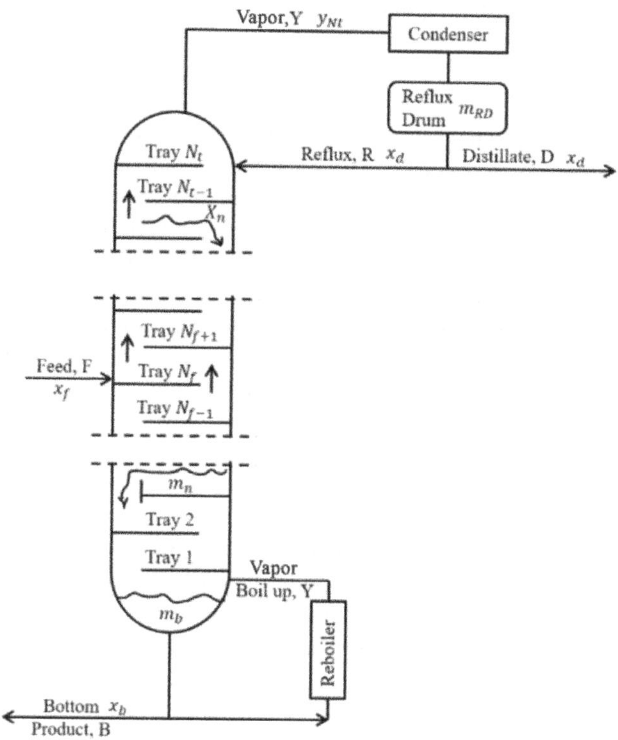

Figure 11.1. *Schematic diagram of ideal simple distillation (prepared for this work)*

At the bottom of the column, a portion of the outlet stream is removed as a bottom product at rate B, which has a composition of x_b. The rest of the content in the stream is sent into a reboiler for vapor buildup. It is also considered that the liquids in the reboiler and at the base of the column are mixed perfectly, which implies that they possess the same composition x_b. The reboiler can also hold a total of m_{RD} moles, similar to the reflux drum. It is also assumed that the vapor leaving the bottom of the column and the vapor entering tray 1 have the same composition, denoted by y_b, and are in equilibrium with the liquid of composition x_b. The column has N_t theoretical trays, each with a liquid holdup, denoted as m_n, including the downcomer. The liquid on each tray is assumed to be perfectly mixed and has a composition of x_n. The vapor holdup is considered negligible throughout the system. Despite the large vapor volume, moles are typically tiny due to the vapor phase's significantly lower density than the liquid phase. It is necessary to note that this assumption may not hold good in high-pressure columns. This assumption means the vapor flow rate through each tray is the same.

$$Y_1 = Y_2 = Y_3 = \ldots = Y_n = \ldots = Y_{Nt-1} = Y_{Nt} = Y$$

Another assumption made is that of equimolal overflow, with negligible temperature differences across each tray. This implies that the vapor and liquid flow rates through the stripping and rectifying sections are constant under steady-state conditions, eliminating the need to consider energy and total balances. However, the liquid rates throughout the column vary dynamically, depending on the fluid mechanics of the tray. Various formulas relate the liquid holdup of the tray (m_n) to the rate of liquid leaving the tray (X_n), ranging from the simple Francis weir formula to more complex relationships involving the effects of vapor rate, densities, compositions and other factors. A more straightforward relation between them is assumed.

$$m_n = f(X_n) \qquad [11.2]$$

Finally, the dynamics of the reboiler and condenser are neglected, and it is assumed that there is no chemical reaction in the column throughout the distillation process. Figure 11.1 represents a simplified ideal case: binary component distillation diagram. The equations describing the system of a distillation process based on the continuity equation (conservation of mass) are as follows:

Reboiler and column base

Overall material balance:

$$\frac{dm_b}{dt} = X_1 - Y - B \qquad [11.3]$$

Component mass balance:

$$\frac{d(m_b x_b)}{dt} = X_1\, x_1 - Y y_1 - B x_b \qquad [11.4]$$

First tray (n = 1)

Overall material balance:

$$\frac{dm_1}{dt} = X_2 - X_1 \qquad [11.5]$$

Component mass balance:

$$\frac{d(m_1 x_1)}{dt} = X_2\, x_2 - X_1\, x_1 + Y(y_b - y_1) \qquad [11.6]$$

Feed tray (n = N_f)

Overall material balance:

$$\frac{dm_{Nf}}{dt} = X_{Nf+1} - X_{Nf} + F \qquad [11.7]$$

Component mass balance:

$$\frac{d(m_{Nf} x_f)}{dt} = X_{2Nf+1}\, x_{Nf+1} - X_{Nf}\, x_{Nf} + Y\left(y_{Nf-1} - y_{Nf}\right) + F x_f \qquad [11.8]$$

Next to top tray (n = N_t – 1)

Overall material balance:

$$\frac{dm_{Nt-1}}{dt} = X_{Nt} - X_{Nt-1} \qquad [11.9]$$

Component mass balance:

$$\frac{d(m_{Nt-1} x_{Nt-1})}{dt} = X_{Nt}\, x_{Nt} - X_{Nt-1}\, x_{Nt-1} + Y(y_{Nt-2} - y_{Nt-1}) \qquad [11.10]$$

Top tray ($n = N_t$)

Overall material balance:

$$\frac{dm_{Nt}}{dt} = R - X_{Nt} \quad [11.11]$$

Component mass balance:

$$\frac{d(m_{Nt}x_{Nt})}{dt} = Rx_d - X_{Nt}\, x_{Nt} + Y(y_{Nt-1} - y_{Nt}) \quad [11.12]$$

Condenser and reflux drum ($n = N_t$)

Overall material balance:

$$\frac{dm_{RD}}{dt} = Y - R - D \quad [11.13]$$

Component mass balance:

$$\frac{d(m_{RD}x_d)}{dt} = Yy_{Nt} - (R + D)x_d \quad [11.14]$$

Generalizing for nth tray

Overall material balance:

$$\frac{dm_n}{dt} = X_{n+1} - X_n \quad [11.15]$$

Component mass balance:

$$\frac{d(m_n x_n)}{dt} = X_{n+1}\, x_{n+1} - X_n\, x_n + Y(y_{n-1} - y_n) \quad [11.16]$$

Like mass balance, each tray has its equilibrium relation [11.1] and hydraulic equations [11.2]. The equations that represent the level controllers at the reflux drum and base are as follows:

$$D = f(m_{RD}) \quad [11.17]$$

$$B = f(m_b) \quad [11.18]$$

The degrees of freedom of this system can be calculated by subtracting the number of equations from the number of variables.

Number of variables:

Compositions on tray (x_n and y_n)	$2N_t$
Liquid flows on the tray (X_n)	N_t
Liquid holdups on the tray (m_n)	N_t
Composition of reflux drum (x_d)	1
Flows of reflux drum (R and D)	2
Foldup of reflux drum (m_{RD})	1
Compositions of base (x_b and y_b)	2
Flows of base (Y and B)	2
Holdup of base (m_b)	1
	$4N_t + 9$

Number of equations:

Component continuity of tray	N_t
Total continuity of tray	N_t
Equilibrium (trays plus base)	$N_t + 1$
Hydraulic	N_t
Level controllers	2
Component continuity of reflux drum	1
Total continuity of reflux drum	1
Component continuity of base	1
Total continuity of base	1
	$4N_t + 7$

$4N_t + 9$ variables describe this system, while only $4N_t + 7$ equations specify the system. Therefore, this system is underspecified. This means there are only two variables to control this system. These variables can be reflux flowrate R and vapor boil-up Y. These variables can be held constant (operated as an open loop system) or varied to set two other variables as constant (operated as a closed loop system).

11.3. Nonideal multicomponent distillation

A multicomponent ideal distillation is a more realistic distillation process than the one considered above. This type of distillation is very similar to the many real-life distillation processes seen in other industries. As previously mentioned, some assumptions will also be considered when studying this system. The feed mixture contains more than two components or N_c components. The distillation column in this system runs under nonideal conditions, unlike in the previous case. All of the trays in the column are assumed to have some inefficiencies, which can be due to many reasons, for example, flooding, weeping, bypassing, fouling, contamination, etc.

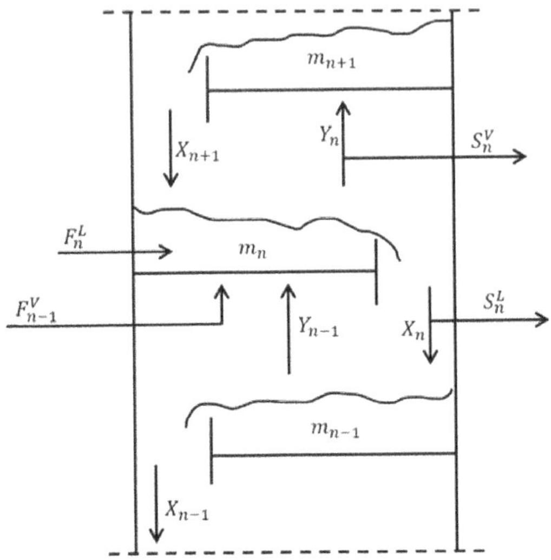

Figure 11.2. *Schematic diagram of nonideal multicomponent distillation (prepared for this work)*

The other assumption considered in this system is that the liquid on the trays is mixed perfectly and incompressible. As with the ideal binary system, the vapor holdups on the tray are negligible. Vapor and liquid in this system are considered in thermal equilibrium but not phase equilibrium. The deviation of the vapor and liquid of this system is described using Murphree vapor phase efficiency.

$$E_{n,i} = \frac{y_{n,i} - y_{n-1,i}}{y_{n,i}^e - y_{n-1,i}} \qquad [11.19]$$

where:

– $E_{n,i}$ is the Murphree vapor efficiency for an ith component on the nth tray;

– $y_{n,i}$ is the actual composition of vapor of ith component exiting the nth tray;

– $y_{n-1,i}$ is the actual composition of vapor of ith component entering the nth tray;

– $y_{n,i}^e$ is the composition of vapor of ith component in phase equilibrium with liquid on the nth tray with composition $x_{n,i}$.

It is further assumed that the multiple feeds in both the liquid and vapor phases and side stream draw-offs in both the liquid and vapor phases are permitted. The dynamics of the column reboiler and condenser are also neglected. Figure 11.2 represents a simplified diagram of a section in a nonideal multicomponent distillation column. Overall material balance:

$$\frac{dm_n}{dt} = X_{n+1} - X_n + F_n^L + F_{n-1}^V + Y_{n-1} - Y_n - S_n^L - S_n^V \qquad [11.20]$$

Component mass balance:

$$\frac{dm_n x_{n,i}}{dt} = X_{n+1}x_{n+1,i} - X_n x_{n,i} + F_n^L x_{n,i}^f + F_{n-1}^V y_{n-1,i}^f + Y_{n-1}y_{n-1,i}$$

$$- Y_n y_{n,i} - S_n^L x_{n,i} - S_n^V y_{n,i} \qquad [11.21]$$

Energy balance:

$$\frac{dm_n h_n}{dt} = X_{n+1}h_{n+1} - X_n h_n + F_n^L h_n^f + F_{n-1}^V H_{n-1}^f + Y_{n-1}H_{n-1}$$

$$- Y_n H_n - S_n^L h_n - S_n^V H_n \qquad [11.22]$$

where:

– h denotes the enthalpy in the liquid phase (energy/mole);

– H denotes the enthalpy in the vapor phase (energy/mole).

Relationships must be used to find the value of appropriate vapor–liquid equilibrium. From the two vapors entering the trays, it can be calculated. Additional equations include physical property relationships, the vapor hydraulic equation and the liquid hydraulic relationship. These are used to determine values for densities and enthalpies, vapor flow rates from known tray pressure drops and liquid flow rates over the weirs from known tray holdups.

By subtracting the number of equations from the number of variables, it was determined that the degree of freedom equals the number of side streams plus two. This means that if there are no side streams or zero side streams, then the degree of freedom of this system is two, which is the same as ideal binary distillation. In general, the multicomponent system will be simplified into a pseudo-binary system by controlling the impurities of the heavy key component in the distillate and the light key component in the bottom product or by controlling the purity of the light key component in the distillate and the heavy key component in the bottom product.

11.4. Batch distillation

The two cases of the distillation process discussed now are continuous processing or continuous distillation, as the real-life applications of distillation in the industry have it. However, the batch distillation process is also in practice, so studying this is equally necessary. This type of distillation seems to be advantageous because of its ability to separate multiple (N_c) components using a single distillation column instead of using N_c-1 columns that run continuously.

However, this is not true in all cases, like in large-volume products, because of the higher energy consumption of batch distillation compared to continuous distillation. Even after the large energy consumption, the batch distillation process has found applications in small-volume products with value as the high energy costs have minimal effect on the economy compared to the high value of the product.

The working of a batch distillation with a multicomponent feed can be explained as follows. The feed mixture with N_c components is fed into the container and heated until the mixture begins to boil. As the boiling occurs in the container, the vapor slowly rises into the condenser and settles in the reflux drum after condensation.

After the contents in the reflux drum reach a certain level, the reflux is pumped back into the column at rate R. As this process continues, the composition of the lightest component or the first component x_{d1} reaches the maximum or the required purity, which is when it is withdrawn at rate D.

As the process continues, with time the composition of the first component in the container decreases, decreasing its purity in the reflux drum. At this juncture, the above-explained whole process starts to occur again for the next heavier component or component two and this cycle continues until all the components are recovered separately. Thus, multiple components can be separated using a single column with the required purity.

Figure 11.3 represents a simplified diagram of batch distillation.

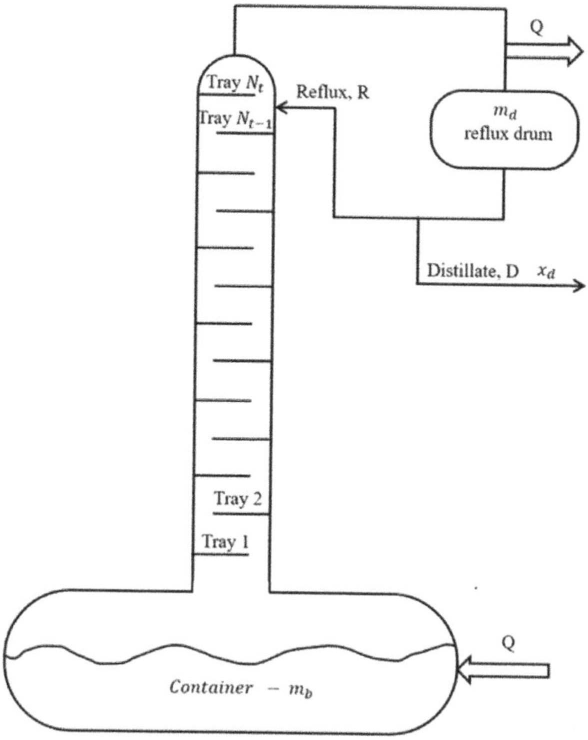

Figure 11.3. *Schematic diagram of batch distillation (prepared for this work)*

Due to the separation of multiple components, which have varied physical and thermal properties in a single column, the column has to operate at various temperatures, pressures and reflux ratios. This raises the need for the optimization of the design and operation of the batch. The number of theoretical trays is assumed to be N_t with constant relative volatilities and equimolal overflow throughout the distillation setup. The total amount of the materials fed into the column is represented as F(moles). These materials can be fresh feed or a feed mixture with slop cuts, whose composition is denoted as $x_{f,i}$. The initial amount of materials in the container is denoted by m_{bo}, which is the same as F. The initial composition of the components in the container is $x_{bo,\, i}$. The composition in the container at any given time, t, will be $x_{b,\, i}$. Likewise, the holdup in the container at any given time t will be mb. The reflux drum, column trays and container are all assumed to be filled with material of composition $x_{bo,\, i}$, initially. The vapor boilup rate is assumed to be constant at Y (moles/hr). Along with that, the holdup capacity of the trays and reflux drum are also assumed to be constant.

Container

$$\frac{dm_b}{dt} = -D \tag{11.23}$$

$$\frac{d(m_b x_{b,i})}{dt} = Rx_{1,i} - Yy_{b,i} \tag{11.24}$$

$$y_{b,i} = \frac{\alpha_i x_{b,i}}{\sum_{j=1}^{N_C} \alpha_j x_{b,j}} \tag{11.25}$$

Nth tray

$$m_n \frac{d(x_{n,i})}{dt} = R(x_{n+1,i} - x_{n,i}) + Y(y_{n-1,i} - y_{n,i}) \tag{11.26}$$

$$y_{n,i} = \frac{\alpha_i x_{n,i}}{\sum_{j=1}^{N_C} \alpha_j x_{n,j}} \tag{11.27}$$

Top tray (n = N_t)

$$m_{Nt} \frac{d(x_{Nt,i})}{dt} = R(x_{d,i} - x_{Nt,i}) + Y(y_{Nt-1,i} - y_{Nt,i}) \tag{11.28}$$

$$y_{Nt,i} = \frac{\alpha_i x_{Nt,i}}{\sum_{j=1}^{N_C} \alpha_j x_{Nt,j}} \tag{11.29}$$

Reflux drum

$$m_d \frac{d(x_{d,i})}{dt} = Yy_{Nt,i} - (R + D)x_{d,i} \tag{11.30}$$

$$R = Y - D \tag{11.31}$$

11.5. Applications and advancements

Binary and multicomponent distillation are critical for separating and purifying chemical mixtures in various industries. In the petrochemical industry, distillation is used for crude oil fractionation and natural gas processing, separating crude oil into fractions like gasoline, kerosene and diesel and natural gas into methane, ethane, propane and butane. Distillation purifies solvents and organic chemicals in the chemical industry, while in the pharmaceutical industry, it ensures the high purity of active pharmaceutical ingredients and intermediates. The food and beverage industry employs distillation in the production of alcoholic beverages and the extraction of flavors and fragrances. Environmental engineering applications include wastewater treatment and solvent recovery. Specific examples include crude oil distillation units in refineries, ethylene plants, pharmaceutical manufacturing for antibiotic synthesis and the production of high-purity ethanol for beverages. Distillation processes, whether binary or multicomponent, are essential for efficient separation and purification across these industries.

Recent advancements in binary and multicomponent distillation have markedly enhanced efficiency, sustainability and process control. Innovations like dividing wall columns (DWC) and heat-integrated distillation have reduced energy consumption and equipment costs. Enhanced control strategies, including model predictive control (MPC) and real-time optimization (RTO), have improved performance and stability. Advanced techniques such as reactive distillation and membrane-assisted distillation provide superior separation capabilities. Energy efficiency is further improved through low-energy configurations, better insulation and advanced materials. Computational advances, including AI and machine learning, optimize maintenance and operations, while modern simulators improve design and troubleshooting. Sustainability initiatives include using ionic liquids for extractive distillation and incorporating distillation into carbon capture and utilization (CCU) schemes. Modular and intensified systems, such as micro distillation and hybrid processes, offer greater flexibility and efficiency, aligning with the trend towards more efficient, adaptable and environmentally friendly distillation processes.

11.6. Conclusion

This chapter has dived into the mathematical modeling methods applied to different feed compositions, encompassing binary distillation (simple distillation), multicomponent distillation and batch and continuous operations, while considering both ideal and nonideal conditions. This was achieved by establishing assumptions, defining variables and formulating mass and energy balances. This approach not only enhances the analytical capabilities but also provides a clear-cut perspective on the distillation process, enabling proper and calculated decision-making in industrial applications.

11.7. References

Agachi, P.S., Nagy, Z.K., Cristea, M.V., Imre-Lucaci, A. (2007). *Model Based Control: Case Studies in Process Engineering*. Wiley-VCH, Weinheim.

Bortz, M. and Asprion, N. (2022). *Simulation and Optimization in Process Engineering: The Benefit of Mathematical Methods in Chemical Industry Applications*. Elsevier, Amsterdam.

Gorak, A. and Sorensen, E. (2014). *Distillation: Fundamentals and Principles*. Academic Press, Cambridge.

Jana, A.K. (2011). *Chemical Process Modeling and Computer Simulation*. Phi Learning, New Delhi.

Lei, Z., Dai, C., Chen, B., Ding, Z. (2021). *Special Distillation Processes*. Elsevier, Amsterdam.

Luyben, W.L. (2014). *Process Modeling, Simulation, and Control for Chemical Engineers*. McGraw Hill, New Delhi.

Mendes, M. (ed.) (2017). *Distillation: Innovative Applications and Modeling*. BoD – Books on Demand, Norderstedt.

Mujtaba, I.M. (2004). *Batch Distillation: Design and Operation*. World Scientific Publishing Company, Singapore.

Pan, W. (1984). Mathematical modeling of distillation operation from non-equilibrium approach. PhD Dissertation, New Jersey Institute of Technology, Newark.

Rasmuson, A. (2014). *Mathematical Modeling in Chemical Engineering*. Cambridge University Press, New York.

Rose, L.M. (1974). *The Application of Mathematical Modelling to Process Development and Design*. Elsevier, Amsterdam.

Steffen, V. and da Silva, E.A. (2017). Steady-state modeling of equilibrium distillation. In *Distillation: Innovative Applications and Modeling*, Mendes, M. (ed.). BoD – Books on Demand, Norderstedt.

Stephanopoulos, G. (1984). *Chemical Process Control*. Prentice Hall, New York.

Taylor, R. and Krishna, R. (1993). *Multicomponent Mass Transfer*. John Wiley & Sons, New York [Online]. Available at: https://www.wiley.com/en-us/Multicomponent+Mass+Transfer-p-9780471574170.

12

Filtration and Membrane Processes

Given the physicochemical properties of membranes, different types of membranes were developed and used for various applications. This chapter explores various membrane fabrication techniques and their applications across industries. The synthesis methods, including phase inversion, sintering, electrospinning and track-etching, offer precise control over membrane properties. Sintering, a critical process in ceramic membrane production, involves presintering, thermolysis and final sintering phases, ensuring membrane integrity. Applications in water and wastewater treatment showcase membrane processes as crucial in addressing global water scarcity, offering advantages such as reduced energy consumption and minimal chemical usage. Challenges, particularly membrane fouling, underscore the need for ongoing research to optimize membrane performance. Desalination applications highlight the significance of precisely controlled membranes in membrane distillation, providing energy-efficient solutions for freshwater scarcity. In the food industry, membrane processing proves advantageous with its gentle treatment, low energy consumption and simultaneous fractionation and concentration capabilities. Membranes play pivotal roles in pharmaceutical and biotechnological domains, contributing to drug delivery, purification and bioseparation processes. Gas separation processes using membranes gain traction due to robust materials and a modular design, while medical applications showcase the versatility of membranes in drug delivery, artificial organs, tissue regeneration and diagnostics. This chapter emphasizes the indispensable role of membrane technology across diverse industries, highlighting its multifaceted contributions to scientific research and its transformative impact on global challenges. Challenges like fouling prompt a continuous quest for mitigation strategies, solidifying membrane technology's standing as a transformative force in various sectors.

Chapter written by Esmaeil ALLAHKARAMI and Ebrahim ALLAHKARAMI.

For a color version of all figures in this chapter, see www.iste.co.uk/kumararya/chemical engineering1.zip

12.1. Introduction

In recent decades, membrane technology has emerged as a prominent separation method, finding widespread applications and establishing itself as an indispensable technology in the field (Birniwa et al. 2024). The inherent advantages of membrane technology, such as operating without the need for additional chemicals, low energy consumption, ease of handling and well-organized process conductions, have contributed to its prominence. Membrane separation demonstrates enhanced efficiency in comparison to traditional techniques, with the efficacy directly tied to the membrane's characteristics (Fattah et al. 2023).

The inspiration for membrane separation processes finds its roots in natural biological systems, where numerous chemical engineering processes occur. A notable example is the human body, wherein the heart serves as a pump for blood circulation within organs, and the lungs function as a blower for gas exchange. Biological cells are enveloped by semipermeable membranes, such as the cell membrane. The skin, the body's largest organ, acts as a protective barrier against harmful environments while facilitating the expulsion of sweat, akin to the principles of reverse osmosis (RO) (Jhaveri and Murthy 2016). The stomach, in the digestive system, absorbs essential substances from consumed food and converts them into important compounds through metabolic processes. The exchange of substances between cells in the stomach and liver can be compared to processes such as microfiltration (MF) or ultrafiltration (UF). Red blood cells (RBCs) facilitate gas exchange between cells and lung alveoli, functioning similarly to a liquid membrane. Alveoli, on the other hand, facilitate the interchange of air components with inhaled air, similar to the process of gas permeation (GP). The renal blood filtration process shares similarities with dialysis, where artificial kidneys, referred to as hemodialysis in medical terminology, use comparable principles (Jhaveri and Murthy 2016).

A membrane functions as a selective barrier that separates two phases by regulating the movement of components across it (Xiang et al. 2022). This is achieved by controlling the motion of several elements. Membranes have undergone substantial advancements since their establishment in the 18th century, rendering them more adaptable for a diverse range of applications. The concept of membranes originates from that specific time period. In essence, a membrane functions as a discrete and slender surface that regulates the permeation of chemical species by means of its pores. Regarding composition and structure, this interface can exhibit either chemical or physical heterogeneity, or it can possess molecular homogeneity, demonstrating complete uniformity in both aspects. For example, it could contain holes or pores with specific diameters, or it could have a multilayer structure. The term "filter" is frequently employed to denote structures with the ability to segregate particle suspensions above a size range of 1–10 μm. Nevertheless, a conventional filter is closely associated with the concept of a membrane.

The separation process is fundamentally straightforward: the membrane functions as a semi-permeable layer situated between two phases, regulating the transport of substances between them. For instance, a filter allows water to permeate the membrane while capturing suspended solids and other substances. There exist diverse methods to facilitate the penetration of substances through a membrane. Membranes exhibit heightened efficiency in the generation of process water from sources such as groundwater, surface water or wastewater (Ersahin et al. 2012).

Synthetic membranes, although possessing less selectivity and energy efficiency compared to biological membranes, provide greater chemical and mechanical stability. Defining the precise structure, function and versatility of membranes is challenging. Generally, membranes serve as barriers that either separate or connect two distinct regions, regulating the exchange of matter and energy between them. Biological membranes, found in living organisms, perform complex and specific transport tasks, often using active transport. In contrast, synthetic membranes, created by humans, typically exhibit passive transport properties.

Membranes have become significantly important in process engineering since the 1970s, having a notable influence on both technical and commercial aspects. The extensive utilization of membrane processes is apparent in three primary domains: applications such as seawater desalination or wastewater purification, where membranes economically compare to alternative processes; applications in the food and drug industry, where membranes provide both technical and commercial benefits; and membrane applications in artificial organs and therapeutic systems, where no feasible alternative exists. The continuous advancement of innovative and effective membranes is propelling the spread of membrane processes beyond their existing boundaries. Additionally, developing technologies such as membrane contactors and membrane reactors are increasingly prominent in the field of process engineering (Baker 2024).

Membrane processes are acknowledged for their energy efficiency in various separation applications, notably in water treatment. While energy requirements are a key cost factor, other considerations include investment and maintenance costs, pre- and post-treatment procedures, required product quality and feed mixture composition. Micro- and ultra-filtration, common in water purification, have low energy demands but compete with even lower-energy options like biological treatment or sand bed filtration. In seawater desalination, reverse osmosis is the primary economical membrane process, but its energy advantage is tempered by the efficiency of pressure-generating pumps. Electrodialysis also plays a role, with energy consumption tied to feed solution concentration (Al-Amshawee and Yunus 2023).

The overall economics of membrane processes involve factors such as investment, operating costs and plant capacity. Reverse osmosis tends to have a cost advantage in seawater desalination, while both electrodialysis and reverse osmosis outperform distillation processes in desalination of brackish water. For surface water desalting, reverse osmosis and ultrafiltration yield higher-quality product water, retaining most contaminants. However, the choice of water purification methods depends on feed water composition and desired product quality, often requiring a combination of processes (Moreira et al. 2022).

Environmentally, membrane processes have a low impact, employing no hazardous chemicals or heat generation. Effluents, typically concentrated brine, pose challenges in brackish water desalination and may require post-treatment. Pressure-driven membrane processes pose no health hazards, and the obtained product is generally of high quality, requiring minimal post-treatment. Chlorination may be used for potable water sterilization, especially in hot climates with extended storage times (Adam et al. 2022).

In various applications such as water desalination, purification, drug and food product separation and medical/pharmaceutical processes, membrane processes compete with traditional water treatment methods. Membrane processes like reverse osmosis offer advantages such as energy efficiency, simplicity and high-quality product outcomes. They operate at ambient temperature, preventing product changes or degradation. However, challenges include potential unreliability in the chemical and petrochemical industry, the need for extensive pretreatment and sensitivity to fouling and concentration polarization. Despite these challenges, recent advancements, particularly in reverse osmosis seawater desalination, have improved membrane performance, stability and resistance to operational errors. Combining membrane processes with conventional methods often results in cost-effective and high-quality water treatment.

Current membrane processes exhibit satisfactory performance in various applications, such as sea and brackish water desalination through reverse osmosis, hemodialysis and hemofiltration. These applications have experienced years of successful operation, with membranes offering adequate flux and salt rejection. However, for processes like micro- and ultra-filtration, while membrane properties are generally satisfactory, further improvements are sought in areas such as process design, control, application knowledge and long-term operating experience. Concentration polarization and membrane fouling are key challenges in these applications, prompting the need for better membrane modules and process design concepts to enhance control and prolong membrane life.

However, pervaporation, gas separation, membrane reactors and fuel cell separators require high-quality membranes, improved process design, extensive application knowledge and long-term experience to be reliable technologies.

In addition to current membrane processes, developing membrane operations such as membrane contactors and membrane reactors are increasingly being used at the industrial level, and are now considered standard unit operations in process engineering. These advancements enhance the overall influence of membrane engineering on industrial manufacturing. All membrane activities are in accordance with the criteria of process intensification and sustainable industrial growth. The principles of chemical process rationalization and miniaturization, which are fundamental to process intensification, are applied to many manufacturing processes. Membranes and processes possess essential characteristics such as high energy efficiency, operational simplicity, transport selectivity, operational flexibility and environmental compatibility. These attributes make them vital tools for advanced molecular separations and chemical transformations, surpassing the constraints of conventional industrial processes.

Given the physicochemical properties of membranes, different types of membranes were developed and used for various applications. This chapter summarizes the techniques used for the fabrication of different membranes. It also presents the application of membrane in different industrial areas, such as food, medicine, pharmacy, biotechnology, industrial effluent and wastewater treatment. Furthermore, this chapter discusses fouling as a significant challenge in membrane-based processes.

12.2. Types of membranes

Membranes can be categorized as either isotropic or anisotropic based on their characteristic properties (Lee et al. 2016). Isotropic membranes exhibit a consistent composition and physical structure. If the material is microporous, its permeation fluxes are comparatively high compared to when it is nonporous (dense), which severely limits its application due to low permeation fluxes. Isotropic microporous membranes are commonly used in microfiltration applications (Fane et al. 2011).

Anisotropic membranes, in contrast, exhibit non-uniformity throughout their surface and consist of distinct layers with varying architectures and compositions. These membranes consist of a thin selective layer that is backed by a larger and highly permeable layer. They are specifically used in reverse osmosis (RO) procedures (Baker 2024).

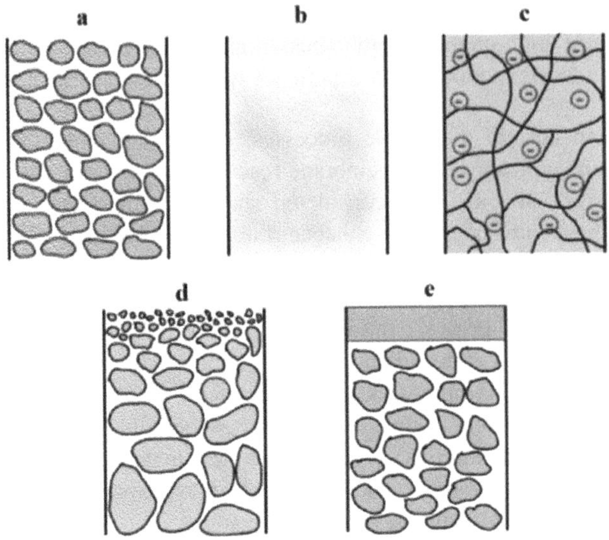

Figure 12.1. *Graphical representation of several membrane classes: isotropic (a, b and c) and anisotropic (d and e) membranes: macroporous (a), nonporous dense (b), electrically charged (c), Loeb-Sourirajan (d), composite (e). (Ahmad 2020)*

Figure 12.1 showcases distinct classes of membranes, distinguishing between isotropic and anisotropic types (Lee et al. 2016). Isotropic membranes exhibit a chemically homogeneous composition throughout their structure. These membranes maintain uniformity in their structure, lacking variations in composition or arrangement (Firouzi and Żur 2023). Anisotropic membranes are characterized by both chemical and structural heterogeneity. These membranes consist of layers with varying structures and compositions, contributing to their non-uniformity. This graphical representation emphasizes the fundamental distinction between isotropic and anisotropic membranes, highlighting their chemical and structural characteristics (Lalia et al. 2013).

In terms of membrane composition, membranes fall into two categories: organic or inorganic. Organic membranes use synthetic organic polymers, with common materials for pressure-driven separation processes (such as reverse osmosis, nanofiltration, ultrafiltration and microfiltration) including polyethylene (PE), polytetrafluoroethylene (PTFE), polypropylene and cellulose acetate, among others. On the other hand, inorganic membranes employ materials like ceramics, metals, zeolites or silica. Renowned for their chemical and thermal stability, inorganic membranes find extensive use in industrial applications like hydrogen separation, ultrafiltration and microfiltration.

Membranes can also be categorized by their geometric shape, including sheet, tubular, spiral-wound and hollow fiber membranes. Spiral-wound configurations are commonly used for reverse osmosis membranes, while hollow fiber membranes are employed in ultrafiltration systems.

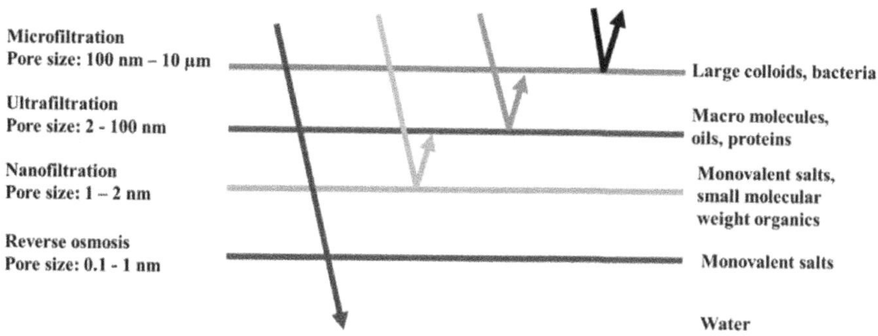

Figure 12.2. *Different membrane processes and retained species*

Generally, there are four different types of membranes based on the type of substance to be separated. These include reverse osmosis, nanofiltration, ultrafiltration and microfiltration, distinguished by the size of their pores (Figure 12.2).

12.2.1. *Microfiltration (MF)*

Microfiltration (MF) closely resembles conventional filtration processes, effectively separating discrete particles from a solution. The boundary between UF and MF processes is not sharply defined. To maintain consistency with industry standards, the upper pore size limit for MF is typically set at less than 10 µm, and the lower size limit is established at 0.1 µm. It is worth noting that MF stands out as the most widely applied process for inorganic membranes in the separation of small insoluble particles, bacteria and yeast cells from aqueous solutions. This versatile application makes MF a crucial technique in various industries for the efficient removal of contaminants and particles from liquids (Wakeman and Williams 2002).

Microfiltration, with the largest membrane pores (ranging from 0.08 to 0.5 microns), is primarily used for separating suspended solids, bacteria and fat globules. Solid particles, bacteria and fat globules are the only substances restricted from passing through microfiltration membranes.

The fundamental role of a microfiltration membrane is to selectively retain components from a liquid feed, essentially functioning as a sieving process (Tanudjaja et al. 2022).

Microfiltration finds extensive use in diverse industries such as pharmaceuticals, food and biotechnology, particularly for the sterilization of heat-sensitive feeds that cannot undergo traditional heat-based sterilization methods. In contemporary applications, microfiltration membranes have evolved to serve advanced purposes and are well suited for developments in healthcare, food processing, textiles and wastewater treatment. The versatility of microfiltration technology makes it a valuable tool across various sectors, contributing to advancements in sterilization, separation and treatment processes.

Microfiltration (MF) is a versatile and critical technology employed across various industries for a multitude of applications. Its primary function is to separate particles ranging from 0.1 to 10 micrometers in size from liquids by passing the fluid through a porous membrane. This process ensures that suspended solids, bacteria and other microorganisms are effectively removed, making MF essential in many sectors.

In the realm of water treatment, MF is extensively used to remove suspended solids, bacteria and microorganisms, ensuring the provision of safe and potable water. Municipal water treatment facilities rely on MF membranes to effectively eliminate pathogens and particulate matter, delivering clean drinking water to communities (Allgeier et al. 2005). The process involves passing untreated water through microfiltration membranes, which trap contaminants while allowing clean water to pass through. The filtered water is then further treated or directly supplied for public consumption, ensuring compliance with health and safety standards.

Additionally, MF plays a crucial role in wastewater treatment, particularly during the tertiary treatment phase. In this stage, MF significantly reduces pathogen levels and suspended solids before the treated effluent is either discharged into the environment or reused. The use of MF in wastewater treatment helps in achieving stringent discharge standards and facilitates water reuse, thus contributing to water conservation efforts (Judd 2010).

The food and beverage industry also significantly benefits from microfiltration technology. MF is essential for clarifying and sterilizing products, thereby enhancing their quality and extending shelf life. In the dairy industry, MF membranes are used to remove bacteria from milk, enabling the production of pasteurized milk, cheese and yogurt. This process not only ensures the safety of dairy products but also improves their texture and taste by removing unwanted microorganisms (Jönsson 2013).

For beverages such as juices, wine and beer, MF assists in the clarification process by removing yeast and other spoilage organisms. The removal of these impurities enhances the stability, taste and overall quality of the final products. For instance, in the wine industry, MF is used to remove tartrate crystals and other particulates, resulting in a clear and visually appealing product (Le et al. 2014). Similarly, in beer production, MF helps in removing yeast cells post-fermentation, which improves the product's shelf life and taste consistency.

In the pharmaceutical sector, ensuring sterility during the production process is of paramount importance, and MF is a key technology in achieving this goal. During drug manufacturing, MF sterilizes liquid formulations by removing bacteria and particulates without compromising the integrity of the active ingredients (Lutz et al. 2013). This is particularly crucial for injectable drugs, where even minimal contamination can lead to severe health risks.

In the production of biologics, MF is integral in filtering cell cultures and fermentation broths, purifying proteins and other biologically active compounds essential for therapeutic applications. The technology ensures that these products meet the required purity and safety standards, which is critical for their efficacy and regulatory approval (Meltzer and Jornitz 2007).

Biotechnology relies heavily on MF for critical processes such as cell harvesting and protein purification. MF is used to separate cells from culture media in the production of vaccines and therapeutic proteins. This separation is essential for obtaining high yields of target cells or proteins, which are then used in further downstream processes (Harrison et al. 2015).

Furthermore, in protein purification, MF clarifies cell lysates to obtain high-purity target proteins. This step is crucial for both research and commercial production, as it ensures that the proteins are free from contaminants that could interfere with their function or stability. The use of MF in these processes enhances the overall efficiency and reliability of biotechnological applications.

The chemical processing industry also employs MF to clarify solutions and recover valuable products, ensuring high quality and efficiency in various chemical processes. MF removes impurities from chemical solutions, enhancing the purity and quality of the final products. For example, in the production of fine chemicals and specialty polymers, MF is used to remove catalyst residues and other particulates, resulting in a purer product (Baker 2023).

Additionally, MF is applied in the recovery of catalysts and valuable compounds from reaction mixtures. This application not only improves the overall efficiency of the process but also contributes to resource conservation and cost reduction.

By recovering and reusing catalysts, chemical manufacturers can reduce their raw material costs and minimize waste (Hussain 2020).

Recent advancements in microfiltration technology have significantly expanded its applications and improved its performance. The development of advanced membrane materials, such as ceramic and polymer composites, has enhanced the robustness and fouling resistance of MF membranes. These materials provide longer service life and better performance in challenging filtration environments, such as those with high levels of particulates or aggressive chemicals (Basile et al. 2015; Gitis and Rothenberg 2016).

Ceramic membranes, for example, offer superior chemical and thermal stability compared to traditional polymeric membranes. This makes them ideal for applications involving harsh chemicals or high temperatures. Their increased durability reduces the frequency of membrane replacement, thus lowering operational costs and downtime.

Polymer composite membranes combine the flexibility and lower cost of polymeric materials with the enhanced performance characteristics of inorganic components. These hybrid membranes exhibit improved fouling resistance and mechanical strength, making them suitable for a wider range of applications. The development of such advanced materials represents a significant step forward in the capabilities of MF technology (Basile and Charcosset 2016).

The integration of MF with other membrane technologies, such as ultrafiltration (UF) and nanofiltration (NF), has led to the creation of high-performance hybrid systems. These systems offer superior separation and purification capabilities, achieving higher purity levels and efficiently removing a broader range of contaminants. For instance, in water treatment, combining MF with UF or NF can enhance the removal of both particulate and dissolved contaminants, providing a higher quality of treated water (Singh 2006).

Such hybrid systems are particularly advantageous in applications requiring stringent quality standards. In the pharmaceutical industry, for example, a combination of MF and UF can be used to ensure the sterility and purity of injectable drugs, meeting both regulatory and safety requirements. These integrated systems provide a more comprehensive solution to complex filtration challenges, offering enhanced performance and reliability.

The advent of smart filtration systems, which incorporate automation and sensor-driven technology, has revolutionized the MF landscape. These smart systems enable real-time monitoring and optimization of filtration processes, allowing for dynamic adjustments to operational parameters. As a result, they maintain optimal performance and efficiency, reduce downtime and lower operational costs (Rudolph et al. 2019; Dansawad et al. 2022).

Smart filtration systems can detect changes in feed water quality or membrane fouling and adjust the filtration parameters accordingly. This capability enhances the reliability and consistency of the filtration process, ensuring that the treated water or product meets the required standards at all times. Additionally, the use of automation reduces the need for manual intervention, further increasing operational efficiency.

Environmental applications of MF have seen significant enhancements, particularly in addressing emerging challenges such as the removal of microplastics and other contaminants from water sources. Advanced MF methods are being developed to improve water quality and protect aquatic ecosystems. For example, new membrane designs and materials are being tested to enhance the removal of microplastics, which are a growing environmental concern due to their persistence and potential impact on marine life (Pereao et al. 2020).

The ability of MF to remove emerging contaminants, such as pharmaceuticals and personal care products, from water sources is also being improved. These contaminants, which are not effectively removed by conventional treatment methods, pose a threat to both human health and the environment. Advanced MF techniques are being developed to address these challenges, highlighting the critical role of MF in environmental conservation efforts (Radwan et al. 2023).

Moreover, innovations in MF technology have led to the creation of low-energy systems, which are designed to reduce operational costs and minimize environmental impact. These energy-efficient MF systems are especially beneficial for large-scale applications where energy consumption is a significant concern. By lowering energy requirements, these systems contribute to sustainable practices and enhance the overall efficiency of filtration processes (Figoli and Criscuoli 2017; Castel and Favre 2018).

Energy-efficient MF systems use advanced membrane materials and optimized process designs to reduce the energy needed for filtration. For example, the use of low-resistance membranes and energy recovery devices can significantly decrease the power consumption of MF systems. These advancements make MF a more sustainable option for a wide range of applications, from municipal water treatment to industrial processes.

In conclusion, microfiltration is a versatile and indispensable technology with widespread applications across various industries. Its ability to purify, clarify and sterilize liquids makes it a critical component in water treatment, food and beverage production, pharmaceuticals, biotechnology and chemical processing. Recent advancements in MF technology continue to push the boundaries of its capabilities, offering improved performance, energy efficiency and environmental benefits. As these innovations progress, the role of MF in ensuring quality and sustainability across different sectors will only grow in significance.

12.2.2. *Ultrafiltration (UF)*

Ultrafiltration membranes have larger pores compared to reverse osmosis and nanofiltration (approximately 0.1 to 0.01 microns). In ultrafiltration, relatively low pressure is applied, allowing salts, sugars, organic acids and small peptides to pass through, while proteins, fats and polysaccharides cannot permeate. This process is effective in removing bacteria, unicellular organisms and some viruses from water (Moreno et al. 2022).

In ultrafiltration (UF) membranes, the mechanism for separating solvent from solute/colloidal particles is akin to that observed in reverse osmosis (RO) and nanofiltration (NF). The interactions between the membrane surface and the solvent/solutes and the porous nature of the membranes have the greatest influence on solute rejection efficiency. Mass transfer resistances within the membrane and at solution boundaries often influence total solvent transfer.

Ultrafiltration (UF) is a membrane-based separation process that operates on the principle of size exclusion. UF membranes typically have pore sizes ranging from 0.01 to 0.1 micrometers, allowing them to selectively remove particles, colloids, macromolecules and microorganisms from liquids such as water. The process involves applying hydraulic pressure to the feed solution, driving water and small solutes through the membrane while retaining larger particles and solutes on the membrane surface or within its pores (Basile et al. 2015).

UF is characterized by several key features:

– *Selective separation*: UF membranes selectively separate substances based on size, with excellent retention of particles and microorganisms.

– *Energy efficiency*: compared to processes like reverse osmosis (RO), UF operates at lower pressures, resulting in reduced energy consumption.

– *Versatility*: UF is applicable across various industries for both concentration and purification purposes due to its scalability and adaptability to different feed streams.

– *Durability*: modern UF membranes are designed to withstand fouling and degradation, ensuring prolonged operational life with proper maintenance (Gitis and Rothenberg 2016).

The efficiency of UF stems from its membrane's ability to selectively permit the passage of molecules and particles based on their size. This molecular sieving effect allows UF to achieve high removal efficiencies for suspended solids, bacteria, viruses, proteins and other contaminants. The membranes are typically made from materials like polysulfone, polyethersulfone or ceramic, chosen for their durability, chemical resistance and pore size distribution tailored to specific applications (Basile and Charcosset 2016).

UF systems operate in crossflow filtration mode, where the feed stream flows tangentially across the membrane surface. This configuration minimizes membrane fouling by continuously sweeping away retained particles and contaminants from the membrane surface, maintaining high flux rates and efficiency over extended periods (Ai et al. 2022).

Ultrafiltration membranes hold a prominent position among various membrane types, being extensively used across diverse applications, ranging from wastewater treatment to biotechnology. Their effectiveness and efficiency make them valuable in a wide array of scenarios, including:

– Drinking water purification (Wu et al. 2023): ultrafiltration membranes are highly efficient in purifying drinking water, removing contaminants and impurities.

In municipal water treatment, UF plays a crucial role in producing safe drinking water by effectively removing pathogens, turbidity and particulate matter from raw water sources. UF membranes provide a physical barrier that ensures high-quality water output, meeting regulatory standards for potable water (Allgeier et al. 2005). Municipalities worldwide use UF to treat surface water and groundwater contaminated with pathogens and suspended solids. This process not only improves water quality but also enhances public health outcomes by preventing waterborne diseases (Judd 2010).

– Sterilization in food (Hu and Wu 2023), pharmaceutical (Ali et al. 2023) and biotechnology (Castro-Muñoz et al. 2022): ultrafiltration membranes are effective in sterilizing products in the food, pharmaceutical and biotechnology sectors, ensuring product safety and quality.

UF is integral to food and beverage processing operations for concentration, clarification and sterilization of products (Dushkova et al. 2022). In dairy processing, UF concentrates milk proteins to produce products like whey protein concentrates and lactose-free milk. UF also clarifies fruit juices by removing pulp and microorganisms, extending shelf life and improving product stability.

– Wastewater treatment (Karray et al. 2022): they are used in wastewater treatment processes, contributing to the removal of pollutants and the production of treated water.

UF systems are employed in industrial wastewater treatment to reclaim water, concentrate valuable products and reduce disposal costs. Textile industries use UF to treat dye effluents, removing colorants and contaminants before discharge or reuse. This process minimizes environmental impact and conserves water resources, aligning with sustainable industrial practices (Scott 1995).

– Healthcare applications (Zhang et al. 2023): ultrafiltration membranes play a vital role in healthcare applications, including hemodialysis, blood oxygenation and drug delivery systems.

The pharmaceutical industry relies on UF for the production of sterile pharmaceuticals and purification of biopharmaceutical products. UF is used in the manufacturing of injectable drugs to remove bacteria, endotoxins and particulates from drug formulations. This ensures product safety and efficacy, meeting stringent regulatory requirements for pharmaceuticals (Meltzer and Jornitz 2007).

In essence, ultrafiltration membranes represent a crucial category of membranes with widespread applications. Their significance extends to current necessities in water treatment, food processing, healthcare and emerging technologies. As a result, ultrafiltration membranes are poised to play a pivotal role in future developments and advancements in the field of membrane science.

Recent developments in UF technology have focused on improving membrane materials, enhancing system efficiency and integrating smart filtration solutions. Researchers have explored new materials such as graphene oxide and nanocomposites to enhance membrane performance, including fouling resistance, permeability and selectivity. The integration of sensor technologies and automation has led to the development of smart UF systems capable of real-time monitoring, adaptive control and predictive maintenance. These systems optimize operational parameters, reduce downtime and improve overall system reliability (Rudolph et al. 2019). Innovations in energy recovery devices and low-pressure membranes have contributed to reducing energy consumption in UF processes, making them more cost-effective and sustainable.

Looking ahead, the future of UF technology is poised for continued advancements driven by emerging trends and research directions. Continued research into nanomaterials and nanotechnology is expected to enhance membrane properties, offering superior selectivity, durability and fouling resistance. The development of integrated UF systems with complementary technologies like nanofiltration and advanced oxidation processes will enable comprehensive treatment solutions for complex water and wastewater challenges. UF will play a pivotal role in addressing emerging contaminants such as microplastics and pharmaceutical residues in water sources, contributing to environmental sustainability and public health protection. Future UF membranes may incorporate responsive materials and functionalities, enabling adaptive filtration properties and enhanced performance under varying operating conditions.

Ultrafiltration technology continues to evolve as a versatile and efficient solution for water purification, industrial processing and pharmaceutical applications worldwide. With ongoing advancements in materials science, system design and smart technologies, UF is poised to play an increasingly critical role in meeting global demands for clean water, sustainable manufacturing practices and safe pharmaceutical products.

12.2.3. *Nanofiltration (NF)*

Nanofiltration is a specialized membrane technology designed to filter particles with a diameter of less than one nanometer (10 Angstroms) (Mohammad et al. 2015). It exhibits selective rejection capabilities, particularly for organic compounds with a molecular weight falling within the range of 200–400 Da. Additionally, dissolved salts experience rejection rates ranging from 20–98%. Notably, monovalent anions (e.g. sodium chloride or calcium chloride) exhibit rejection rates of 20–80%, while divalent anions (e.g. magnesium sulfate) demonstrate higher rejection rates ranging from 90–98%. This technology finds application in various scenarios, including:

– Color and Total Organic Carbon (TOC) removal (Yuksekdag et al. 2023): nanofiltration is employed to remove color and TOC from surface water.

– Elimination of hardness or radium (Maroufi and Hajilary 2023): it is used to eliminate hardness or radium from well water.

– Total Dissolved Solids (TDS) reduction (Gholami et al. 2023): nanofiltration contributes to the overall reduction of TDS in water.

– Separation in specialty food and wastewater applications (Yadav et al. 2022; Siddique 2023): it is used for the separation of organic from inorganic particles in specialty food and wastewater applications.

Typical operating conditions for nanofiltration involve transmembrane pressures ranging from 3.5 to 16 bar (50 to 225 PSI). The versatility of nanofiltration makes it a valuable tool in water treatment processes, addressing various challenges associated with water quality and composition.

Nanofiltration (NF) is a membrane filtration process that operates between ultrafiltration (UF) and reverse osmosis (RO) in terms of its selectivity and pore size. NF membranes typically have pore sizes ranging from approximately 1 to 10 nanometers (nm), which allows them to selectively reject particles, ions and dissolved compounds based on both size and charge (Agboola et al. 2020).

– *Dissolved ions*: NF membranes effectively reject divalent ions like calcium (Ca^{2+}) and magnesium (Mg^{2+}), while allowing monovalent ions such as sodium (Na^{+}) to pass through.

In water softening and desalination processes, NF membranes effectively remove divalent ions (e.g. Ca^{2+}, Mg^{2+}) from feedwater streams. This selective rejection allows for targeted treatment without removing essential monovalent ions, resulting in improved water quality and reduced scaling potential (Mulder 2012).

– *Organic compounds*: NF membranes can reject larger organic molecules, including pesticides, herbicides and natural organic matter (NOM), based on their molecular weight and size.

NF is effective in removing organic compounds such as pesticides and humic substances from drinking water sources. The rejection of these contaminants improves water safety and aesthetic qualities, ensuring compliance with drinking water standards (Khulbe and Matsuura 2021).

NF membranes typically reject organic compounds with molecular weights ranging from approximately 200 to 1000 daltons (Da) or higher, depending on the membrane material and pore size distribution. This range encompasses a variety of contaminants found in water and wastewater streams, making NF suitable for applications where selective removal of specific contaminants is required while retaining essential components.

Recent advancements in NF technology have expanded its applications across various industries, demonstrating its effectiveness in water treatment, food processing and industrial applications.

NF membranes are used in drinking water treatment plants to remove organic contaminants, color and unpleasant tastes from surface water and groundwater sources. For example, NF is employed in the treatment of water sourced from rivers and lakes, where it effectively removes dissolved organic matter and improves water clarity (Rehman et al. 2020).

NF plays a crucial role in concentrating and fractionating components in food processing applications. For instance, in juice production, NF membranes are used to concentrate fruit juices by removing water while retaining flavor compounds and nutrients. In dairy processing, NF helps in the fractionation of milk components to produce concentrated proteins and lactose-free products (Lipnizki et al. 2020).

NF is employed in wastewater treatment within the textile industry to remove dyes and other contaminants from effluent streams. NF membranes effectively reject dye molecules, allowing for the reuse of treated water and minimizing environmental impact (Adak et al. 2023).

Moving forward, the future of NF technology is expected to focus on enhanced selectivity, energy efficiency and integration with advanced treatment technologies. Continued research into membrane materials and surface modifications will enhance NF membranes' selectivity for specific contaminants, improving overall efficiency and reducing operational costs. Innovations in energy recovery devices and membrane design will contribute to reducing the energy footprint of NF processes, making them more sustainable and economically viable. NF will increasingly be integrated with other advanced treatment technologies such as membrane bioreactors (MBRs) and advanced oxidation processes (AOPs) to tackle emerging contaminants and enhance overall treatment efficacy (Lipnizki et al. 2020).

Nanofiltration (NF) represents a critical advancement in membrane technology, offering selective removal of ions and organic compounds based on size and charge. Its ability to treat water and wastewater streams effectively has led to widespread adoption across diverse industries, from drinking water purification to industrial wastewater treatment. With ongoing advancements and applications in various sectors, NF continues to play a pivotal role in addressing water quality challenges and advancing sustainable practices globally.

12.2.4. *Reverse osmosis (RO)*

First observed in 1748 by the French scientist Jean-Antoine Nollet, reverse osmosis is renowned for its efficiency in allowing only water molecules to pass through, making it the most effective membrane for liquid separation. In this process, the densest membrane possible is used to separate liquids. Water is the only substance that can permeate, while bacteria, spores, fats, proteins, salts, sugars, minerals, etc., are prevented from passing through. The pore size of reverse osmosis filters is 0.0001 microns, allowing only water molecules to pass (Hailemariam et al. 2020).

Reverse osmosis is commonly used for desalination of seawater, turning brackish water into drinkable water and industrial purposes. It is also employed in wastewater treatment and reclaiming used water. Additionally, it can eliminate phosphates, calcium, heavy metals and other substances (Biesheuvel et al. 2022).

Recent advancements have focused on developing high-flux and high-rejection membranes capable of efficiently removing a wide range of contaminants, including salts, pathogens and organic compounds (Cai et al. 2023). Thin-film composite (TFC) membranes have become the standard in RO due to their superior performance and durability compared to earlier generations of cellulose acetate membranes. They offer enhanced salt rejection and longer operational lifespan (Kadhom 2023).

Energy recovery devices (ERDs) such as pressure exchangers and turbochargers have revolutionized RO systems by significantly reducing energy consumption. These devices recover energy from the concentrate stream and transfer it to the feedwater, thereby lowering the overall energy requirements of the process (Alawad et al. 2023).

Advances in membrane materials and system design have enabled the development of low-energy RO (LER) systems that operate at lower pressures while maintaining high water recovery rates. This reduces operational costs and environmental impact (Harby et al. 2024).

RO systems are increasingly being designed as modular and compact units, allowing for flexibility in installation and operation. Modular systems facilitate scalability and ease of maintenance, catering to varying water treatment needs from small-scale applications to large industrial plants (Luong et al. 2023).

Integration of automation and sensor technologies enables real-time monitoring of RO performance parameters such as flow rates, pressure and water quality. Smart systems optimize operational efficiency, detect fouling or membrane degradation early and enable proactive maintenance (Shahid et al. 2023).

RO technology has diverse applications across various industries and sectors. RO plays a pivotal role in desalinating seawater and brackish water to produce potable water for coastal regions and arid areas where freshwater resources are limited (Almasoudi and Bassam 2024). RO systems are used in municipal water treatment plants for removing contaminants such as salts, nitrates and dissolved organic compounds from drinking water sources. RO is essential in industrial applications such as food and beverage processing, pharmaceutical manufacturing and electronics production where high-purity water is required for operations (Guan et al. 2023).

Continued research into advanced membrane materials, nanotechnology and surface modifications aims to improve RO membrane performance, including fouling resistance, selectivity and durability. Innovations in energy recovery, membrane cleaning techniques and brine management will enhance the sustainability of RO processes by reducing energy consumption and minimizing environmental impact (Werber et al. 2016). RO systems integrated with renewable energy sources such as solar and wind power offer potential solutions for decentralized water treatment in remote areas, contributing to water and energy sustainability (Amy et al. 2017).

Reverse osmosis (RO) technology continues to evolve with advancements in membrane materials, energy efficiency and system design, expanding its applications across diverse sectors globally. Innovations in high-performance membranes, energy recovery devices and intelligent monitoring systems have enhanced RO's capability to provide safe and reliable water purification solutions. As the demand for clean water grows amid increasing water scarcity and environmental challenges, RO technology plays a crucial role in ensuring access to sustainable water resources worldwide.

12.3. Membrane fabrication techniques

Diverse membrane fabrication methodologies were employed in the preparation of synthetic membranes designed for applications in microfiltration and ultrafiltration. The membrane synthesis processes encompassed techniques such as phase inversion, sintering, electrospinning, track-etching and stretching. Additionally, certain membrane variants were synthesized through the integration of disparate fabrication procedures (Ismail et al. 2015).

Phase inversion constitutes a demixing process wherein an initially uniform polymer solution undergoes a meticulously controlled transition from a liquid to a solid state (Drioli and Giorno 2009). This transformative progression can be actualized through diverse methodologies (Mulder 1996), delineated as follows (Lalia et al. 2013):

(a) Immersion precipitation: this approach involves immersing the polymer solution in a coagulation bath that mainly consists of a non-solvent – often water. The demixing and precipitation process begins when the solvent from the polymer solution and the non-solvent from the coagulation bath interact, necessitating the compatibility of these two substances.

(b) Thermally induced phase separation: this method is based on the observation that the effectiveness of the solvent usually decreases as the temperature falls. After the separation of components, the solvent is removed by methods such as extraction, evaporation or freeze drying.

(c) Evaporation-induced phase separation: the polymer solution, composed of a solvent or a volatile non-solvent, undergoes evaporation in a controlled manner, leading to the formation of solid particles or separation of different components. This procedure is also known as the solution casting method.

(d) Vapor-induced phase separation: in this technique, the polymer solution is placed in an environment filled with a non-solvent, usually water, until it becomes saturated; the presence of the non-solvent causes the polymer to separate or precipitate.

Immersion precipitation and thermally induced phase separation are the main procedures used in the synthesis of polymeric membranes. These techniques produce membranes with various morphologies (Pinnau and Freeman 2000; Liu et al. 2011).

The utilization of the sintering technique enables the fabrication of symmetric membranes, primarily employed for the preparation of ceramic or metallic membranes intended for applications in ultrafiltration (UF) and microfiltration (MF). The first stage in the critical phases of sintering in ceramic membrane production is presintering, which initiates the heating process, focusing on eliminating water from the membrane precursor to prevent the formation of cracks. This phase is crucial in ensuring the membrane precursor becomes water-free, achieved by reaching temperatures surpassing 200°C. Presintering plays a vital role in eliminating various forms of water, such as residual liquid from forming and drying, or moisture acquired during transport and setting. The prevention of uneven stress during the vaporization process is emphasized to avoid detrimental effects on the membrane precursor (Purkait and Singh 2018).

The second phase, known as thermolysis, involves the removal of organic components like binders, plasticizers or dispersants from the membrane precursor. Controlled heating is essential in managing the formation of defects that can adversely affect membrane performance. Factors influencing thermolysis include binder composition, membrane precursor size and shape, heating rate and gas composition. Successful thermolysis requires a suitable binder and controlled heating to generate internal gas pressure effectively released during burnout and vaporization of organic components. The concentration of the binder significantly influences the speed of this process, impacting pore channel opening (Purkait and Singh 2018).

The final stage, termed final sintering, unfolds in three steps: initial, intermediate and final. These steps involve distinct movements and features of the membrane precursor and particles, including full densification, grain coarsening and pore closure. Temperature regulation during final sintering is contingent upon the membrane precursor and material type. Lower temperatures facilitate rapid densification with minimal grain growth, while higher temperatures result in rapid grain growth and subdued densification. Surface diffusion is the primary mode of transport during final sintering, leading to membrane surface smoothing, particle fusion and pore rounding, ultimately resulting in the production of the final ceramic membrane (Purkait and Singh 2018).

The electrospinning technique is a method employed to produce micro to nanoscale fibers by using electrostatic repulsion forces in the presence of high-intensity electric fields (Burke et al. 2015). In a typical electrospinning setup, a controlled rate feeds a solution through a conductive capillary, often a metallic needle, maintaining a droplet at its end. This capillary, connected to a high-voltage power supply, acts as an electrode to charge the solution, while a grounded collector, positioned at a distance, subjects the highly charged droplet to a potent electrostatic force. Figure 12.3 illustrates the schematic representation of electrospinning apparatus for two different types of setups (vertical and horizontal setups) (Isaac et al. 2021).

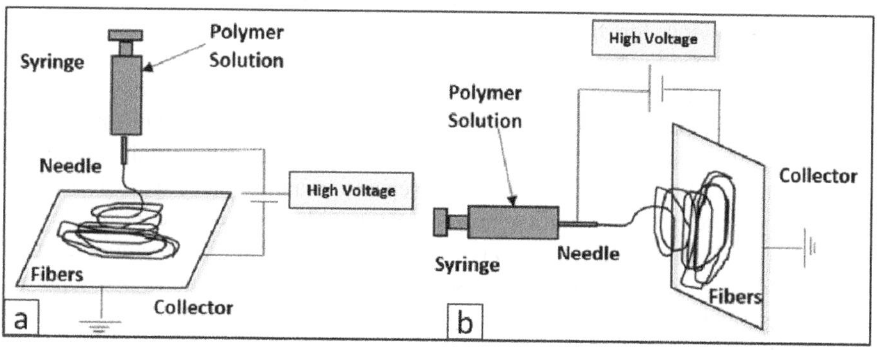

Figure 12.3. *Graphical representation of electrospinning method setup, (a) vertical setup and (b) horizontal setup (Isaac et al. 2021)*

As the voltage increases, repulsion forces overcome surface tension, causing the droplet to form a cone known as the Taylor cone (Taylor 1969). At a critical voltage, material expulsion from the cone's tip occurs, resulting in the formation of fibers

(Yarin et al. 2001). If the solution's viscosity prevents droplet breakup, a consistent jet forms, undergoing a whipping instability before depositing randomly oriented fibers on the grounded collector (Deitzel et al. 2001). The electrospinning process creates charged polymer jets that solidify into fibers as the solvent evaporates during their travel.

Despite its simplicity, the electrospinning setup allows precise control over fiber properties by adjusting various parameters such as acceleration, voltage, humidity, temperature, solution concentration and composition (Baji et al. 2010). However, limitations include the random deposition of nanofibers, the use of single or homogeneous materials, and a rate limitation determined by the solution flow rate, which can be constraining for fast-drying or highly viscous polymers (Bhardwaj and Kundu 2010).

The track-etching method is used for fabricating membranes, particularly in the domain of polymer membranes. In this process, a polymeric film sheet, often made of materials like polycarbonate or polyethylene, is exposed to high-energy particles in the presence of a radiation source. The areas bombarded by these particles undergo chemical alteration and degradation. Following irradiation, the polymer film is subjected to an etching process in a bath containing either alkaline or hydrogen peroxide, depending on the specific material used. During etching, the damaged regions along the paths traced by the high-energy particles are selectively removed, resulting in the formation of well-defined pores (Ma et al. 2020).

The track-etching method offers precise control over the creation of a porous membrane, allowing for the manipulation of pore size and distribution. After fabrication, the membrane undergoes thorough rinsing and characterization, often using techniques like scanning electron microscopy to assess its pore structure. Post-treatment steps may be employed based on the intended application, and stringent quality control measures ensure consistency and reliability in the fabricated membranes. Widely applicable in filtration, separation and sensing applications, the method is favored for its ability to produce membranes with controlled pore characteristics, high reproducibility and uniformity. The resulting membranes find utility in diverse fields where precise pore control is essential for optimal performance, making it valuable for both scientific research and industrial processes (Ivanova et al. 2021).

Microporous membranes used in microfiltration (MF), ultrafiltration (UF) and membrane distillation (MD) are commonly produced through the extrusion followed by the stretching technique (Lalia et al. 2013). Originating in the 1970s and held under proprietary ownership, this method involves heating polymers above their melting points, extruding them into thin sheets and subsequently stretching to induce porosity (Sarada et al. 1983). The procedure is well suited for highly crystalline polymers. It involves cold stretching to create micropores and hot stretching to further enhance the pore structure of the final membrane (Sadeghi et al. 2007).

The outcome and properties of these membranes are influenced by the material's physical properties and processing parameters. Carreau's research group, for example, focused on polypropylene (PP) membranes, identifying molecular weight as a critical factor in controlling membrane structure. They developed blend membranes with improved lamellae thickness and porosity by incorporating long-branched PP (Sadeghi et al. 2008). Similarly, studies on polyvinylidene fluoride (PVDF) and polytetrafluoroethylene (PTFE) demonstrated the impact of blending different melt rheologies and stretching operations on water vapor permeability and fibril connections in the membranes (Sadeghi et al. 2009).

Additional research explored blending different molecular weights of PP, revealing that higher molecular weight species led to more uniform pores and improved connectivity without significant effects on mechanical properties (Tabatabaei et al. 2008). Annealing's influence on crystallite size and crystallinity in precursor PP hollow fibers was also studied, emphasizing the crucial role of pre-annealing for porous structure formation (Kim et al. 2008). Overall, this extrusion and stretching technique allows for precise control over membrane properties by manipulating material composition and processing parameters.

12.4. Application of membrane processes

Membrane technology holds great importance in multiple industrial areas, such as food, medicine, pharmacy, biotechnology, industrial effluent and wastewater treatment, and chemicals (Baker 2024). Furthermore, this technique has also spread into households through the use of drinking water filtration systems. The advantages of this process compared to traditional separation techniques include its high capacity for removal, operational flexibility and cost-effectiveness. Furthermore, the reduced energy needs and the easy accessibility of membrane materials are factors that contribute to the increasing demand for membrane-based processes.

The chemical industry faces significant challenges in the separation, concentration and purification of molecular mixtures. Traditional methods such as ion exchange, adsorption, precipitation, extraction, crystallization and distillation have been complemented by the integration of membrane processes. These processes use semipermeable membranes as separation barriers. Membranes find widespread application in various sectors, including the production of potable water from sea and brackish water, the purification of industrial effluents, the recovery of valuable constituents, the concentration of macromolecular mixtures in the food and drug industries and the separation of gases and vapors in petrochemical processes. Their utility extends to critical roles in chemical reactors, energy conversion and storage systems, drug delivery devices and artificial organs (Abdelrasoul 2020).

12.4.1. *Water and wastewater treatment*

The connection between human activities and water is fundamental, and with the continuous growth of the global population, significant amounts of wastewater are produced daily in household, industrial and agricultural settings. Regrettably, freshwater resources fail to replenish rapidly enough to meet the rising demands. This leads to increased competition and an uneven distribution of the limited freshwater resources among different sectors, resulting in a critical problem of insufficient access to clean water. This is especially pronounced in developing countries, affecting millions of people (Rezai and Allahkarami 2021).

The consequences also affect agricultural activities, where limited water resources impede continuous irrigation and livestock production on farms. This crisis is particularly evident in various regions, including the Middle East, Africa, Asia and Latin America. The harsh realities emphasize the scale of the issue, with 2.1 billion individuals worldwide lacking access to safe drinking water in their homes. Furthermore, almost four billion people face severe water scarcity for at least one month each year. The urgent need for sustainable water management practices and innovative solutions to ensure fair access to this essential resource for everyone is highlighted by the pressing challenges associated with water scarcity and uneven distribution (Sonune and Ghate 2004).

Over time, various efforts have been made to introduce different technologies for treating wastewater, such as traditional filtration, coagulation–flocculation and biological treatment systems. Simultaneously, continuous work aims to improve existing technologies to meet current discharge or reuse standards. Among the technologies that have seen notable progress, membrane technology stands out. Its growth in recent decades is attributed to the compelling advantages it offers to water and wastewater treatment processes.

Membrane technology boasts several benefits, including a significant reduction in equipment size, energy requirements and capital costs. This makes it a promising option in wastewater treatment, as emphasized by Singh and Hankins, who argue that membrane technology can bridge the economic and sustainability gap. Features like the potential for minimal or zero chemical usage, environmental friendliness and widespread accessibility contribute to making membrane technology an increasingly favored choice in contemporary wastewater treatment.

Technologies like microfiltration (MF), ultrafiltration (UF), nanofiltration (NF) and reverse osmosis (RO) are of interest due to their inherent advantages. These include selective separation capabilities, continuous and automatic operation, well-organized process conductions, chemical-free purification, straightforward scale-up procedures and minimal space requirements. As a result, membrane techniques, especially membrane bioreactors (MBRs), have found broad application in advanced water and wastewater treatment.

Membranes based on reverse osmosis, nanofiltration and ultrafiltration are proficient in removing various water impurities, including turbidity, bacteria, viruses, color, taste, hardness, sulfates, total dissolved solids (TDS), nitrates, fluoride, arsenic and salts.

Notably, reverse osmosis membranes excel due to their minute pore size, enabling the removal of a broader range of impurities.

Nevertheless, despite their many advantages, membranes for water and wastewater treatment face certain limitations. Operating costs are affected by factors like power needs, labor expenses, material costs, membrane cleaning, scale inhibition and membrane lifespan and replacement.

A notable challenge is the decrease in permeate flux over time due to membrane fouling, presenting a significant hurdle to the efficiency of reverse osmosis (RO), nanofiltration (NF), ultrafiltration (UF) and microfiltration (MF) technologies. Tackling membrane fouling remains a pivotal concern in progressing these membrane-related technologies.

12.4.2. *Desalination*

In membrane distillation, the availability of membranes with precisely controlled properties is a pivotal factor. The effectiveness of the process hinges directly on the structural and physicochemical attributes inherent in the membranes employed. This consideration becomes particularly crucial when using advanced membrane technology, which demands meticulously structured and functionalized membranes (Drioli et al. 2015).

The structural properties of the membrane film, such as its porosity, pore distribution, mean pore size, thickness and shape, naturally affect the performance of membrane distillation. The capacity of the membrane to interface two mediums without producing phase dispersion into the other is crucial to the effective performance of the process (Abdel-Karim et al. 2021). The membrane must also be able to resist liquid penetration into its holes while integrating high volumetric mass transfer.

For membrane contactor applications in membrane distillation, the ideal membranes should possess specific characteristics. They need to be porous to facilitate mass transfer, hydrophobic to repel liquid intrusion, exhibit good thermal stability and showcase excellent chemical resistance to withstand exposure to various feed solutions. These membrane attributes collectively contribute to the efficiency and reliability of the membrane distillation process.

Freshwater scarcity has become a pressing challenge in the contemporary era, exacerbated by factors such as population growth, improved living standards, a thriving agricultural sector and industrialization (Eckardt et al. 2009). This global issue is underscored by the staggering estimate that over one billion people lack access to clean freshwater. Compounding the problem, conventional energy sources and freshwater reservoirs are depleting rapidly, necessitating the development of less energy-intensive and environmentally friendly water purification techniques.

While the overall quantity of freshwater in reservoirs may be adequate for current needs, the global challenge lies in its uneven distribution. Approximately 97% of global water reserves are in the ocean, leaving only about 3% as freshwater. Most of this is locked in glaciers and ice caps (2.06%), with a smaller portion as groundwater (0.90%) and surface/other freshwater resources (0.03%). In response to this challenge, desalination techniques for seawater and brackish water have gained popularity as alternatives (Shiklomanov 1991; Shirazi et al. 2012). Second-generation desalination technologies, primarily based on membrane operations like reverse osmosis (RO), have seen significant growth in the last two decades, constituting 60% of global desalination installations due to their energy efficiency (Greenlee et al. 2009).

However, desalination technologies face limitations such as the disposal of brine and the urgent requirement for additional reductions in energy usage to promote sustainable development. To address these constraints, current research is centered on third-generation desalination methods, including membrane distillation (MD), forward osmosis and capacitive deionization. MD has gained notice for its notable benefits.

Within the context of MD, a microporous hydrophobic membrane creates a temperature gradient, which enables the concentration of solutions to reach their saturation point without a noticeable decrease in flux. The utilization of waste heat sources, such as solar energy, geothermal energy and low-temperature industrial streams, can provide the necessary power for this procedure. The MD membrane exhibits selective permeability, allowing only vapors to flow through while trapping any nonvolatile impurities on the retentate side. As a result, the product obtained is theoretically free from any solid or nonvolatile pollutants, ensuring a purity level of 100% (Camacho et al. 2013). MD's qualities make it a viable component in third-generation desalination methods, reducing the inherent disadvantages of current reverse osmosis (RO) procedures (Khayet 2011).

12.4.3. *Membrane processing in the food industry*

Membrane processing presents numerous advantages for the food industry, offering a range of benefits that contribute to substantial quality improvements (Cuperus and Nijhuis 1993). Some of the key advantages include:

– Gentle treatment: membrane processing allows for the gentle treatment of products at ambient temperatures, which is particularly advantageous for maintaining the quality of sensitive food products.

– Low energy consumption: the process is characterized by low energy consumption, making it a more energy-efficient option compared to some traditional methods.

– Simultaneous fractionation and concentration: membrane processing enables simultaneous fractionation and concentration of food products, providing a versatile approach to modifying the composition of various components.

– Demineralization: membranes can be employed for demineralization purposes, allowing for the removal of specific minerals from food products.

– Increased yield: the technology contributes to increased yield, ensuring more efficient utilization of raw materials and reducing waste.

– Simple plant layout and operating mechanisms: membrane processing systems are known for their simple plant layout and operating mechanisms, which enhance ease of operation and maintenance.

However, despite these advantages, membrane processing in the food industry is not without challenges (Daufin et al. 2001). Special problems limit its widespread use, with fouling being a primary concern. Fouling can lead to a decrease in flux and, in some cases, a change in separation characteristics, affecting the overall efficiency of the process. Additionally, the maximum concentration factor achievable in certain membrane processes may be restricted by factors such as viscosity and osmotic pressure (Li and Chen 2010). Despite these challenges, ongoing research and technological advancements aim to address these limitations, making membrane processing an increasingly valuable tool in the food industry.

Membrane processing technologies have significantly advanced the food industry by providing efficient, non-thermal methods for separation, concentration and purification. These processes preserve the quality of food products while enhancing operational efficiency and sustainability. This section explores specific applications, their benefits, real-world examples and the challenges associated with membrane processing, along with potential solutions. Membrane processing finds extensive applications across various sectors of the food industry:

a) *Dairy industry*:

– Ultrafiltration (UF) is used to concentrate whey proteins from whey, a by-product of cheese production. Ultrafiltration increases protein content, creating whey protein concentrates (WPC) and isolates (WPI) which are high-value products used in nutritional supplements and food formulations. For example, Arla Foods Ingredients uses UF to produce high-quality whey protein concentrates that retain functional properties such as solubility and emulsification (Smithers 2008).

b) *Fruit juice industry*:

– Microfiltration (MF) and reverse osmosis (RO) are employed to clarify and concentrate fruit juices, respectively. MF removes suspended solids and microbial contaminants without altering flavor, while RO concentrates the juice by removing water, enhancing the natural sweetness and reducing transportation costs. For example, Tropicana uses MF to produce clear, shelf-stable orange juice that maintains a fresh taste and bright color, followed by RO to concentrate the juice for efficient shipping (Cheryan 1998).

c) *Vegetable oil processing*:

– Ultrafiltration (UF) and nanofiltration (NF) are used to remove phospholipids and free fatty acids from crude vegetable oils. These processes improve oil quality by enhancing its stability and reducing the need for chemical refining steps, leading to a more natural product. For example, Cargill employs UF and NF technologies in the refining of soybean oil, resulting in higher purity oil with better sensory properties (Oladosu et al. 2017).

12.4.4. *Pharmaceutical and biotechnology*

Membranes are integral in various facets of the pharmaceutical industry. They play a vital role in understanding drug absorption and cellular interaction, influencing bioavailability. In manufacturing, membrane filtration is employed for purification, while membrane-based drug delivery systems, such as transdermal patches and liposomes, enable controlled release. Membrane chromatography enhances the purification of biomolecules, contributing to pharmaceutical product quality. Membranes also feature prominently in dialysis machines, filtering toxins during hemodialysis. Additionally, they serve as barriers in packaging, safeguarding drugs from external elements to maintain stability. Biological membrane studies, including cell culture models, contribute insights into drug transport and toxicity. Overall, membranes are indispensable for ensuring the effectiveness, safety and quality of pharmaceutical products throughout various stages of development and production (Charcosset 2012).

Filtration and membrane technology are crucial in the fermentation process, particularly in the production of antibiotics and enzymes. These technologies enhance production efficiency by facilitating the separation of cells or biomass from the fermentation broth, leading to improved rates and yields. The implementation of membrane filtration systems automates and streamlines the harvesting process, reducing manual workload for operators and increasing operational efficiency. Additionally, membrane systems contribute to lower maintenance costs due to their durability and longevity. Membranes have become standard components in industrial production lines, particularly for concentrating enzymes before subsequent processing steps, further optimizing efficiency in enzyme-related applications. In summary, filtration and membrane technology play vital roles in optimizing the fermentation process, offering benefits such as improved efficiency, cost savings and operational enhancements in industrial production lines (Van Reis and Zydney 2001).

12.4.5. *Gas separation*

In recent years, membrane processes for gas separation have seen increased acceptance in the industry, presenting a competitive edge against established methods like pressure swing absorption and cryogenic distillation. The successful expansion of membrane applications into challenging and harsh environments, such as petrochemistry, relies on the development of robust, high-performance materials (Bernardo et al. 2009).

The growing importance of membrane processes in gas separation is attributed to several key factors. The development of new, durable and high-performance materials is critical, especially in industries like petrochemistry, to broaden the applications of membrane processes in challenging environments. The modular design inherent to membrane operations aligns with process intensification principles, promoting flexibility and adaptability for diverse gas separation applications. Membrane processes naturally lend themselves to process intensification strategies, enhancing efficiency while minimizing resource consumption. The versatility of membrane processes is a significant driver for their widespread adoption, allowing them to effectively address diverse separation challenges and become increasingly competitive in the market (Chakrabarty et al. 2022).

As the industry continues to seek more sustainable, energy-efficient and cost-effective solutions, membrane processes offer a compelling alternative for gas separation. The ongoing development of resilient materials and the inherent advantages of modularity and versatility position membrane processes as a promising and competitive technology in the gas separation landscape.

12.4.6. *Medical applications*

Membrane technology holds immense significance in medical applications, playing a critical role in several life-saving treatment methods (Baker 2024). Membranes are used in drug delivery systems, ensuring controlled and targeted release of pharmaceutical substances within the body. Membranes play a crucial role in the development of artificial organs, contributing to advancements in organ transplant technologies. Membranes are also employed in tissue regeneration applications, facilitating the repair and replacement of damaged tissues (Liu et al. 2020). Membranes are incorporated into diagnostic devices, aiding in various medical tests and analyses. Membranes serve as coatings for various medical devices, enhancing their functionality and biocompatibility. Membranes are used in bioseparation processes, enabling the isolation and purification of biomolecules (Bittel et al. 2021).

The production of membranes for medical applications is substantial, with the total membrane area used in medical contexts almost equivalent to all industrial membrane applications combined. Notably, in fiscal terms, the value of medical membrane products surpasses that of all other applications combined (Baker 2024). For example, in the United States alone, the medical membrane market is estimated to approach 1.5 billion dollars annually and continues to experience steady growth. This underscores the critical role that membrane technology plays in advancing medical treatments, diagnostics and therapeutic interventions, contributing significantly to the healthcare industry's continuous development (Stamatialis et al. 2008).

12.5. Membrane fouling

Fouling, in a broad sense, refers to the undesired accumulation of deposits either on the surface or within the pores of a membrane, leading to a decline in permeation flux and salt rejection performance (Gryta 2007; He et al. 2008). This issue poses a significant challenge in membrane-based processes, particularly in pressure-driven technologies like reverse osmosis (RO), nanofiltration (NF) and ultrafiltration (UF), where fouling can adversely affect desalination and purification processes.

Typically, foulants exhibit a colloidal nature and interact either among themselves or with the membrane surface, resulting in the formation of deposits. These deposits can obstruct the flow of water through the membrane, diminishing its efficiency over time. Effectively managing fouling is crucial for sustaining the performance of membrane-based processes. Ongoing research is dedicated to developing strategies and materials to mitigate fouling effects and extend the operational life of membranes, ensuring the continued effectiveness of these essential separation technologies (Tijing et al. 2015).

The fouling characteristics affecting microfiltration (MF) and ultrafiltration (UF) membranes differ significantly from those impacting nanofiltration (NF) and reverse osmosis (RO) membranes. Membrane fouling manifests as a consequence of intricate and multifaceted phenomena, involving electrostatic and hydrophobic interactions between the membrane surface and the constituents present in the feed water (Tanudjaja et al. 2022). The principal categories of fouling encompass organic, inorganic, biofouling and colloidal/particulate fouling (Alkhatib et al. 2021) (Figure 12.4).

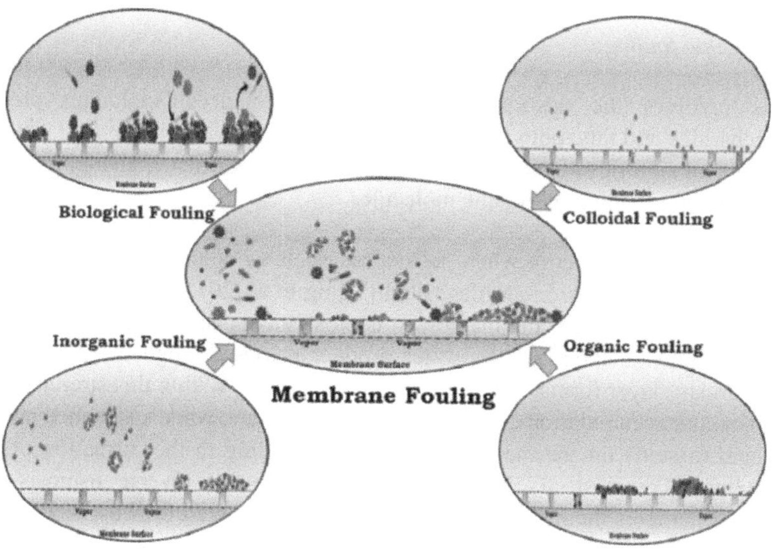

Figure 12.4. *Different classifications of membrane fouling (Alkhatib et al. 2021)*

Fouling refers to the accumulation of unwanted materials on the surface or within the pores of a membrane, resulting in decreased permeability and efficiency. The primary mechanisms of fouling include:

a) *Particulate fouling*:

This is caused by the deposition of suspended solids, such as proteins, fats and colloidal particles, on the membrane surface. This type of fouling reduces the flow rate and increases the pressure required to maintain desired permeate flux.

b) *Scaling*:

This involves the precipitation and deposition of inorganic salts, like calcium carbonate and calcium sulfate, due to supersaturation at the membrane surface. Scaling can lead to severe permeability loss and membrane damage if not adequately controlled.

c) *Biofouling*:

This results from the growth and accumulation of microorganisms, such as bacteria and algae, forming biofilms on the membrane surface. Biofouling not only reduces membrane performance but also poses hygiene risks, especially in food processing applications.

d) *Organic fouling*:

This involves the adsorption of organic substances, including proteins, polysaccharides and fats, onto the membrane surface and within its pores. This type of fouling can significantly reduce membrane flux and is often challenging to clean due to the strong binding of organic molecules.

Inorganic fouling occurs when substances such as silica, calcium, aluminum, sulfate, phosphorus and iron are deposited (Tijing et al. 2015). The fouling process occurring at the surface of the membrane is closely connected to the concentration polarization phenomenon. As the water product passes through the membrane, a dense boundary layer forms on the surface of separation. Within this stratum, higher levels of solute accumulation cause the salts to precipitate, while suspended particles may begin to settle on the membrane's exterior, leading to the formation of scale and fouling (Tijing et al. 2014). Scale deposition on reverse osmosis (RO) membranes hampers the hydrodynamic conditions of the feed flow by accumulating on and inside the membranes. If the circumstances that cause fouling are not properly regulated, scaling can develop into a self-perpetuating phenomenon (Al-Amoudi and Farooque 2005). During cases of high polarization, many problems may arise, including reduced reverse osmosis (RO) efficiency, channeling and membrane surface degradation.

Unlike organic and inorganic fouling, which include non-living chemicals as foulants, biofouling presents a more complex issue. In the process of biofouling, living microorganisms attach themselves to the surface of the membrane, causing rapid growth and obtaining nutrients from the surrounding environment. This biological activity leads to the creation of a consolidated "biofilm" that has the ability to inhabit almost the whole surface (Cho et al. 2004; Tijing et al. 2008). During their metabolic processes, these microbes often emit significant quantities of extracellular polymeric substances (EPS). The EPS have a crucial function in enhancing the attachment of bacteria to the surface of the membrane and provide resistance against biocides and toxins found in the solution (Griebe and Flemming 1998).

Polysaccharides (as carbs) and proteins are thought to be the main sources of fouling. Thus, determining the concentration of EPS or SMP is mostly dependent on analyzing carbs and proteins. Proteins, which are more hydrophobic than carbs, and carbohydrates, which are hydrophilic, both contribute to increased fouling propensity, with carbohydrate fractions from both EPS and SMP being particularly remarkable in this context (Yigit et al. 2008; Li et al. 2012).

Colloids are available in a wide range of sizes, from a few nanometers to several micrometers. Water colloids include clay, colloidal silica, silt, sulfur, iron and aluminum compounds, and solidified crystals generating through incomplete treatment processes. According to Yiantsios and Karabelas (1998), some high molecular weight organic substances, such as polysaccharides, peptidoglycans, proteins and humic aggregates, along with their clusters that look like cell debris, are likely to be colloidal foulants because they behave similarly to inorganic colloidal particles. Membrane fouling by particles and colloids is usually caused by initially blocked pores and then the formation of a cake layer (Lim and Bai 2003).

The potential for pore blockage in filtration processes is influenced by the size of colloids in relation to membrane pores. Colloids closely approaching the size of membrane pores can lead to pore blockage, while those significantly larger may result in the formation of a more easily removable cake (Huang et al. 2008). The transport of colloids and fine particles to the membrane surface involves various mechanisms, including gravitational settling, shear-induced diffusion, Brownian diffusion, convection, lateral migration and inertial lift.

Pore blocking in filtration processes can manifest in various forms. Standard pore blocking occurs when the pore opening constricts due to the deposition of particles around the pore entry. In cases of complete pore blocking, also known as pore sealing, a pore becomes entirely plugged. Intermediate pore blocking represents a combination of the two aforementioned scenarios, wherein the accumulation of deposited particles starts to bridge the pore openings. Understanding these different modes of pore blocking is crucial for optimizing filtration systems and addressing specific challenges associated with the deposition of particles during the filtration process (Guo et al. 2012).

Several strategies have been developed and refined to mitigate the effects of fouling in membrane processes. These include:

1) *Membrane modification*:

– Surface modification: the application of hydrophilic coatings, grafting of antifouling polymers or incorporation of nanoparticles to enhance membrane surface properties. For example, membranes modified with hydrophilic polymers such as polyethylene glycol (PEG) exhibit reduced protein adsorption and improved resistance to organic fouling (Cassano et al. 2015).

– Antifouling membranes: development of membranes with inherent antifouling properties, such as those incorporating silver nanoparticles or titanium dioxide, which possess antimicrobial and self-cleaning capabilities (Zhang et al. 2016).

2) *Optimized operation and cleaning protocols*:

– Operational parameters: adjusting operating conditions like crossflow velocity, transmembrane pressure and feed concentration can minimize the rate of fouling. Operating at higher crossflow velocities helps to reduce the deposition of foulants by maintaining turbulent flow near the membrane surface (Meng et al. 2017).

– Advanced cleaning techniques: employing chemical cleaning agents, enzymatic treatments or physical methods such as backwashing and air sparging to remove fouling layers effectively. Enzymatic cleaners are particularly effective in breaking down organic foulants without damaging the membrane (Abdullah et al. 2023).

3) *Pretreatment processes*:

– Coagulation and flocculation: pre-treating feed water with coagulants and flocculants to aggregate fine particles and reduce the fouling load on membranes.

– Pre-filtration: using microfiltration or ultrafiltration as a pretreatment step to remove larger particulates and reduce the burden on the downstream membrane processes (Hamedi et al. 2019).

4) *Real-time monitoring and control systems*:

– Sensors and automation: integration of sensors for real-time monitoring of fouling indicators such as pressure drop, permeate flux and conductivity. Automated control systems can adjust operational parameters in response to fouling conditions, optimizing membrane performance and cleaning schedules (Niu et al. 2022).

– Predictive maintenance: using data analytics and machine learning to predict fouling events and schedule maintenance proactively. Predictive models can help in identifying trends and patterns associated with fouling, allowing for timely interventions (Goh et al. 2016).

Research continues to focus on developing more effective fouling mitigation strategies and innovative membrane technologies. Research is exploring the use of graphene oxide and carbon nanotubes in membrane fabrication. These materials offer superior mechanical strength, increased permeability and enhanced antifouling properties. For instance, graphene oxide membranes exhibit high hydrophilicity and smooth surfaces, which reduce the tendency for foulants to adhere (Soren et al. 2023). In addition, development of membranes inspired by natural systems, such as mussel-inspired coatings that mimic the adhesive and antifouling properties of mussel foot proteins, offer a novel approach to creating fouling-resistant surfaces.

Furthermore, combining membrane technologies with other separation techniques, such as adsorption and filtration, enhances fouling control and overall process efficiency. For example, integrating membrane filtration with activated carbon adsorption can effectively remove organic foulants and improve membrane lifespan (Bamasag et al. 2023).

Additionally, development of smart membranes that respond to environmental stimuli, such as pH and temperature changes, to dynamically alter their properties and minimize fouling. These membranes can self-clean or adjust their surface characteristics in response to fouling conditions, maintaining optimal performance (Geise et al. 2010). Research into self-healing materials aims to create membranes that can repair minor damages autonomously, extending their operational life and reducing maintenance needs. These materials incorporate dynamic bonds or microcapsules containing healing agents that activate upon damage (Kammakakam and Lai 2023).

In addition, development of biodegradable membranes made from renewable resources reduces environmental impact. These membranes offer a sustainable alternative to traditional synthetic membranes and align with the industry's move toward greener technologies (Baker 2023). Research into environmentally friendly cleaning agents focuses on developing solutions that are effective at removing fouling while ensuring they do not damage the membrane or pose any harm to the environment. Natural enzymes and biodegradable chemicals are being explored as alternatives to harsh chemical cleaners (Goh et al. 2016).

Fouling is a pervasive challenge in membrane processing, impacting efficiency, product quality and operational costs. However, advancements in membrane materials, innovative fouling mitigation strategies and ongoing research into new technologies offer promising solutions. By understanding the mechanisms of fouling and implementing effective mitigation techniques, industries can optimize membrane performance, enhance sustainability and improve overall process efficiency. The future of membrane technology looks bright with continuous innovation and development aimed at overcoming the challenges posed by fouling.

12.6. Conclusion

In summary, membrane processes play crucial roles in water treatment, industrial applications and environmental protection, offering diverse solutions for liquid separation and purification.

Understanding the diverse types and applications of water filtration membranes is crucial for selecting the most suitable technology for specific water treatment needs. Further research and development in membrane filtration technologies continue to enhance their efficiency and applicability in various industrial and environmental contexts.

The current landscape of membrane processes and applications exhibits varying levels of satisfaction. In micro- and ultra-filtration, the current status of membrane properties is generally acceptable. However, there is a necessity for progress in application knowledge, control, long-term operating experience and process design, especially in chemical and food industry applications or wastewater treatment. Challenges like concentration polarization and membrane fouling remain significant, prompting a need for innovative membrane modules and process design concepts to enhance fouling control and extend membrane life.

Beyond traditional membrane processes and applications, there is a growing trend toward adopting emerging membrane operations such as membrane contactors and membrane reactors in industrial settings. These innovative approaches are increasingly recognized as standard unit operations in process engineering, playing a significant role in shaping the landscape of membrane engineering across various industrial productions. The integration of membrane contactors and reactors is contributing to process intensification, enhancing efficiency and enabling sustainable industrial development. It is crucial to underscore that the implementation of membrane operations must align with the principles of process intensification and sustainable practices, ensuring their effectiveness and positive impact on industrial processes. This emphasis on aligning membrane operations with broader industrial goals reflects a commitment to advancing both technology and environmental responsibility in tandem.

Fouling poses a significant challenge in membrane-based processes, particularly in pressure-driven technologies like reverse osmosis (RO), nanofiltration (NF) and ultrafiltration (UF), where fouling can adversely affect desalination and purification processes. The principal categories of fouling encompass organic, inorganic, biofouling and colloidal/particulate fouling. Inorganic fouling occurs when substances such as silica, calcium, aluminum, sulfate, phosphorus and iron are deposited. The fouling process occurring at the surface of the membrane is closely connected to the concentration polarization phenomenon. Membranes can easily be fouled by a range of dissolved inorganic and organic substances. In the process of biofouling, living microorganisms attaching to the surface of the membrane have the ability to inhabit almost the whole surface of the membrane. During their metabolic processes, they often emit significant quantities of extracellular polymeric substances that have a crucial function in enhancing the attachment of bacteria to the surface of the membrane and provide resistance against biocides and toxins found in the solution.

However, the intrinsic characteristics of membranes, including high energy efficiency, operational simplicity, transport selectivity, operational flexibility and environmental compatibility, establish them as pivotal tools for advanced molecular separations and chemical transformations. Membranes and the associated processes they enable are instrumental in surmounting constraints inherent in traditional industrial approaches. They contribute significantly to the objectives of chemical process rationalization, miniaturization and sustainable industrial development. By harnessing the unique features of membranes, industries can achieve more streamlined and efficient processes, aligning with contemporary goals of resource optimization and environmental responsibility. The versatile applications of membranes underscore their importance in advancing the frontiers of chemical engineering and facilitating a transition toward more sustainable and efficient industrial practices.

12.7. References

Abdel-Karim, A., Leaper, S., Skuse, C., Zaragoza, G., Gryta, M., Gorgojo, P. (2021). Membrane cleaning and pretreatments in membrane distillation – a review. *Chemical Engineering Journal*, 422, 129696.

Abdelrasoul, A. (2020). *Advances in Membrane Technologies*. IntechOpen, London.

Abdullah, R., Astira, D., Widiyanto, A.R., Dharma, H.N.C., Hidayat, A.R.P., Santoso, L., Sulistiono, D.O., Rahmawati, Z., Gunawan, T., Jaafar, J. (2023). Recent development of mixed matrix membrane as a membrane bioreactor for wastewater treatment: A review. *Case Studies in Chemical and Environmental Engineering*, 8, 100485.

Adak, M.K., Kumar, K., Das, A., Dhak, D. (2023). Nano-filtration application in the textile industry for wastewater treatment. In *Nanofiltration Membrane for Water Purification*, Ahmad, A. and Alshammari, M.B. (eds). Springer, Berlin.

Adam, M.R., Othman, M.H.D., Kurniawan, T.A., Puteh, M.H., Ismail, A.F., Khongnakorn, W., Rahman, M.A., Jaafar, J. (2022). Advances in adsorptive membrane technology for water treatment and resource recovery applications: A critical review. *Journal of Environmental Chemical Engineering*, 10, 107633.

Agboola, O., Sanni, S.E., Sadiku, R., Popoola, P., Fasiku, V.O. (2020). Advanced nanofiltration membranes for wastewater treatment. In *Contaminants and Clean Technologies*, Chowdhary, P. and Raj, A. (eds). CRC Press, Boca Raton.

Ahmad, H.-B. (2020). Ceramic membranes: New trends and prospects (short review). *Water and Water Purification Technologies. Scientific and Technical News*, 27, 4–31.

Ai, X., Li, Y.-H., Li, Y.-W., Gao, T., Zhou, K.-G. (2022). Recent progress on the smart membranes based on two-dimensional materials. *Chinese Chemical Letters*, 33, 2832–2844.

Al-Amoudi, A.S. and Farooque, A.M. (2005). Performance restoration and autopsy of NF membranes used in seawater pretreatment. *Desalination*, 178, 261–271.

Al-Amshawee, S.K.A. and Yunus, M.Y.B.M. (2023). Electrodialysis membrane desalination with diagonal membrane spacers: A review. *Environmental Science and Pollution Research*, 1–25.

Alawad, S.M., Mansour, R.B., Al-Sulaiman, F.A., Rehman, S. (2023). Renewable energy systems for water desalination applications: A comprehensive review. *Energy Conversion and Management*, 286, 117035.

Ali, S.S., Sabry, R., Ali, H.M., Gadallah, H., Mansor, E.S., Abdallah, H., Shalaby, M., Shaban, A.M. (2023). Integrated system of anoxic/activated sludge and ultrafiltration membrane for zero liquid discharge of pharmaceutical industrial wastewater treatment. *Journal of Environmental Chemical Engineering*, 11, 109068.

Alkhatib, A., Ayari, M.A., Hawari, A.H. (2021). Fouling mitigation strategies for different foulants in membrane distillation. *Chemical Engineering and Processing - Process Intensification*, 167, 108517.

Allgeier, S., Alspach, B., Vickers, J. (2005). *Membrane Filtration Guidance Manual*. United States Environmental Protection Agency, Washington D.C.

Almasoudi, S.M. and Bassam, J. (2024). Desalination technologies and their environmental impacts: A review. *Sustainable Chemistry One World*, 100002.

Amy, G., Ghaffour, N., Li, Z., Francis, L., Linares, R.V., Missimer, T., Lattemann, S. (2017). Membrane-based seawater desalination: Present and future prospects. *Desalination*, 401, 16–21.

Baji, A., Mai, Y.-W., Wong, S.-C., Abtahi, M., Chen, P. (2010). Electrospinning of polymer nanofibers: Effects on oriented morphology, structures and tensile properties. *Composites Science and Technology*, 70, 703–718.

Baker, R.W. (2023). *Membrane Technology and Applications*. John Wiley & Sons, Hoboken.

Baker, R.W. (2024). *Membrane Technology and Applications*. John Wiley & Sons, Hoboken.

Bamasag, A., Almatrafi, E., Alqahtani, T., Phelan, P., Ullah, M., Mustakeem, M., Obaid, M., Ghaffour, N. (2023). Recent advances and future prospects in direct solar desalination systems using membrane distillation technology. *Journal of Cleaner Production*, 385, 135737.

Basile, A. and Charcosset, C. (2016). *Integrated Membrane Systems and Processes*. John Wiley & Sons, Hoboken.

Basile, A., Cassano, A., Rastogi, N.K. (2015). *Advances in Membrane Technologies for Water Treatment: Materials, Processes and Applications*. Elsevier, Amsterdam.

Bernardo, P., Drioli, E., Golemme, G. (2009). Membrane gas separation: A review/state of the art. *Industrial & Engineering Chemistry Research*, 48, 4638–4663.

Bhardwaj, N. and Kundu, S.C. (2010). Electrospinning: A fascinating fiber fabrication technique. *Biotechnology Advances*, 28, 325–347.

Biesheuvel, P.M., Porada, S., Elimelech, M., Dykstra, J.E. (2022). Tutorial review of reverse osmosis and electrodialysis. *Journal of Membrane Science*, 647, 120221.

Birniwa, A.H., Habibu, S., Abdullahi, S.S.A., Mohammad, R.E.A., Hussaini, A., Magaji, H., Al-Dhawi, B.N.S., Noor, A., Jagaba, A.H. (2024). Membrane technologies for heavy metals removal from water and wastewater: A mini review. *Case Studies in Chemical and Environmental Engineering*, 9, 100538.

Bittel, M., Reichert, P., Sarfati, I., Dressel, A., Leikam, S., Uderhardt, S., Stolzer, I., Phu, T.A., Ng, M., Vu, N.K. (2021). Visualizing transfer of microbial biomolecules by outer membrane vesicles in microbe-host-communication in vivo. *Journal of Extracellular Vesicles*, 10, e12159.

Burke, L., Keshvari, A., Wright, C.J., Hilal, N. (2015). Electrospinning: A practical approach for membrane fabrication. In *Membrane Fabrication*, Hilal, N., Ismail, A.F., Wright, C. (eds). CRC Press, Boca Raton.

Cai, Y., Wu, J., Shi, S.Q., Li, J., Kim, K.-H. (2023). Advances in desalination technology and its environmental and economic assessment. *Journal of Cleaner Production*, 397, 136498.

Camacho, L.M., Dumée, L., Zhang, J., Li, J.-D., Duke, M., Gomez, J., Gray, S. (2013). Advances in membrane distillation for water desalination and purification applications. *Water*, 5(1), 94–196.

Cassano, A., Basile, A., Rastogi, N.K. (2015). *Advances in Membrane Technologies for Water Treatment: Materials, Processes and Applications*. Woodhead Publishing, Cambridge.

Castel, C. and Favre, E. (2018). Membrane separations and energy efficiency. *Journal of Membrane Science*, 548, 345–357.

Castro-Muñoz, R., Serna-Vázquez, J., García-Depraect, O. (2022). Current evidence in high throughput ultrafiltration toward the purification of monoclonal antibodies (mAbs) and biotechnological protein-type molecules. *Critical Reviews in Biotechnology*, 42, 827–837.

Chakrabarty, T., Giri, A.K., Sarkar, S. (2022). Mixed-matrix gas separation membranes for sustainable future: A mini review. *Polymers for Advanced Technologies*, 33, 1747–1761.

Charcosset, C. (2012). *Membrane Processes in Biotechnology and Pharmaceutics*. Elsevier, Amsterdam.

Cheryan, M. (1998). *Ultrafiltration and Microfiltration Handbook*. CRC Press, Boca Raton.

Cho, Y.I., Fridman, A.F., Lee, S.H., Kim, W.T. (2004). Physical water treatment for fouling prevention in heat exchangers. In *Advances in Heat Transfer, Volume 38*, Greene, G.A., Harnett, J.P., Bar-Cohen, A., Cho, Y.I. (eds). Elsevier, Amsterdam.

Cuperus, F.P. and Nijhuis, H.H. (1993). Applications of membrane technology to food processing. *Trends in Food Science & Technology*, 4, 277–282.

Dansawad, P., Yang, Y., Li, X., Shang, X., Li, Y., Guo, Z., Qing, Y., Zhao, S., You, S., Li, W. (2022). Smart membranes for oil/water emulsions separation: A review. *Advanced Membranes*, 2, 100039.

Daufin, G., Escudier, J.P., Carrère, H., Bérot, S., Fillaudeau, L., Decloux, M. (2001). Recent and emerging applications of membrane processes in the food and dairy industry. *Food and Bioproducts Processing*, 79, 89–102.

Deitzel, J.M., Kleinmeyer, J., Harris, D.E.A., Tan, N.C.B. (2001). The effect of processing variables on the morphology of electrospun nanofibers and textiles. *Polymer*, 42, 261–272.

Drioli, E. and Giorno, L. (eds) (2009). *Membrane Operations: Innovative Separations and Transformations*. Wiley-VCH, Weinheim.

Drioli, E., Ali, A., Macedonio, F. (2015). Membrane distillation: Recent developments and perspectives. *Desalination*, 356, 56–84.

Dushkova, M., Mihalev, K., Dinchev, A., Vasilev, K., Georgiev, D., Terziyska, M. (2022). Concentration of polyphenolic antioxidants in apple juice and extract using ultrafiltration. *Membranes*, 12, 1032.

Eckardt, N.A., Cominelli, E., Galbiati, M., Tonelli, C. (2009). The future of science: Food and water for life. *The Plant Cell*, 21, 368–372.

Ersahin, M.E., Ozgun, H., Dereli, R.K., Ozturk, I., Roest, K., Van Lier, J.B. (2012). A review on dynamic membrane filtration: Materials, applications and future perspectives. *Bioresource Technology*, 122, 196–206.

Fane, A.G., Wang, R., Jia, Y. (2011). Membrane technology: Past, present and future. *Membrane and Desalination Technologies*, 1–45.

Fattah, I.M.R., Farhan, Z.A., Kontoleon, K.J., Kianfar, E., Hadrawi, S.K. (2023). Hollow fiber membrane contactor based carbon dioxide absorption – stripping: A review. *Macromolecular Research*, 31, 299–325.

Figoli, A. and Criscuoli, A. (2017). *Sustainable Membrane Technology for Water and Wastewater Treatment*. Springer, Berlin.

Firouzi, N. and Żur, K.K. (2023). On the generalized nonlinear mechanics of compressible, incompressible, isotropic, and anisotropic hyperelastic membranes. *International Journal of Solids and Structures*, 264, 112088.

Geise, G.M., Lee, H.S., Miller, D.J., Freeman, B.D., Mcgrath, J.E., Paul, D.R. (2010). Water purification by membranes: The role of polymer science. *Journal of Polymer Science Part B: Polymer Physics*, 48, 1685–1718.

Gholami, F., Ghanizadeh, G., Zinatizadeh, A.A., Zinadini, S., Masoumbeigi, H. (2023). Arsenic and total dissolved solids removal using antibacterial/antifouling nanofiltration membranes modified by functionalized graphene oxide and copper ferrodioxide. *Water Environment Research*, 95, e10902.

Gitis, V. and Rothenberg, G. (2016). *Ceramic Membranes: New Opportunities and Practical Applications*. John Wiley & Sons, Hoboken.

Goh, P.S., Matsuura, T., Ismail, A.F., Hilal, N. (2016). Recent trends in membranes and membrane processes for desalination. *Desalination*, 391, 43–60.

Greenlee, L.F., Lawler, D.F., Freeman, B.D., Marrot, B., Moulin, P. (2009). Reverse osmosis desalination: Water sources, technology, and today's challenges. *Water Research*, 43, 2317–2348.

Griebe, T. and Flemming, H.-C. (1998). Biocide-free antifouling strategy to protect RO membranes from biofouling. *Desalination*, 118, 153–156.

Gryta, M. (2007). Influence of polypropylene membrane surface porosity on the performance of membrane distillation process. *Journal of Membrane Science*, 287, 67–78.

Guan, K., Fang, S., Zhou, S., Fu, W., Li, Z., Gonzales, R.R., Xu, P., Mai, Z., Hu, M., Zhang, P. (2023). Thin film composite membrane with improved permeance for reverse osmosis and organic solvent reverse osmosis. *Journal of Membrane Science*, 688, 122104.

Guo, W., Ngo, H.-H., Li, J. (2012). A mini-review on membrane fouling. *Bioresource Technology*, 122, 27–34.

Hailemariam, R.H., Woo, Y.C., Damtie, M.M., Kim, B.C., Park, K.-D., Choi, J.-S. (2020). Reverse osmosis membrane fabrication and modification technologies and future trends: A review. *Advances in Colloid and Interface Science*, 276, 102100.

Hamedi, H., Ehteshami, M., Mirbagheri, S.A., Rasouli, S.A., Zendehboudi, S. (2019). Current status and future prospects of membrane bioreactors (MBRs) and fouling phenomena: A systematic review. *The Canadian Journal of Chemical Engineering*, 97, 32–58.

Harby, K., Emad, M., Benghanem, M., Abolibda, T.Z., Almohammadi, K., Aljabri, A., Alsaiari, A., Elgendi, M. (2024). Reverse osmosis hybridization with other desalination techniques: An overview and opportunities. *Desalination*, 117600.

Harrison, R.G., Todd, P., Rudge, S.R., Petrides, D.P. (2015). *Bioseparations Science and Engineering*. Oxford University Press, Oxford.

He, F., Gilron, J., Lee, H., Song, L., Sirkar, K.K. (2008). Potential for scaling by sparingly soluble salts in crossflow DCMD. *Journal of Membrane Science*, 311, 68–80.

Hu, Y. and Wu, W. (2023). Application of membrane filtration to cold sterilization of drinks and establishment of aseptic workshop. *Food and Environmental Virology*, 15, 89–106.

Huang, H., Spinette, R., O'Melia, C.R. (2008). Direct-flow microfiltration of aquasols: I. Impacts of particle stabilities and size. *Journal of Membrane Science*, 314, 90–100.

Hussain, C.M. (2020). *Handbook of Functionalized Nanomaterials for Industrial Applications*. Elsevier, Amsterdam.

Isaac, B., Taylor, R.M., Reifsnider, K. (2021). Mechanical and dielectric properties of aligned electrospun fibers. *Fibers*, 9, 4.

Ismail, A.F., Chandra Khulbe, K., Matsuura, T., Ismail, A.F., Khulbe, K.C., Matsuura, T. (2015). Membrane fabrication/manufacturing techniques. *Gas Separation Membranes: Polymeric and Inorganic*, 193–220.

Ivanova, N.M., Filippova, E.O., Tverdokhlebov, S.I., Levkovich, N.V., Apel, P.Y. (2021). Preparation, structure, and properties of track-etched membranes based on polylactic acid. *Membranes and Membrane Technologies*, 3, 282–290.

Jhaveri, J.H. and Murthy, Z.V.P. (2016). A comprehensive review on anti-fouling nanocomposite membranes for pressure driven membrane separation processes. *Desalination*, 379, 137–154.

Jönsson, A.S. (2013). Microfiltration, ultrafiltration and diafiltration. *Separation and Purification Technologies in Biorefineries*, 205–231.

Jornitz, M. W. (ed.) (2019). *Filtration and Purification in the Biopharmaceutical Industry*. CRC Press, Boca Raton.

Judd, S. (2010). *The MBR Book: Principles and Applications of Membrane Bioreactors for Water and Wastewater Treatment*. Elsevier, Amsterdam.

Kadhom, M. (2023). A review on the polyamide thin film composite (TFC) membrane used for desalination: Improvement methods, current alternatives, and challenges. *Chemical Engineering Research and Design*, 191, 472–492.

Kammakakam, I. and Lai, Z. (2023). Next-generation ultrafiltration membranes: A review of material design, properties, recent progress, and challenges. *Chemosphere*, 316, 137669.

Karray, R., Elloumi, W., Ali, R.B., Loukil, S., Chamkha, M., Karray, F., Sayadi, S. (2022). A novel bioprocess combining anaerobic co-digestion followed by ultra-filtration and microalgae culture for optimal olive mill wastewater treatment. *Journal of Environmental Management*, 303, 114188.

Khayet, M. (2011). Membranes and theoretical modeling of membrane distillation: A review. *Advances in Colloid and Interface Science*, 164, 56–88.

Khulbe, K.C. and Matsuura, T. (2021). Membrane modification. In *Nanotechnology in Membrane Processes*. Springer, Cham. doi: 10.1007/978-3-030-64183-2_4.

Kim, J., Kim, S.S., Park, M., Jang, M. (2008). Effects of precursor properties on the preparation of polyethylene hollow fiber membranes by stretching. *Journal of Membrane Science*, 318, 201–209.

Lalia, B.S., Kochkodan, V., Hashaikeh, R., Hilal, N. (2013). A review on membrane fabrication: Structure, properties and performance relationship. *Desalination*, 326, 77–95.

Le, T.T., Cabaltica, A.D., Bui, V.M. (2014). Membrane separations in dairy processing. *Journal of Food Technology Research*, 2, 1–14.

Lee, A., Elam, J.W., Darling, S.B. (2016). Membrane materials for water purification: Design, development, and application. *Environmental Science: Water Research & Technology*, 2, 17–42.

Li, H. and Chen, V. (2010). Membrane fouling and cleaning in food and bioprocessing. In *Membrane Technology*, Cui, Z.F. and Muralidhara, H.S. (eds). Elsevier, Amsterdam.

Li, J., Yang, F., Liu, Y., Song, H., Li, D., Cheng, F. (2012). Microbial community and biomass characteristics associated severe membrane fouling during start-up of a hybrid anoxic–oxic membrane bioreactor. *Bioresource Technology*, 103, 43–47.

Lim, A.L. and Bai, R. (2003). Membrane fouling and cleaning in microfiltration of activated sludge wastewater. *Journal of Membrane Science*, 216, 279–290.

Lipnizki, F., Thuvander, J., Rudolph, G. (2020). Membrane processes and applications for biorefineries. In *Current Trends and Future Developments on (Bio-) Membranes*, Figoli, A., Li, Y., Basile, A. (eds). Elsevier, Amsterdam.

Liu, F., Hashim, N.A., Liu, Y., Abed, M.R.M., Li, K. (2011). Progress in the production and modification of PVDF membranes. *Journal of Membrane Science*, 375, 1–27.

Liu, X., He, X., Jin, D., Wu, S., Wang, H., Yin, M., Aldalbahi, A., El-Newehy, M., Mo, X., Wu, J. (2020). A biodegradable multifunctional nanofibrous membrane for periodontal tissue regeneration. *Acta Biomaterialia*, 108, 207–222.

Luong, V.T., Kurz, E.E.C., Hellriegel, U., Dinh, D.N., Tran, H.T., Figoli, A., Gabriele, B., Luu, T.L., Hoinkis, J. (2023). Modular desalination concept with low-pressure reverse osmosis and capacitive deionization: Performance study of a pilot plant in Vietnam in comparison to seawater reverse osmosis. *Journal of Environmental Management*, 329, 117078.

Lutz, H., Wilkins, R., Carbrello, C. (2013). Sterile filtration: Principles, best practices and new developments. In *Sterile Product Development: Formulation, Process, Quality and Regulatory Considerations*, Kolhe, P., Shah, M., Rathore, N. (eds). Springer, New York, 431–459.

Ma, T., Janot, J.M., Balme, S. (2020). Track-etched nanopore/membrane: From fundamental to applications. *Small Methods*, 4, 2000366.

Maroufi, N. and Hajilary, N. (2023). Nanofiltration membranes types and application in water treatment: A review. *Sustainable Water Resources Management*, 9, 142.

Meltzer, T.H. and Jornitz, M.W. (2007). Concerning mechanisms of particle removal by filters. In *Filtration and Purification in the Biopharmaceutical Industry*, Jornitz, M.J., Jornitz, M.W., Meltzer, T.H. (eds). CRC Press, Boca Raton, 105–174.

Meng, F., Zhang, S., Oh, Y., Zhou, Z., Shin, H.-S., Chae, S.-R. (2017). Fouling in membrane bioreactors: An updated review. *Water Research*, 114, 151–180.

Mohammad, A.W., Teow, Y.H., Ang, W.L., Chung, Y.T., Oatley-Radcliffe, D.L., Hilal, N. (2015). Nanofiltration membranes review: Recent advances and future prospects. *Desalination*, 356, 226–254.

Moreira, V.R., Lebron, Y.A.R., De Souza Santos, L.V., Amaral, M.C.S. (2022). Low-cost recycled end-of-life reverse osmosis membranes for water treatment at the point-of-use. *Journal of Cleaner Production*, 362, 132495.

Moreno, M., Mazur, L.P., Weschenfelder, S.E., Regis, R.J., De Souza, R.A.F., Marinho, B.A., Da Silva, A., De Souza, S.M.A.G.U., De Souza, A.A.U. (2022). Water and wastewater treatment by micellar enhanced ultrafiltration – A critical review. *Journal of Water Process Engineering*, 46, 102574.

Mulder, M. (1996). *Basic Principles of Membrane Technology*. Springer Science & Business Media, Berlin.

Mulder, M. (2012). *Basic Principles of Membrane Technology*. Springer Science & Business Media, Berlin.

Niu, C., Li, X., Dai, R., Wang, Z. (2022). Artificial intelligence-incorporated membrane fouling prediction for membrane-based processes in the past 20 years: A critical review. *Water Research*, 216, 118299.

Oladosu, W.A., Manan, Z.A., Alwi, S.R.W. (2017). Recovery of vegetable oil from spent bleaching earth: State-of-the-art and prospect for process intensification. *Chemical Engineering Transactions*, 56, 133–138.

Pereao, O., Opeolu, B., Fatoki, O. (2020). Microplastics in aquatic environment: Characterization, ecotoxicological effect, implications for ecosystems and developments in South Africa. *Environmental Science and Pollution Research*, 27, 22271–22291.

Pinnau, I. and Freeman, B.D. (1999). Formation and modification of polymeric membranes: Overview. In *Membrane Formation and Modification*. American Chemical Society, Washington, D.C.

Purkait, M.K. and Singh, R. (2018). *Membrane Technology in Separation Science*. CRC Press, Boca Raton.

Radwan, E.K., Abdel Ghafar, H.H., Ibrahim, M.B.M., Moursy, A.S. (2023). Recent trends in treatment technologies of emerging contaminants. *Environmental Quality Management*, 32, 7–25.

Rehman, Z.U., Khojah, B., Leiknes, T., Alsogair, S., Alsomali, M. (2020). Removal of bacteria and organic carbon by an integrated ultrafiltration–nanofiltration desalination pilot plant. *Membranes,* 10, 223.

Rezai, B. and Allahkarami, E. (2021). Wastewater treatment processes – Techniques, technologies, challenges faced, and alternative solutions. In *Soft Computing Techniques in Solid Waste and Wastewater Management*, Karri, R.R., Ravindran, G., Dehghani, M.H. (eds). Elsevier, Amsterdam.

Rudolph, G., Virtanen, T., Ferrando, M., Güell, C., Lipnizki, F., Kallioinen, M. (2019). A review of in situ real-time monitoring techniques for membrane fouling in the biotechnology, biorefinery and food sectors. *Journal of Membrane Science*, 588, 117221.

Sadeghi, F., Ajji, A., Carreau, P.J. (2007). Analysis of microporous membranes obtained from polypropylene films by stretching. *Journal of Membrane Science*, 292, 62–71.

Sadeghi, F., Ajji, A., Carreau, P.J. (2008). Microporous membranes obtained from polypropylene blends with superior permeability properties. *Journal of Polymer Science Part B: Polymer Physics*, 46, 148–157.

Sadeghi, F., Tabatabaei, S.H., Ajji, A., Carreau, P.J. (2009). Effect of PVDF characteristics on extruded film morphology and porous membranes feasibility by stretching. *Journal of Polymer Science Part B: Polymer Physics*, 47, 1219–1229.

Sarada, T., Sawyer, L.C., Ostler, M.I. (1983). Three dimensional structure of celgard® microporous membranes. *Journal of Membrane Science*, 15, 97–113.

Scott, K. (1995). *Handbook of Industrial Membranes*. Elsevier, Amsterdam.

Shahid, M.K., Mainali, B., Rout, P.R., Lim, J.W., Aslam, M., Al-Rawajfeh, A.E., Choi, Y. (2023). A review of membrane-based desalination systems powered by renewable energy sources. *Water*, 15, 534.

Shiklomanov, I.A. (1991). *The World's Water Resources*. UNESCO, Paris.

Shirazi, M.M.A., Kargari, A., Shirazi, M.J.A. (2012). Direct contact membrane distillation for seawater desalination. *Desalination and Water Treatment*, 49, 368–375.

Siddique, J.A. (2023). Nanofiltration membrane use for separation of heavy metals from wastewater. In *Emerging Techniques for Treatment of Toxic Metals from Wastewater*, Ahmad, A., Kumar, R., Jawaid, M. (eds). Elsevier, Amsterdam.

Singh, R. (2006). *Hybrid Membrane Systems for Water Purification: Technology, Systems Design and Operations*. Elsevier, Amsterdam.

Smithers, G.W. (2008). Whey and whey proteins – From 'gutter-to-gold'. *International Dairy Journal*, 18, 695–704.

Sonune, A. and Ghate, R. (2004). Developments in wastewater treatment methods. *Desalination*, 167, 55–63.

Soren, S., Panda, P., Chakroborty, S. (2023). Nanotechnology in water and wastewater treatment. In *Agricultural and Environmental Nanotechnology: Novel Technologies and their Ecological Impact*, Fernandez-Luqueno, F. and Patra, J.K. (eds). Springer, Berlin.

Stamatialis, D.F., Papenburg, B.J., Gironés, M., Saiful, S., Bettahalli, S.N.M., Schmitmeier, S., Wessling, M. (2008). Medical applications of membranes: Drug delivery, artificial organs and tissue engineering. *Journal of Membrane Science*, 308, 1–34.

Tabatabaei, S.H., Carreau, P.J., Ajji, A. (2008). Microporous membranes obtained from polypropylene blend films by stretching. *Journal of Membrane Science*, 325, 772–782.

Tanudjaja, H.J., Anantharaman, A., Ng, A.Q.Q., Ma, Y., Tanis-Kanbur, M.B., Zydney, A.L., Chew, J.W. (2022). A review of membrane fouling by proteins in ultrafiltration and microfiltration. *Journal of Water Process Engineering*, 50, 103294.

Taylor, G.I. (1969). Electrically driven jets. *Proceedings of the Royal Society of London. A. Mathematical and Physical Sciences*, 313, 453–475.

Tijing, L.D., Pak, B.C., Lee, D.H., Cho, Y.I. (2008). Heat-treated titanium balls for the mitigation of mineral fouling in heat exchangers. *Experimental Heat Transfer*, 21, 115–132.

Tijing, L.D., Woo, Y.C., Johir, M.A.H., Choi, J.-S., Shon, H.K. (2014). A novel dual-layer bicomponent electrospun nanofibrous membrane for desalination by direct contact membrane distillation. *Chemical Engineering Journal*, 256, 155–159.

Tijing, L.D., Woo, Y.C., Choi, J.-S., Lee, S., Kim, S.-H., Shon, H.K. (2015). Fouling and its control in membrane distillation – A review. *Journal of Membrane Science*, 475, 215–244.

Van Reis, R. and Zydney, A. (2001). Membrane separations in biotechnology. *Current Opinion in Biotechnology*, 12, 208–211.

Wakeman, R.J. and Williams, C.J. (2002). Additional techniques to improve microfiltration. *Separation and Purification Technology*, 26, 3–18.

Werber, J.R., Osuji, C.O., Elimelech, M. (2016). Materials for next-generation desalination and water purification membranes. *Nature Reviews Materials*, 1, 1–15.

Wu, Q., Chen, C., Zhang, Y., Tang, P., Ren, X., Shu, J., Liu, X., Cheng, X., Tiraferri, A., Liu, B. (2023). Safe purification of rural drinking water by biological aerated filter coupled with ultrafiltration. *Science of the Total Environment*, 868, 161632.

Xiang, H., Min, X., Tang, C.-J., Sillanpää, M., Zhao, F. (2022). Recent advances in membrane filtration for heavy metal removal from wastewater: A mini review. *Journal of Water Process Engineering*, 49, 103023.

Yadav, D., Karki, S., Ingole, P.G. (2022). Nanofiltration (NF) membrane processing in the food industry. *Food Engineering Reviews*, 14, 579–595.

Yarin, A.L., Koombhongse, S., Reneker, D.H. (2001). Taylor cone and jetting from liquid droplets in electrospinning of nanofibers. *Journal of Applied Physics*, 90, 4836–4846.

Yiantsios, S.G. and Karabelas, A.J. (1998). The effect of colloid stability on membrane fouling. *Desalination*, 118, 143–152.

Yigit, N.O., Harman, I., Civelekoglu, G., Koseoglu, H., Cicek, N., Kitis, M. (2008). Membrane fouling in a pilot-scale submerged membrane bioreactor operated under various conditions. *Desalination*, 231, 124–132.

Yuksekdag, A., Korkut, S., Kaya, R., Pasaoglu, M.E., Turken, T., Agtas, M., Ersahin, M.E., Ozgun, H., Koyuncu, I. (2023). Upgrading of conventional water treatment plant by nanofiltration for enhanced organic matter removal. *Separation and Purification Technology*, 325, 124766.

Zhang, R., Liu, Y., He, M., Su, Y., Zhao, X., Elimelech, M., Jiang, Z. (2016). Antifouling membranes for sustainable water purification: Strategies and mechanisms. *Chemical Society Reviews*, 45, 5888–5924.

Zhang, Q., Wang, N., Nie, L., Lu, C., Chen, H., He, W., Li, M., Wang, Y., Zhao, J., Xiong, J. (2023). Visit-to-visit ultrafiltration volume variability predicts all-cause mortality in patients receiving hemodialysis. *Renal Failure*, 45, 2194439.

13

Pervaporation

During the last few decades, the membrane separation process has become one of the emerging technologies to resolve several industrial issues. Membrane processes like ultrafiltration, micro-filtration, nano-filtration and reverse osmosis have well-established domestic and industrial applications. Owing to its nondestructive and energy-efficient nature, advanced membrane processes such as pervaporation have recently received huge interest. The separation depends on the preferential transport of one of the components through a specific membrane. Apart from the common applications such as the dehydration of organic solvents and removal of dilute organic compounds from aqueous or organic mixtures, the pervaporation process can break an azeotropic mixture without adding any other components (as required in the azeotropic distillation process). This chapter will cover the basics of membrane processes with an emphasis on pervaporation for the separation of azeotropic mixture.

13.1. Introduction

Separation processes play a crucial role in the chemical industry by purifying raw materials, recovering products with desired purity and mitigating pollution by treating waste streams released into the environment. Membrane separation processes, known for their significance in water desalination, effluent treatment, food production, pharmaceuticals, gas separations and liquid mixture separation, encompass various methods classified based on membrane usage, driving forces and feed solution characteristics (Zhou 2005). Pressure-driven processes like reverse osmosis (RO), ultrafiltration (UF), nanofiltration (NF) and microfiltration (MF) are well established, while newer techniques such as gas separation and pervaporation (PV) have emerged (Mandal and Pangarkar 2002).

Chapter written by Tapas PALAI, Abhishek KUMAR and Leela Manohar AESHALA.

Pervaporation stands out as a promising alternative to several separation processes due to its energy-saving one-step nature. When the appropriate membrane material is chosen, PV can effectively separate azeotropic and close-boiling mixtures, addressing challenges faced by traditional conventional processes (Mandal and Pangarkar 2003).

Membranes act as molecular-level barriers that allow the preferential passage of one component over another. Various materials like polymers, carbon molecular sieves and zeolites can be employed to create membranes.

While standard separation methods for organic compounds from aqueous streams are available, the industry faces challenges in separating organic–organic systems, particularly in polar–nonpolar mixtures (Greenlaw et al. 1977; Kulprathipanja et al. 1992; Okada and Matsuura 1991), aromatic/aliphatic combinations (Razzak et al. 1999; Devi et al. 2006; Guo et al. 2006) and the separation of aromatic/alicyclic mixtures (Mansur et al. 2004; Lewandowska 2005). The main focus of this chapter will be on PV process.

However, basic membrane processes are summarized in the subsequent section for a better understanding of membrane processes.

13.2. Basics of membrane separation processes

A membrane is a physical barrier (may be solid or liquid), placed between two phases or media, that separates selectively various components from a mixture. The stream passing through the membrane is known as permeate and the rejected stream is known as retentate as shown in Figure 13.1. Polymers are most widely used as the membrane material. However, in recent times, ceramic, glass, metal and liquid materials are also being used as membrane materials based on their specific uses. Membranes can be neutral or charged materials.

Separation in the membrane occurs due to the differences in the transport rates of chemical species through it. The driving forces for membrane processes may be concentration gradient, pressure gradient, temperature and electrical potential difference. In some cases, the transport of molecules may occur because of more than one driving forces. Hence, the chemical potential gradient is the driving force.

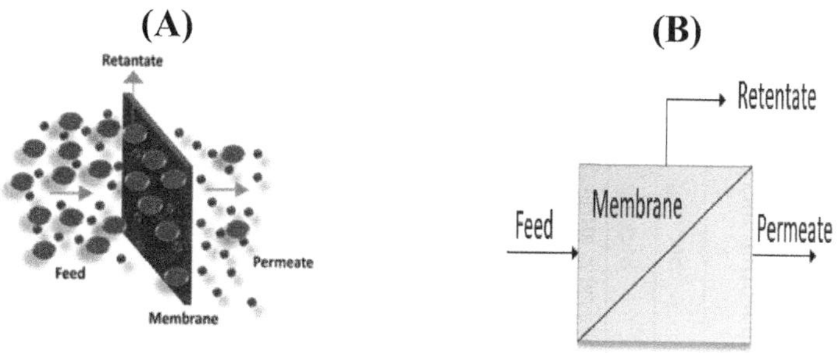

Figure 13.1. *(A) Process diagram of membrane separation. (B) Schematic presentation of membrane separation. For a color version of this figure, see www.iste.co.uk/kumararya/chemicalengineering1.zip*

Membrane processes have various advantages and disadvantages compared to other conventional processes. Some of these are presented in Table 13.1.

Advantages	Disadvantages
1. Inherently simple process. 2. Works at ambient or moderate temperature. 3. Suitable for heat-sensitive materials. 4. Efficient and faster process. 5. Nondestructive process, i.e. components are not damaged or chemically altered. 6. Cleaner and more energy-efficient route for high-quality products. 7. Easy to scale up.	1. Membranes are prone to fouling and flux is gradually declined. 2. Limited chemical resistance to a wide range of pH and organic solvents (polymer). 3. Lower life of the membrane.

Table 13.1. *Advantages and disadvantages of membrane processes (Saleh and Gupta 2016)*

13.2.1. *Classification of membrane processes*

The selectivity of a membrane process depends on the membrane pore size, driving forces, transport mechanism, etc. Based on the pore size, it can be classified as microfiltration (MF), ultrafiltration (UF), nanofiltration (NF) and reverse osmosis (RO). Furthermore, based on the driving forces, membrane processes may be classified as pressure-driven (e.g. RO, UF, NF, MF), concentration-driven (e.g. dialysis, liquid membrane), temperature-driven (e.g. membrane distillation) and electrical potential-driven (e.g. electro-dialysis). A summary of various membrane processes is given in the Figure 13.2 and Table 13.2.

Membrane process	Membrane type	Pore size	Driving force	Separation mechanism
Microfiltration	Microporous	> 100 nm	Pressure difference (< 2 bar)	Sieving
Ultrafiltration	Microporous	2–100 nm	Pressure difference (1–10 bar)	Sieving
Nanofiltration	Charged or neutral	1–2 nm	Pressure difference (3–20 bar)	Sieving, attraction or repulsion
Reverse osmosis	Homogeneous/dense	< 1 nm	Pressure difference (10–80 bar)	Solution-diffusion-desorption
Dialysis	Homogeneous/microporous	0.1–10 nm	Concentration difference	Sieving, solution-diffusion
Gas separator	Sintered	500–10000 nm	Pressure difference	Sorption, diffusion, Knudsen diffusion
Pervaporation	Homogeneous/dense	0.1–10 nm	Activity differences	Solution-diffusion-desorption
Electrodialysis	Ion exchange membrane	-	Electrical potential gradient	Migration
Liquid membrane	-	-	Concentration gradient	Solution-diffusion-desorption

Table 13.2. *Summary of various membrane processes (Dutta 2007)*

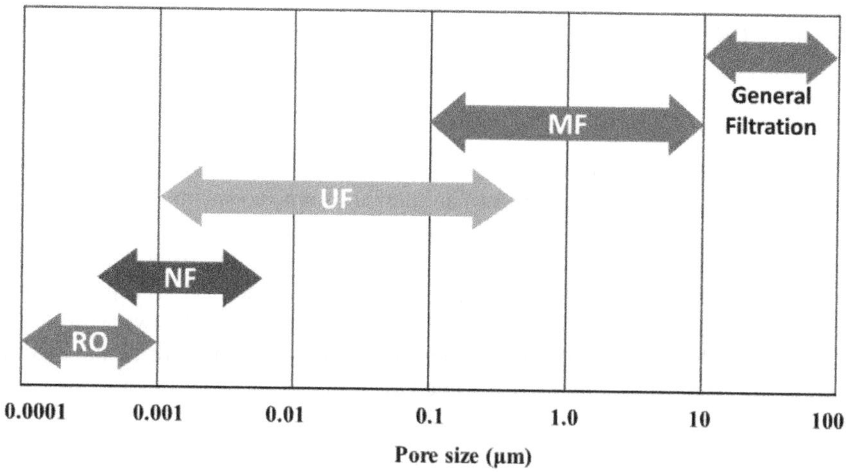

Figure 13.2. *Membrane process classification based on pore size. For a color version of this figure, see www.iste.co.uk/kumararya/chemicalengineering1.zip*

13.2.2. *Conventional membrane separation processes*

The main methods used in many industries to separate and purify components from mixtures, sometimes including fluids or gases, are conventional membrane separation procedures. These processes make use of semipermeable membranes, which limit or allow some compounds to flow through depending on their size, shape or chemical makeup. By acting as barriers, these membranes make it possible to separate components in a mixture without the need for a lot of heat or chemical processes.

Subsequent sections outline a comprehensive overview of conventional membrane separation processes (Kavitha et al. 2022).

13.2.2.1. *Reverse osmosis (RO)*

One of the well-known membrane separation methods that is usually used for desalination and water purification is RO. Osmosis is a natural phenomenon where the solvent molecules (water) pass through a semi-permeable membrane from side A to side B, as shown in Figure 13.3, because of the higher chemical potential of water and rejects salts, bigger pollutants, ions and minerals. Hence, the level on the solution side (side B) keeps on rising till the osmotic equilibrium is achieved. The osmotic pressure (π) is given by the Vant-Hoff equation for a dilute solution and the Virial-type equation for the concentrated solution as per equations [13.1] and [13.2], respectively (De 2011).

$$\pi = CRT \qquad [13.1]$$

$$\pi = B_1C + B_2C^2 + B_3C^3 \qquad [13.2]$$

where C is the molar concentration of salt, R is the ideal gas constant, T is the absolute temperature, and B_1, B_2 and B_3 are the virial-type constants.

If an external pressure slightly higher than the equilibrium osmotic pressure is applied on the solution side (side B), the water molecules get transported from side B to A leading to the purification of water. The process is known as RO. The RO is exclusively used for the desalination of seawater to produce drinking water in places where freshwater supplies are limited (Tawalbeh et al. 2023).

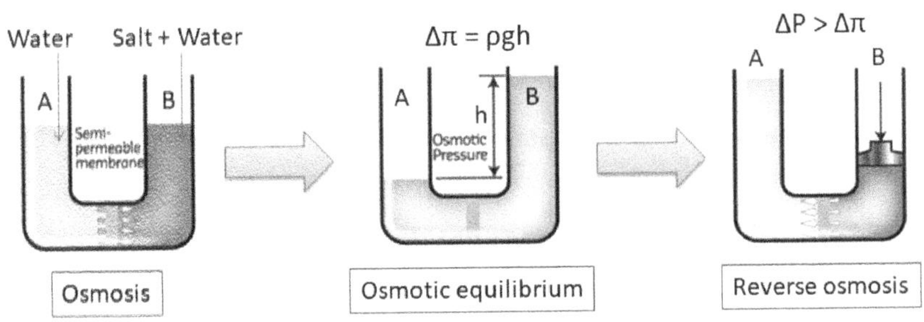

Figure 13.3. *Schematic presentation of osmosis and reverse osmosis processes. For a color version of this figure, see www.iste.co.uk/kumararya/chemicalengineering1.zip*

13.2.2.2. *Ultrafiltration (UF)*

Ultrafiltration (UF) is a pressure-driven membrane process that is used to separate dissolved solids and colloidal particles. The membranes are porous having pore size 2–100 nm. UF works on the principle of sieving action or diffusion based on the size of the pores. The mechanism of solute transport through the UF membrane depends on the pore size.

a) Diffusive ultrafiltration:

In this UF process, the pore size of the membrane is 10–20°A. The transport of solute mainly occurs by diffusion mechanism. The solute diffusion rate through the membrane is much smaller than that of the solvent. The pore flow model (Hagen–Poiseuille equation) may be applicable for calculating the permeate flux as given in equation [13.3]

$$J_p = \frac{\varepsilon_m d_p^2 \Delta P}{32\bar{V}\mu l} \quad [13.3]$$

where J_p is the permeate flux, ε_m is the membrane porosity, D_p is the average pore diameter, ΔP is the transmembrane pressure, μ is the viscosity, $\bar{V}$ is the specific molar volume, l is the average pore length and L_m is the membrane permeability.

b) Microporuous ultrafiltration:

Microporous UF, on the other hand, works on the principle of a sieving mechanism. The pore size of the membrane is higher than the diffusive UF. The osmotic pressure model can be useful for the estimation of permeate flux, as per equation [13.4].

$$J_p = \frac{1}{\mu R_m}(\Delta P - \Delta\pi) \quad [13.4]$$

where R_m is the membrane hydraulic resistance and $\Delta\pi$ is the osmotic pressure difference between feed and permeate side.

In general, $\Delta\pi$ is negligible in UF as the solutes are generally macromolecular in nature and have very little contribution to the osmotic pressure.

Major applications of UF include recovery of paint, latex, polyvinyl acetate (PVA), lignosulfates from spent sulfite liquor in the paper industry, reduction of color in caustic bleach effluent in pulp and paper industries, and separation of oil-water emulsion. Application of UF in the food industry is very common. The concentration of milk and recovery of protein from cheese whey, separation and recovery of protein and enzymes, potato starch, high-value products, wastewater treatment, etc. are a few examples. More details are available in other studies (for example Cheryan 1998; Dutta 2007).

Proteins and colloidal particles are among the bigger molecules that are separated by UF. It is used in the food and beverage sector to remove microorganisms from beverages and concentrate the proteins in dairy products (Kammakakam and Lai 2023).

13.2.2.3. *Microfiltration (MF)*

Microfiltration (MF) is a pressure-driven membrane process that is used to separate dissolved solids and colloidal particles where membrane pore size is 100–1000 nm. Larger pore diameters, which are characteristic of microfiltration, enable the separation of particles like bacteria and suspended particulates. MF works on the principle of a sieving mechanism. For laminar flow through pores, the flux may be estimated using the Hagen–Poiseuille equation as given by equation [13.3]. Details are available in other studies (see, for example, Dutta 2007).

MF finds its application in food and beverage, dairy, pharmaceutical industries and wastewater treatment. It is used for removing contaminants and clarifying liquids in the biotechnology and pharmaceutical sectors (Tanudjaja et al. 2022).

13.2.2.4. *Gas separation*

Traditional membrane separation techniques are not limited to liquid separations and may also be applied to gas separations. Some membranes have exceptional gas selectivity, making them useful for carbon dioxide collection, hydrogen purification and natural gas processing.

Common materials for membranes are polymer, ceramic and metallic materials, which are selected for their durability, affordability and adaptability. The problem of fouling is solved by using cutting-edge materials and cleaning methods.

Constant advancements employing molecular-level tailoring and nanomaterials aim to improve durability, permeability and selectivity. Traditional membrane systems use less energy than alternatives, which is in line with the world's sustainability objectives, widely used in food and beverage production for liquid concentration and purification, petrochemicals for gas separation and pharmaceuticals for medication purification essential to the recycling and reuse of wastewater, supporting environmental stewardship. Process control will be improved by integration with automation and sophisticated sensing. Inspired by natural systems, biomimetic membranes have the potential to achieve significant improvements in selectivity and efficiency.

Traditional membrane separation techniques provide sophisticated answers to separation problems in a variety of sectors. The combination of cutting-edge technology and biomimetic methodologies shows promise for even more effective and long-lasting applications as research continues (Sidhikku Kandath Valappil 2021).

13.2.3. *Advanced membrane separation processes*

Membrane separation is currently being researched across the Globe. Apart from the conventional processes, several membrane processes have been developed, such as dialysis, electro-dialysis, vapor permeation, PV, membrane distillation, membrane chromatography, membrane contactor, liquid membrane, etc. The subsequent section focus on pervaporation (PV).

13.3. Pervaporation

Pervaporation (PV) is a combination of two processes: selective permeation of one component through a membrane followed by evaporation ((Per)meation + E(vaporation) = Pervaporation). PV was first observed accidentally by Philip Adolph Kober in 1917 during a dialysis experiment (Kober 1917). However, PV received huge interest in the late 1990s. PV is a membrane separation process where feed is taken on the upstream side of the membrane and the downstream side is kept under vacuum. With its unique method of separating liquid mixtures, PV works based on selective vapor transport across specific membranes. The secret to PV's effectiveness is its capacity to take advantage of variations in vapor pressure among liquid mixture constituents. The more volatile component can selectively evaporate and pass through a dense membrane, which functions as a molecular sieve, leaving the less volatile component behind.

The thickness, porosity and surface chemistry of the membrane, as well as its design, play a major role in this process. Tailored affinities in thin, dense membranes increase the effectiveness of selective permeation and enable accurate component separation (Castro-Muñoz 2020).

PV has several advantages over conventional membrane separation processes (Pabby et al. 2009):

1. The azeotropic mixture can easily be separated by using PV with an appropriate membrane.

2. Entrainers are not required as in the case of azeotropic distillation. Hence, downstream separation is not required. This decreases the operating cost as well as environmental pollution.

3. There are lower energy requirements as compared to other conventional separation processes.

4. It is easy to operate and requires less floor space.

5. There is flexibility in operation and easy scale-up.

6. There are properties of the membrane that can be adjusted as per the requirement.

7. PV can easily be combined with other separation processes to make a more economical hybrid membrane process.

13.3.1. *Theory of pervaporation*

Overall PV process consists of several transport and physical processes. It involves a sequence of three major steps:

- selective sorption of the desired component from the mixture;
- selective diffusion through the membrane;
- desorption into a vapor phase on the permeate side.

The process of sorption, mass transport and desorption due to evaporation is complicated. No such universally accepted mechanism exists to explain the mechanism. Furthermore, the permeation flux depends on the properties of the membranes, such as thickness and porosity, affinity for the targeted components, diffusivity of permeate components across membranes, etc. (Peng et al. 2021). Based on the mass transfer mechanism, several models have been proposed to explain the PV process such as physical mechanism (molecular sieving or capillary flow, solution-diffusion) and chemical mechanism (facilitated transport). Some of them are discussed in the subsequent sections.

13.3.1.1. *Facilitated transport mechanism*

Facilitated transport, also known as carrier-mediated transport, is a chemical mechanism. Membrane pores are filled with carriers. The carrier may be classified into three categories: a) free moving, b) semi-mobile and c) fixed site. The preferential permeation of a species across the membrane occurs due to a reversible chemical reaction between the permeating species and the carrier, diffusion of the reacted species and desorption of the target species via reverse reaction. Thus, the carrier remains in the membrane and the targeted species gets transported to the other side of the membrane (Peng et al. 2021). In PV, the reversible chemical reactions occur due to π-complexation, π–π interaction and host–guest complexation (Song et al. 2019).

Figure 13.4 depicts the facilitated transport mechanism during PV. The target molecule (A) present in the feed mixture (containing A & B) gets preferentially sorbed on the membrane and reacts reversibly with the carrier (C) to form reacted species/complex (AC) as per the following reaction equation [13.5].

$$A + C \rightleftharpoons AC \tag{13.5}$$

The molecule AC is then diffused to the downstream side of the membrane because of the concentration gradient. On the downstream side of the membrane, the reverse reaction occurs and A is desorbed from the membrane.

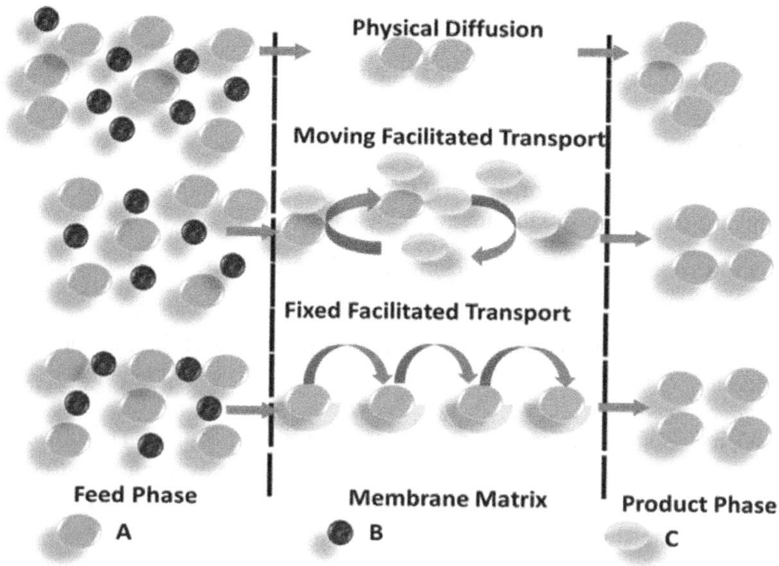

Figure 13.4. *Different mass transport mechanisms during the PV process. For a color version of this figure, see www.iste.co.uk/kumararya/chemicalengineering1.zip*

13.3.1.2. *Preferential sorption–capillary flow or molecular sieving model*

The model was proposed by Sourirajan and Shiyao in 1987 (Pabby et al. 2009). This is a physical transport mechanism where the membrane acts as a porous barrier. The extent of separation of rapidly permeating molecules depends on the effective sizes of the permeant, pore size and distribution of the membrane. The separation occurs on the basis of the molecular sieving mechanism. The rapidly permeating solute is first preferentially sorbed on the upstream section followed by flow through the membrane matrix (capillaries) and finally gets evaporated at the downstream side of the membrane.

The model assumes the pores are straight and cylindrical. The target molecule first preferentially sorbed on the membrane surface due to the interaction between the membrane and the permeant molecule. The molecules having a size lower than the pore size get transported through the cylindrical pores and desorbed at the pore's outlet by vaporization. A vapor–liquid equilibrium is established at the pore's outlet as shown in Figure 13.5A.

13.3.1.3. *Solution–diffusion mechanism*

The solution–diffusion mechanism is the most widely accepted for the transport of mass across the dense membrane during PV. The overall process occurs in three successive steps (Pabby et al. 2009):

a) Preferential sorption of the target molecule from the feed liquid mixture in the membrane matrix (upstream side).

b) Diffusion of the sorbed solutes through the membrane under a concentration gradient.

c) Desorption of the permeants at the downstream side of the membrane in the vapor phase.

The first two steps generally control the effective separation. The downstream side is generally kept under a vacuum which facilitates the evaporation of permeate at a lower temperature than the normal boiling point of the permeant. Figure 13.5B depicts the mechanism of the PV process.

The mass transport of permeant (component A) through the membrane during PV is generally described by Fick's law of diffusion (equation [13.6])

$$J_A = -D_{Am}\frac{dC_{Am}}{dz} \qquad [13.6]$$

where D_{Am} is the diffusion coefficient of A in the membrane, J_A is the flux of A and C_{Am} is the molar concentration of A in the membrane. Diffusion coefficients of penetrating species through polymers, in general, depend strongly on the state of swelling of the polymer. There are several relationships to describe the concentration dependency of the diffusion coefficient (equations [13.7]–[13.9]).

a) Constant diffusion coefficient

$$D_A = D_A^0 \qquad [13.7]$$

b) Linear dependent diffusion coefficient

$$D_A = D_A^0(A_{AA}C_{Am} + A_{AB}C_{Bm}) \qquad [13.8]$$

c) Exponential dependent diffusion coefficient

$$D_A = D_A^0 \exp(A_{AA}C_{Am} + A_{AB}C_{Bm}) \qquad [13.9]$$

As the membrane is very perm-selective (specifically separates one molecule from the mixture), convective flux through the membrane can be neglected compared to the diffusional flux. So, the mass transport of component A through the membrane for a binary system of A and B can be expressed as per equation [13.10].

$$N_A = J_A = -D_{Am}\frac{dC_{Am}}{dz} \qquad [13.10]$$

Substituting equation [13.9] into equation [13.10], we have

$$J_A = -D_{Am}^0 \exp\left[A_{AA}C_{Am} + A_{AB}C_{Bm}\right]\frac{dC_{Am}}{dz} \qquad [13.11]$$

$$J_A dz = -D_{Am}^0 \exp\left[A_{AA}C_{Am} + A_{AB}C_{Bm}\right]dC_{Am} \qquad [13.12]$$

On the integration of equation [13.12] across the membrane thickness (δ)

$$J_A\int_0^{\delta} dz = -D_{Am}^0 \exp\left(A_{AB}C_{jm}^F\right)\int_{C_{Am}^F}^{C_{Mm}^P} \exp\left(A_{AA}C_{Am}\right)dC_{Am} \qquad [13.13]$$

$$J_A\delta = -D_{Am}^0 \exp\left(A_{AB}C_{Bm}^F\right)\left[\frac{\exp\left(A_{AA}C_{Am}\right)}{A_{AA}}\right]_{C_{Am}^F}^{C_{Am}^P} \qquad [13.14]$$

$$J_A = \frac{-D_{Am}^0 \exp\left(A_{AB}C_{Bm}^F\right)}{A_{AA}\delta}\left[\exp\{A_{AA}(C_{Am}^P - C_{Am}^F)\}\right] \qquad [13.15]$$

The superscript “F” denotes the feed side and “P” denotes the permeate side of the membrane.

The activity coefficients in the liquid phase and the membrane phase may be written as in equations [13.16] and [13.17], respectively.

$$\gamma_{Am} C_{Am} = \gamma_A x_A \text{ for component A} \quad [13.16]$$

$$\gamma_{Bm} C_{Bm} = \gamma_B x_B \text{ for component B} \quad [13.17]$$

where, γ_A and γ_B are the activity coefficients of A and B, respectively, in the liquid phase, γ_{Am} and γ_{Bm} are in the membrane phase and x is the mol fraction of the respective component.

These coefficients are assumed to remain constant across the membrane (ideality of swelling). Interface equilibria are given by equations [13.18]–[13.21] (Smitha et al. 2004; Shao and Huang 2007).

$$\gamma_{Am} C_{Am}^F = \gamma_A^F x_A^F \quad [13.18]$$

$$\gamma_{Bm} C_{Bm}^F = \gamma_B^F x_B^F \quad [13.19]$$

$$\gamma_{Am} C_{Am}^P = \frac{P_A^P}{P_A^{Sat}} \exp\left[-\frac{\bar{V}_A \left(P^F - P_A^{Sat} \right)}{RT} \right] \quad [13.20]$$

$$\gamma_{Bm} C_{Bm}^P = \frac{P_B^P}{P_B^{Sat}} \exp\left[-\frac{\bar{V}_B \left(P^F - P_B^{Sat} \right)}{RT} \right] \quad [13.21]$$

$\left(P^F - P_A^{Sat} \right)$ for the upstream is very less as no extra pressure is applied. Hence, the exponential terms are close to 1. Thus, exponential terms may be neglected. However, it cannot be neglected for the permeate side. The flux may be written as equation [13.22].

$$J_A = -\frac{D_{Am}^0 \exp\left(A_{AB} C_{Am}^F \right)}{A_{AA} \delta} \left[\exp\left\{ -\frac{A_{AA}}{\gamma_{Am}} \left(\gamma_A^F x_A^F - \frac{P_A^P}{P_A^{Sat}} \right) \right\} \right] \quad [13.22]$$

Since $P_A^P << P_A^{Sat}$, the term $\left(P_A^P / P_A^{Sat} \right)$ may be neglected.

$$J_A = -\frac{D_{Am}^0 \exp\left(A_{AB} C_{Bm}^F\right)}{A_{AA}\delta} . \exp\left\{ -\frac{A_{AA}}{\gamma_{Am}} \gamma_A^F x_A^F \right\} \qquad [13.23]$$

$$J_A = -\frac{D_{Am}^0}{A_{AA}\delta} . \exp\left[A_{AB} \frac{\gamma_B^F x_B^F}{\gamma_{Bm}} - A_{AA} \frac{\gamma_A^F x_A^F}{\gamma_{Am}} \right] \qquad [13.24]$$

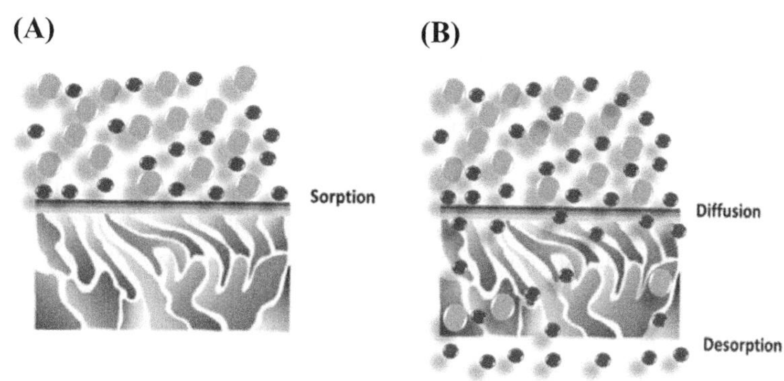

Figure 13.5. *Schematic of solution–diffusion mechanism: (A) sorption and (B) diffusion and desorption. For a color version of this figure, see www.iste.co.uk/kumararya/chemicalengineering1.zip*

13.3.2. *Selectivity of pervaporation*

The selectivity of PV is represented in terms of the separation factor. It is defined as given in equation [13.25] (Smitha et al. 2004).

$$\alpha_{AB} = \frac{x_A^P / x_B^P}{x_A^F / x_B^F} \qquad [13.25]$$

The value of α may vary from 0 to infinity. Higher values of α indicate better selectivity of the membrane. It can also be represented in terms of the component permeate fluxes as per equation [13.26].

$$\alpha_{AB} = \frac{J_A x_B^F}{J_B x_A^F} \qquad [13.26]$$

The performance of PV can be expressed by the pervaporation separation index (PSI) defined as the product of normalized flux and selectivity above 1 and given by equation [13.27] (Mandal et al. 2008, 2011).

$$PSI = J_A(\alpha_{AB} - 1) \qquad [13.27]$$

13.3.3. *Pervaporation laboratory scale set-up*

A setup was designed and developed for PV experimental investigations as shown in Figure 13.6 by Thiyagarajan et al. (2011). The test cell is made of glass having provision to house the membrane on a porous steel support that divides the cell into two halves. The feed is taken on the upper side of the cell, and the permeate is collected from the lower side (downstream side).

The whole cell is externally heated to maintain isothermal conditions. The pressure of the upstream side is maintained at 1 atm. and the downstream side is under vacuum with a vacuum pump. Permeated vapor is condensed in a Dewar flask containing liquid nitrogen.

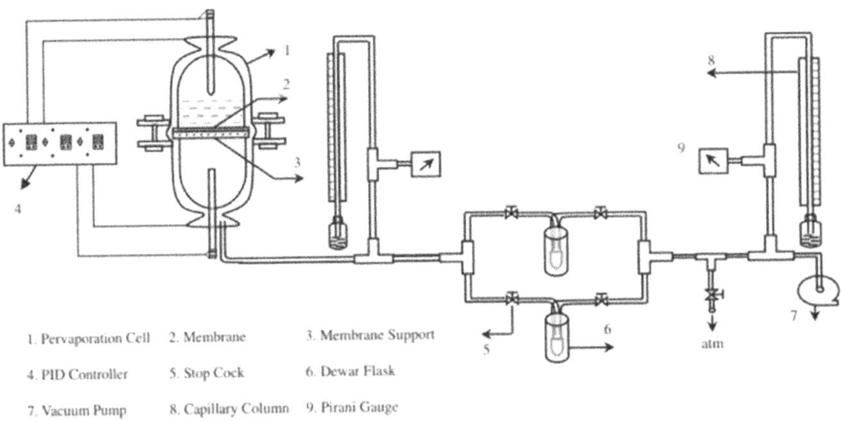

Figure 13.6. *Schematic of pervaporation laboratory set-up. Reproduced with permission (Thiyagarajan et al. 2011)*

13.3.4. *Applications*

Pervaporation has several useful and significant uses as shown in Figure 13.7. One noteworthy use is in the dehydration of organic solvents, a process that presents difficulties for more conventional techniques like distillation. The process of PV is very useful in sectors like chemicals and pharmaceuticals since it effectively removes water from solvents, guaranteeing high-quality goods. Additionally, the method is useful for recovering volatile organic compounds (VOCs) from liquid streams, which reduces emissions and permits resource reuse, both of which promote environmental sustainability. The principles behind PV's operation highlight its potential as a creative and effective molecular separation technology, providing a sophisticated response to the problems encountered by traditional separation techniques. Research is still being done to improve membrane design, increase selectivity and stability, and increase the range of industrial uses for PV (Liu and Jin 2021).

With its exceptional capacity to extract specific constituents from liquid mixtures, PV has found numerous and significant uses in a wide range of sectors. This membrane separation technology is a transformational solution in numerous important fields due to its accuracy and efficiency.

13.3.4.1. *Azeotropic mixture separation*

PV's accuracy overcomes azeotropic mixture, which are difficult to separate using traditional separation techniques. The technique is a useful tool in sectors like chemicals and petrochemicals where azeotropic problems are common since it can transport the more volatile component across a membrane selectively (Yang Kong et al. 2022).

A strong and accurate way for resolving the issues presented by azeotropic mixtures (with constant boiling temperatures) that obstruct separation using conventional techniques: PV. PV's special capacity to move a single component across a membrane in a selective manner depending on variations in vapor pressure has made it a game-changer for businesses struggling with azeotropic separations.

The idea of selective vapor movement is fundamental to azeotropic separation by PV. The constant boiling temperatures of the constituents make it difficult to separate the components in azeotropic mixtures using conventional techniques like distillation. However, in PV, a membrane with particular affinity properties is used to take advantage of the mixture's vapor–liquid balance. The more volatile component preferentially permeates through this membrane, which functions as a molecular filter, keeping the less volatile component out of sight. The interactions between the membrane and the mixture's constituents affect this selective permeation, highlighting the significance of membrane material selection and design. PV's extraordinary adaptability in azeotropic separations is one of its many benefits. It can be used in a variety of sectors, including environmental engineering, medicines, chemicals and petrochemicals. For instance, separating azeotropic mixtures is a typical problem in the chemical industry while producing high-purity compounds. An effective substitute for distillation's energy-intensive needs is PV, which guarantees the intended separation, and likewise, in the petrochemical industry, the hydrocarbon mixture separation.

The correct selection of membrane materials is one of the most important factors impacting the effectiveness of PV in azeotropic separation. For the targeted components, the membrane must have the ideal ratio of permeability to selectivity. By reducing mass transfer resistance and enabling the preferential transport of the more volatile component, a thin, and dense membranes with customized surface qualities improve separation efficiency (Zhang et al. 2022).

PV also has advantages over separation effectiveness. It uses less energy and costs less to run since it operates at lower temperatures than conventional techniques. The process may also be designed in a compact and modular manner, which makes it appropriate for both small- and large-scale industrial applications (Castro-Muñoz et al. 2022).

There are noticeable effects of PV in azeotropic separation on the environment. Comparing this process to traditional separation techniques results in a smaller carbon footprint due to the decreased energy consumption. Recovering and reusing useful components from azeotropic mixtures supports sustainable practices by encouraging waste reduction and resource conservation.

To sum up, PV is a revolutionary concept in the field of azeotropic separation. Because of its accuracy, effectiveness and adaptability, it is a priceless instrument for sectors dealing with complicated mixes that have continuous boiling points. The use of PV in azeotropic separations is set to grow as research into membrane technology and process parameter optimization continues, providing fresh opportunities for improving separation procedures and spurring innovation in a range of industrial domains (Dong et al. 2020).

13.3.4.2. *Recovery of volatile organic compounds (VOCs)*

By making it easier to extract volatile organic compounds (VOCs) from liquid streams, PV is a critical component of environmental stewardship. This application helps industries like paint and coating manufacture, which allows for the recycling of important resources and a reduction in emissions (Fatima et al. 2022). VOCs present in the food and beverages are to be separated for better quality and shelf life. The most common VOCs are methanol, ethanol, propanol, ethyl acetate, acetonitrile, phenol, acetaldehyde, acetone, etc. (Du et al. 2021). Removal of these valuable VOCs from the aqueous phase can be made by various conventional separation techniques such as adsorption, solvent extraction, distillation, etc. However, most of the methods have several limitations. PV could be a good alternative for selective removal and recovery of VOCs more specifically from very dilute solutions. Polydimethylsiloxane (PDMS) is the most investigated polymer membrane for separation of VOCs. Besides this, polyvinyl alcohol (PVA), polydivinylbenzene (PDVB)-coated PDMS, zeolite ZSM-5 and polyether block amide (PEBA) are the other polymers that have been tested for VOC separation (Du et al. 2021).

13.3.4.3. *Wastewater treatment*

PV can be used for the dehydration of organic solvents, removal of organic compounds from aqueous solutions, removal of VOCs such as benzene, toluene, xylene, phenols, styrene, etc. and separation of organic–organic mixture. The concentration and recovery of important components in wastewater treatment is greatly aided by PV. In line with sustainability objectives, it helps lower the amount of wastewater produced and permits the repurposing of treated water in industrial operations (Dutta 2007).

Currently, multistage flash (MSF) distillation and RO are the most widely used desalination technologies for commercial purposes. However, these techniques have high-power consumption and low-heat transfer efficiency. PV is very promising in this regard due to its higher salt rejection and selectivity (Lakshmy et al. 2022). PVA, silica/silicate, graphene oxide (GO) and zeolite are the most studied materials for the desalination of water using PV (Li et al. 2023). The major concern with PV technology is its lowest permeation rate through membranes. Recent research has revealed that inorganic–organic hybrid membranes and TFC membranes have better performance. However, further research is needed to explore the membranes that can show better flux (Lakshmy et al. 2022).

13.3.4.4. *Essential oil extraction*

Essential oils are concentrated liquids extracted from plants (leaves, flowers, branches) and epicarps (peel) of citrus fruits that contain natural chemicals that give the plant its smell and flavor. Essential oils and their components have wide use in chemicals, cosmetics, food and pharmaceutical sectors. These oils have the potential use as a pest-combating agent, an additive in biodegradable films, in food preservation and in active packaging. Essential oils extracted from natural resources have essential oils that contain several terpenes such as limonene, linalool, myrcene, camphor, etc. Hence, suitable separation techniques are to be applied. Conventionally, liquid–liquid extraction and distillation are commonly used for extraction and purification purposes (Silvestre et al. 2019). The use of the

PV process could be an efficient technique for extracting essential oils. It provides a gentle and effective way to acquire high-quality essential oils used in cosmetics, aromatherapy and the food sector by selectively penetrating the volatile components via a membrane (Castro-Muñoz et al. 2023). PDMS, POMS, PVDF, polyvinylpyrrolidone (PVP) and PEBA are the most common polymer membranes tested for essential oil extraction by PV (Silvestre et al. 2019).

13.3.4.5. *Biofuel manufacturing*

Biofuels are the components derived from biomass, such as plant, algae or animal wastes. Alcohols are the major components used as biofuels. The use of microbial species can produce alcohols (such as ethanol and butanol). However, the concentration of alcohol in the fermentation broth is very low. Hence, recovery of such alcohols from dilute fermentation broth is essential for bio-alcohol production (Zheng et al. 2019). Owing to the selectivity of the membranes, PV has received a great deal of attention in recent years. It increases the effectiveness of the methods used to produce bioethanol by assisting in the achievement of the necessary ethanol concentrations for fuel consumption. Hydrophobic polymers are found to be the most versatile and potential membrane materials for ethanol recovery from the dilute broth. Many polymeric membranes have been reported in the literature such as PDMS, poly(1-trimethylsilyl-1-propyne) (PTMSP), polytetrafluoroethylene (PTFE), polyamide, PVDF, polypropylene, styrene-butadiene block copolymer (SBS), etc. (Peng et al. 2021).

In summary, PV finds use in a variety of sectors and offers customized solutions for intricate separation problems. Because of its adaptability, accuracy and energy efficiency, this technology is useful in a variety of industries and helps to enhance operations, lessen environmental effects and promote industrial practices. The potential uses for PV are expected to grow as research into bettering membrane materials and process parameters proceeds, paving the way for new developments in this cutting-edge separation method (Garg et al. 2023).

13.3.5. *Models*

As a sophisticated membrane separation technique, PV depends on complex models to maximize its effectiveness, especially when it comes to the difficult task of separating azeotropic mixtures. PV presents a strong challenge to azeotropes, which are combinations with constant boiling temperatures that are resistant to conventional separation methods because of their accuracy in molecular separation.

Transport models analyze the kinetics of mass transfer across a membrane, considering factors such as membrane porosity and thickness. These models are crucial for predicting the selectivity and separation efficiency of the process. Challenges become more pronounced when dealing with azeotropic mixtures, necessitating the utilization of more intricate models. These specialized models are essential because azeotropic systems exhibit nonideal behaviors and phase equilibria. Constructing prediction models that encompass both thermodynamic and transport aspects is vital. Additionally, advancements in computational techniques, such as molecular dynamics simulations, offer insights into the molecular-level interactions between components of azeotropic mixtures and the membrane (Wang et al. 2020; Yong and Zhang 2021).

The application of these models extends beyond theoretical understanding, providing valuable guidance for designing membranes optimized to address azeotropic challenges in material characteristics. Furthermore, by facilitating process optimization, these models empower engineers to enhance operations and maximize separation efficiency.

Apart from the transport mechanism (discussed in section 13.3.1), several other models (modified solution-diffusion model, thermodynamics model, Maxwell–Stefan model, computational model, etc.) have been proposed in the literature to explain the mass transport mechanism through the membrane for the PV process. Details of these models are available in the literature (Mujiburohman 2020).

In summary, PV models play a crucial role in navigating the complexities of azeotropic mixture separation. By integrating mass transfer, thermodynamics and membrane properties, these models contribute to the advancement of PV as a precise and efficient technique. Ongoing research into these models suggests the potential for PV to overcome challenges posed by azeotropic mixtures, offering a promising avenue for improving separation procedures in various industrial applications (Li et al. 2023).

13.4. Selection of membrane

The selection of a membrane is very crucial for the PV process. The selection is driven by the hydrophilic/hydrophobic or organophilic/organophobic properties of the materials and the nature of the target component (Figoli et al. 2015). Membrane materials include a polymer (PVA, PDMS, PVDF, etc.), inorganic materials (zeolite, metal oxime framework, silica, etc.), mixed-matrix membrane (MMM) (prepared by incorporating inorganic fillers into polymeric matrix) and composite materials (PVA-PVDF, PVA-Cellulose acetate, PVA-graphene oxide-PVDF, etc.). Optimizing PV for the separation of mixtures requires careful selection of the membrane. To ensure effective and selective separation, several considerations must be made while selecting a membrane for this particular application

(Liu and Jin 2021; Li et al. 2023). Some of the properties are discussed in the subsequent sections.

13.4.1. *Affinity for target components*

According to the solution–diffusion model, the PV process depends on the preferential sorption (solubility) and diffusion of the targeted component through the membrane matrix. The membrane needs to show a particular affinity for the targeted component from the mixture. The most common interactions are dipole–dipole, H-bonding and ion–dipole. This allows the target component to pass through the membrane selectively. Achieving good separation selectivity requires an understanding of the interaction between the mixture's constituents and the membrane material. However, a very high affinity may lead to swelling in the membrane which in turn decreases the selectivity and the durability of the membrane (Figoli et al. 2015).

13.4.2. *Permeability and selectivity*

Selectivity in PV is defined as the ratio of two components of a liquid mixture in the permeate to the same components in the feed. It dictates how the targeted components may pass through the membrane compared to other components. For higher selectivity, the membrane matrix should have active centers capable of interacting with the targeted molecule preferentially (Chaudhari et al. 2019).

The two properties of a perfect membrane should be balanced. Selectivity ensures that just the desired component penetrates preferentially, whereas permeability controls the rate at which molecules can flow across the membrane. To attain the best separation efficiency, the membrane parameters must be tailored to improve both features.

It is found in the literature that inorganic membranes have exceptionally high permeability and selectivity as well as thermal and chemical stabilities. However, there membranes suffer from limitations such as high costs and low mechanical stability during the manufacturing process and the low reproducibility of membrane properties (Figoli et al. 2015).

13.4.3. *Stability and durability*

PV membranes undergo severe operating conditions. High-stability and high-durability membranes are required for challenging working circumstances such as high temperatures and severe chemical environments. The membrane material should not degrade during operation. The membrane materials should have very high chemical and mechanical stability. By selecting the proper material, the long-term operational stability of the membrane can be obtained. The development of PV membranes with long-term operational stability with very high performance is still a challenge. The durability of the PV process is guaranteed by materials that can endure these circumstances without suffering appreciable deterioration. Inorganic membranes show exceptionally high thermal and chemical stabilities. However, these are very costly and show low mechanical stability during the manufacturing process (Figoli et al. 2015). PVA membranes are most widely used in PV. However, research shows that crosslinking of PVA with glutaraldehyde enhances the thermo-mechanical stability of the membrane and shows better PV selectivity (Guo et al. 2008).

13.4.4. *Material compatibility*

The membrane material should be compatible with the solvent and other components present in the feed. For the membrane to function consistently over time and to prevent unintended changes in its characteristics, it should not chemically react with the mixture's constituent parts or dissolve in the solvent.

13.4.5. *Thickness*

The membrane's thickness and porosity affect mass transfer and, in turn, the effectiveness of the separation process. As per the solution–diffusion model, the targeted solute should selectively adsorb and diffuse through the membrane during PV. The membrane resistance is given by equation [13.28] (Hyder et al. 2008).

$$R_{m,i} = \frac{L}{D_i S_i} \qquad [13.28]$$

where L is the membrane thickness, D_i is the diffusivity and S_i is the solubility coefficient of component i in the membrane.

As the membrane thickness increases, the resistance also increases, and subsequently, the flux decreases. However, the selectivity remains nearly unaffected. Reduced mass transfer resistance and regulated porosity in thin, well-designed membranes can let the target component penetrate more easily. However, the issue with very thin membranes is the mechanical strength. Hence, asymmetric membranes with a very thin active layer (known as skin) and porous support will provide better mechanical strength and higher flux.

13.4.6. *Resistance to fouling*

Fouling is a critical issue in any membrane process. In the PV process, fouling occurs when inorganic or organic contaminants present in the feed solution get deposited on the surface of the membrane or inside the pores. This increases the membrane resistance and reduces the permeate flux. Fouling depends on the surface roughness, surface charge or surface free energy (SFE) of the membrane. The SFE is commonly used to describe the membrane surface fouling by the contaminants. Lifshitz–van der Waals acid–base (LW-AB) approach is based on this technique that can evaluate the fouling tendency.

The LW-AB approach assumes that the SFE of two phases is due to two components:

a) Lifshitz–van der Waals (γ_{12}^{LW});

b) Lewis acid–base (γ_{12}^{AB}).

The Lifshitz–van der Waals (γ_{12}^{LW}) and Lewis acid–base (γ_{12}^{AB}) can be calculated using equations [13.29] and [13.30], respectively (Li et al. 2023).

$$\gamma_{12}^{LW} = \left(\sqrt{\gamma_1^{LW}} - \sqrt{\gamma_2^{LW}}\right)^2 \qquad [13.29]$$

$$\gamma_{12}^{AB} = 2\left(\sqrt{\gamma_1^{+}\gamma_1^{-}} + \sqrt{\gamma_2^{+}\gamma_2^{-}} - \sqrt{\gamma_1^{+}\gamma_2^{-}} - \sqrt{\gamma_1^{-}\gamma_2^{+}}\right) \qquad [13.30]$$

where γ^+ and γ^- (mJ m^{-2}) are the electron acceptor and electron donor parameters, respectively. These parameters can be evaluated by measuring the sessile drop contact angle using various polar and dispersive liquids, e.g. water, glycerol, ethylene glycol, diiodomethane, etc. (Li et al. 2023).

γ_{12} can be calculated using equation [13.31]

$$\gamma_{12} = \gamma_{12}^{LW} + \gamma_{12}^{AB} \qquad [13.31]$$

The total interaction energy for three phases (fouling solute (s) – water (w) – membrane (m)) can be given by equations [13.32]–[13.34].

$$\Delta G_{swm} = \Delta G_{swm}^{LW} + \Delta G_{swm}^{AB} \qquad [13.32]$$

$$\Delta G_{swm}^{LW} = \gamma_{sm}^{LW} - \gamma_{sw}^{LW} - \gamma_{wm}^{LW} \qquad [13.33]$$

$$\Delta G_{swm}^{AB} = \gamma_{sm}^{AB} - \gamma_{sw}^{AB} - \gamma_{wm}^{AB} \qquad [13.34]$$

The value of ΔG_{swm} is an indication of the fouling tendency of the solute. The larger negative value indicated a higher tendency of membrane fouling (Li et al. 2023).

To avoid fouling, the feed solution is to be pretreated for the removal of contaminants. Moreover, selecting membrane materials having lower SFE values can decrease the fouling tendency. Among the most commonly used polymers, the PVA membrane shows higher fouling and higher fouling resistance, whereas PVDF has a lower fouling resistance (Li et al. 2023).

13.4.7. *Cost and scalability*

Membrane materials should have a production cost commensurate with the profitability of the separation process, and they should be scalable for industrial applications. For practical implementation, performance and cost issues must be balanced. However, the cost is a major issue in most of the membrane processes due to membrane cost and membrane life cycle.

In summary, careful consideration of the material's qualities, compatibility with the working environment, and affinity for the target components are necessary when choosing a membrane for PV in the separation of azeotropic mixtures. By taking these factors into account, scientists and engineers will be able to better design membranes to the unique problems that azeotropic mixtures present, increasing the efficiency of PV in industrial settings (Salahshoori et al. 2021).

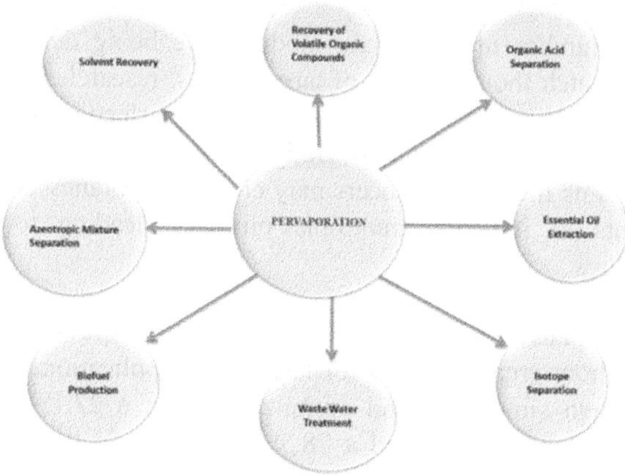

Figure 13.7. *Applications of pervaporation*

13.5. Recent developments

Recent years have seen major advancements in PV, a sophisticated membrane separation technique that provides creative solutions for the separation of azeotropic mixtures and other applications. These developments have broadened the use of PV in a variety of industrial applications, motivated by the convergence of materials science, engineering and sustainability concerns.

13.5.1. *Advanced membrane materials*

The selection of suitable and high-performance membrane materials remains a challenge for PV for various applications. The development of sophisticated materials specifically designed for PV separation has been at the forefront of recent advancements in membrane technology. Membranes that are nanocomposite and contain nanomaterials such as graphene or zeolites demonstrate improved permeability and selectivity. These materials solve issues that have traditionally prevented the broad use of PV and open the door to membranes with greater stability and fouling resistance under demanding operating circumstances. Most of the earlier studies focused on the use of polymeric membranes. High selectivity and flux can be achieved by modifying polymers by blending, composites, carbon nanotubes (CNTs), MOF and surface modification (Nouri et al. 2024). Currently, composite, MOF, and ceramic membranes, MMM, are gaining more interest in the PV process. The use of nanocomposite membranes and inorganic nanofillers in MMMs has been explored for PV recently (Abdali et al. 2024).

13.5.2. *Customized membrane module design*

An increasing number of membrane modules are being designed for specific PV processes. To match the properties of target mixes, researchers are investigating the optimization of membrane thickness, porosity and surface chemistry. Membrane tailoring improves selectivity and overall separation efficiency in particular applications. With this method, engineers may create membranes that precisely bind to the constituents of intricate azeotropic mixtures, leading to more efficient separation procedures.

Owing to high packing density and flexibility, the spiral wound membrane (SWM) modules show great potential for large-scale applications. Lu et al. (2024) demonstrated through simulation and optimization that a 22.1% improvement in ethanol mass transfer coefficient and a 78.5% reduction in the module's specific energy consumption can be achieved using the SWM module. A study conducted by Vatankhah et al. (2021) showed that approximately a 97% reduction in the membrane area can be achieved by using a multi-stage PV design for isopropanol–water separation. More research is needed for efficient module design.

13.5.3. *Integration with different technologies*

One noteworthy recent advancement is the hybrid processes that are produced when PV is combined with different separation technologies, adopting the advantages of each approach. For example, combining PV with distillation or absorption may result in more economical and efficient azeotropic separations. The goal of hybrid techniques is to get beyond the drawbacks of individual separation. An azeotropic mixture of ethanol–water can be separated into absolute ethanol by integrating distillation and PV as shown in Figure 13.8(A). The ethanol is concentrated in the distillate received from the distillation unit. The distillate is then subjected to PV to obtain the absolute ethanol (Purkait et al. 2020).

Haelssig et al. (2012) developed a single-unit membrane delegation process for the separation of an ethanol–water azeotrope. The schematic diagram is shown in Figure 13.8(B). Tubular hydrophilic zeolite membranes were used inside the column that allows the water vapor to pass through. This integrated distillation–PV process improves the efficiency and total energy requirement (Liu et al. 2022).

Boontawan (2012) developed a hybrid system consisting of a continuous stirred tank reactor (CSTR)–distillation–PV system for the esterification of succinic acid into diethyl succinate. A fractionating column is fitted with the CSTR that continuously removes water and ethanol. It increases the conversion rate. Ethanol is then dehydrated by the PV process and recycled to the esterification reactor.

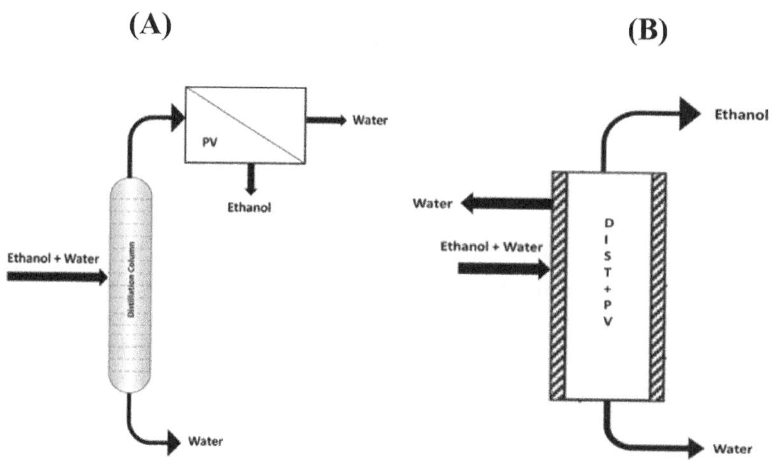

(C)

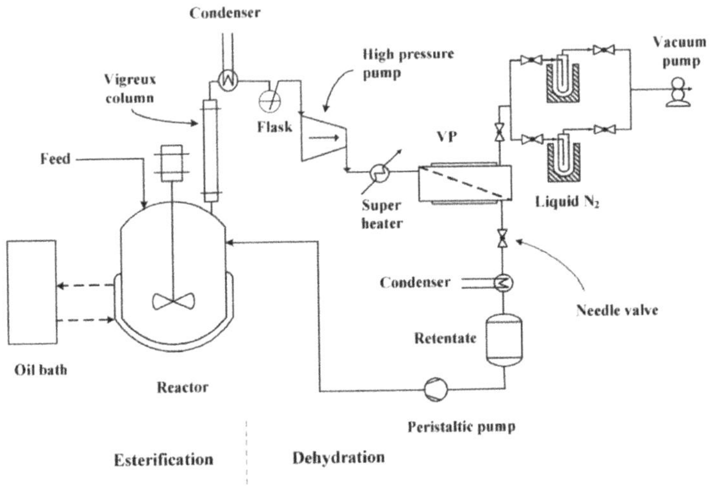

Figure 13.8. *Configurations of various integrated distillation–pervaporation systems: (A) integrated distillation–pervaporation system; (B) single unit distillation–PV system and (C) continuous stirred tank reactor–distillation–PV system (reproduced with permission from Liu et al. (2022))*

13.5.4. *Scalability and process optimization*

It is critical to guarantee that PV processes can be scaled up or down for use in industrial settings. To meet the requirements of large-scale industrial operations, recent innovations have concentrated on improving process parameters, membrane size and operating conditions. Scalability is essential for the effective use of PV in a variety of sectors, including petrochemicals and chemicals.

13.6. Future scope and emerging trends

Research is now being conducted to overcome obstacles and explore new applications for PV in azeotropic separation, which bodes well for the technology's future. Important developments include an emphasis on the selective separation of difficult azeotropes, process optimization for higher energy efficiency and possible uses in developing sectors like the manufacturing of biofuel and renewable energy. Because of its adaptability, PV is a dynamic, developing technology that has the potential to completely transform the way that materials are separated in a variety of industries.

Although encouraging results on ethanol separation using PV have been observed in the literature, it lacks extensive industrial applications due to the unavailability of suitable PV membrane modules. Further research is needed to explore economic industrial module design.

Conclusively, the growing field of PV technology and its latest advancements highlight its significance as a sophisticated membrane technique for azeotropic mixture separation. PV is anticipated to significantly improve efficiency as research into membrane materials and process optimization progresses.

13.7. References

Abdali, A., Eskandarabadi, S.M., Mahmoudian, M., Hakimi Kuranabadi, S. (2024). Inorganic nanofillers in mix matrix membranes for pervaporation process: A review. *Polymer*, 312, 127575.

Ahmed, S.F., Mehejabin, F., Momtahin, A., Tasannum, N., Faria, N.T., Mofijur, M., Hoang, A.T., Vo, D.-V.N., Mahlia, T.M.I. (2022). Strategies to improve membrane performance in wastewater treatment. *Chemosphere*, 306, 135527.

Boontawan, A. (2012). Purification of succinic acid from synthetic solution using vapor permeation-assisted esterification coupled with reactive distillation. *Advanced Materials Research*, 550–553, 3008–3011.

Castro-Muñoz, R. (2020). Breakthroughs on tailoring pervaporation membranes for water desalination: A review. *Water Research*, 187, 116428.

Castro-Muñoz, R., Msahel, A., Galiano, F., Serocki, M., Ryl, J., Hamouda, S.B., Hafiane, A., Boczkaj, G., Figoli, A. (2022). Towards azeotropic MeOH-MTBE separation using pervaporation chitosan-based deep eutectic solvent membranes. *Separation and Purification Technology*, 281, 119979.

Castro-Muñoz, R., Ahmad, M.Z., Cassano, A. (2023). Pervaporation-aided processes for the selective separation of aromas, fragrances and essential (AFE) solutes from agro-food products and wastes. *Food Reviews International*, 39(3), 1499–1525.

Chaudhari, S., Kwon, Y., Shon, M., Nam, S., Park, Y. (2019). Stability and pervaporation characteristics of PVA and its blend with PVAm membranes in a ternary feed mixture containing highly reactive epichlorohydrin. *RSC Advances*, 9(11), 5908–5917.

Cheryan, M. (1998). *Ultrafiltration and Microfiltration Handbook*. CRC Press, Boca Raton.

De, S. (2011). NPTEL lecture on novel separation processes [Online]. Available at: https://nptel.ac.in/courses/103105060.

Devi, D.A., Smitha, B., Sridhar, S., Aminabhavi, T.M. (2006). Dehydration of 1,4-dioxane through blend membranes of poly(vinyl alcohol) and chitosan by pervaporation. *Journal of Membrane Science*, 280(1), 138–147.

Dong, G., Nagasawa, H., Yu, L., Wang, Q., Yamamoto, K., Ohshita, J., Kanezashi, M., Tsuru, T. (2020). Pervaporation removal of methanol from methanol/organic azeotropes using organosilica membranes: Experimental and modeling. *Journal of Membrane Science*, 610, 118284.

Du, C., Du, J.R., Feng, X., Wang, J. (2021). Green extraction of perilla volatile organic compounds by pervaporation. *Separation and Purification Technology*, 261, 118281.

Dutta, B.K. (2007). *Principles of Mass Transfer and Seperation Processes*. PHI Learning, Delhi.

Fatima, S., Govardhan, B., Kalyani, S., Sridhar, S. (2022). Extraction of volatile organic compounds from water and wastewater by vacuum-driven membrane process: A comprehensive review. *Chemical Engineering Journal*, 434, 134664.

Figoli, A., Santoro, S., Galiano, F., Basile, A. (2015). Pervaporation membranes: Preparation, characterization, and application. In *Pervaporation, Vapour Permeation and Membrane Distillation*, Basile, A., Figoli, A., Khayet, M. (eds). Woodhead Publishing, Oxford.

Garg, S., Behera, S., Ruiz, H.A., Kumar, S. (2023). A review on opportunities and limitations of membrane bioreactor configuration in biofuel production. *Applied Biochemistry and Biotechnology*, 195(9), 5497–5540.

Greenlaw, F., Shelden, R., Thompson, E. (1977). Dependence of diffusive permeation rates on upstream and downstream pressures: II. Two component permeant. *Journal of Membrane Science*, 2, 333–348.

Guo, R., Hu, C., Pan, F., Wu, H., Jiang, Z. (2006). PVA–GPTMS/TEOS hybrid pervaporation membrane for dehydration of ethylene glycol aqueous solution. *Journal of Membrane Science*, 281(1), 454–462.

Guo, R., Fang, X., Wu, H., Jiang, Z. (2008). Preparation and pervaporation performance of surface crosslinked PVA/PES composite membrane. *Journal of Membrane Science*, 322(1), 32–38.

Haelssig, J.B., Tremblay, A.Y., Thibault, J. (2012). A new hybrid membrane separation process for enhanced ethanol recovery: Process description and numerical studies. *Chemical Engineering Science*, 68(1), 492–505.

Hyder, M.N., Huang, R.Y.M., Chen, P. (2008). Effect of selective layer thickness on pervaporation of composite poly(vinyl alcohol)–poly(sulfone) membranes. *Journal of Membrane Science*, 318(1), 387–396.

Kammakakam, I. and Lai, Z. (2023). Next-generation ultrafiltration membranes: A review of material design, properties, recent progress, and challenges. *Chemosphere*, 316, 137669.

Kavitha, E., Poonguzhali, E., Nanditha, D., Kapoor, A. (2022). Current status and future prospects of membrane separation processes for value recovery from wastewater. *Chemosphere*, 291(Part 2), 132690.

Kober, P.A. (1917). Pervaporation, perstillation and percrystallization. *Journal of the American Chemical Society*, 39(5), 944–948.

Kulprathipanja, S., Neuzil, R.W., Li, N.N. (1992). Separation of gases by means of mixed matrix membranes. United States Patents, 5,127,925.

Lakshmy, K.S., Lal, D., Nair, A., Babu, A., Das, H., Govind, N., Dmitrenko, M., Kuzminova, A., Korniak, A., Penkova, A., Tharayil, A., Thomas, S. (2022). Pervaporation as a successful tool in the treatment of industrial liquid mixtures. *Polymers (Basel)*, 14(8), 1604.

Lewandowska, K. (2005). The miscibility of poly(vinyl alcohol)/poly(N-vinylpyrrolidone) blends investigated in dilute solutions and solids. *European Polymer Journal*, 41(1), 55–64.

Li, Y., Thomas, E.R., Molina, M.H., Mann, S., Walker, W.S., Lind, M.L., Perreault, F. (2023). Desalination by membrane pervaporation: A review. *Desalination*, 547, 116223.

Liu, G. and Jin, W. (2021). Pervaporation membrane materials: Recent trends and perspectives. *Journal of Membrane Science*, 636, 119557.

Liu, S., Li, H., Kruber, B., Skiborowski, M., Gao, X. (2022). Process intensification by integration of distillation and vapor permeation or pervaporation – An academic and industrial perspective. *Results in Engineering*, 15, 100527.

Lu, X., Huang, J., Pinelo, M., Chen, G., Wan, Y., Luo, J. (2024). Optimizing spiral-wound pervaporation membrane modules through simulation: Unraveling the permeate spacer structure. *AIChE Journal*, 70(10), e18523.

Mandal, S. and Pangarkar, V.G. (2002). Separation of methanol–benzene and methanol–toluene mixtures by pervaporation: Effects of thermodynamics and structural phenomenon. *Journal of Membrane Science*, 201(1), 175–190.

Mandal, S. and Pangarkar, V.G. (2003). Development of co-polymer membranes for pervaporative separation of methanol from methanol–benzene mixture – A solubility parameter approach. *Separation and Purification Technology*, 30(2), 147–168.

Mandal, M.K., Dutta, S., Bhattacharya, P.K. (2008). Characterization of blended polymeric membranes for pervaporation of hydrazine hydrate. *Chemical Engineering Journal*, 138(1), 10–19.

Mandal, M.K., Sant, S.B., Bhattacharya, P.K. (2011). Dehydration of aqueous acetonitrile solution by pervaporation using PVA–iron oxide nanocomposite membrane. *Colloids and Surfaces A: Physicochemical and Engineering Aspects*, 373(1), 11–21.

Mansur, H.S., Oréfice, R.L., Mansur, A.A.P. (2004). Characterization of poly(vinyl alcohol)/poly(ethylene glycol) hydrogels and PVA-derived hybrids by small-angle X-ray scattering and FTIR spectroscopy. *Polymer*, 45(21), 7193–7202.

Mujiburohman, M. (2020). Computational modeling of pervaporation process. In *Polymer Nanocomposite Membranes for Pervaporation*, Thomas, S., George, S.C., Jose, T. (eds). Elsevier, Amsterdam.

Nouri, M., Poorkhalil, A., Farrokhzad, H. (2024). Modification methods categorization of polymeric membranes for pervaporation: A review. *Polymer Reviews*, 1–28. doi: 10.1080/15583724.2024.2396853.

Okada, T. and Matsuura, T. (1991). A new transport model for pervaporation. *Journal of Membrane Science*, 59(2), 133–149.

Pabby, A.K., Rizvi, S.S.H., Sastre, A.M. (eds) (2009). *Handbook of Membrane Separations*. CRC Press, Boca Raton.

Peng, P., Lan, Y., Liang, L., Jia, K. (2021). Membranes for bioethanol production by pervaporation. *Biotechnology for Biofuels*, 14(1), 10.

Purkait, M.K., Singh, R., Mondal, P., Haldar, D. (2020). Pervaporation. In *Thermal Induced Membrane Separation Processes*, Purkait, M.K., Singh, R., Mondal, P., Haldar, D. (eds). Elsevier, Amsterdam.

Razzak, M.T., Zainuddin, Erizal, Dewi, S.P., Lely, H., Taty, E., Sukirno (1999). The characterization of dressing component materials and radiation formation of PVA–PVP hydrogel. *Radiation Physics and Chemistry*, 55(2), 53–165.

Salahshoori, I., Seyfaee, A., Babapoor, A. (2021). Recent advances in synthesis and applications of mixed matrix membranes. *Synthesis and Sintering*, 1(1), 1–27.

Saleh, T.A. and Gupta, V.K. (2016). An overview of membrane science and technology. In *Nanomaterial and Polymer Membranes*, Saleh, T.A. and Gupta, V.K. (eds). Elsevier, Amsterdam.

Shao, P. and Huang, R.Y.M. (2007). Polymeric membrane pervaporation. *Journal of Membrane Science*, 287(2), 162–179.

Sidhikku Kandath Valappil, R., Ghasem, N., Al-Marzouqi, M. (2021). Current and future trends in polymer membrane-based gas separation technology: A comprehensive review. *Journal of Industrial and Engineering Chemistry*, 98, 103–129.

Silvestre, W.P., Livinalli, N.F., Baldasso, C., Tessaro, I.C. (2019). Pervaporation in the separation of essential oil components: A review. *Trends in Food Science & Technology*, 93, 42–52.

Smitha, B., Suhanya, D., Sridhar, S., Ramakrishna, M. (2004). Separation of organic–organic mixtures by pervaporation – A review. *Journal of Membrane Science*, 241(1), 1–21.

Song, Y., Pan, F., Li, Y., Quan, K., Jiang, Z. (2019). Mass transport mechanisms within pervaporation membranes. *Frontiers of Chemical Science and Engineering*, 13(3), 458–474.

Tanudjaja, H.J., Anantharaman, A., Ng, A.Q.Q., Ma, Y., Tanis-Kanbur, M.B., Zydney, A.L., Chew, J.W. (2022). A review of membrane fouling by proteins in ultrafiltration and microfiltration. *Journal of Water Process Engineering*, 50, 103294.

Tawalbeh, M., Qalyoubi, L., Al-Othman, A., Qasim, M., Shirazi, M. (2023). Insights on the development of enhanced antifouling reverse osmosis membranes: Industrial applications and challenges. *Desalination*, 553, 116460.

Thiyagarajan, R., Ravi, S., Bhattacharya, P.K. (2011). Pervaporation of methyl–ethyl ketone and water mixture: Determination of concentration profile. *Desalination*, 277(1), 178–186.

Vatankhah, F., Moheb, A., Mehrabani-Zeinabad, A. (2021). A study on the effects of feed temperature and concentration on design of a multi-stage pervaporation system for isopropanol-water separation using commercial available modules with inter-stage heating. *Journal of Membrane Science*, 618, 118717.

Wang, L., Wang, Y., Wu, L., Wei, G. (2020). Fabrication, properties, performances, and separation application of polymeric pervaporation membranes: A review. *Polymers*, 12(7), 1466.

Yang Kong, Z., Yeh Lee, H., Sunarso, J. (2022). The evolution of process design and control for ternary azeotropic separation: Recent advances in distillation and future directions. *Separation and Purification Technology*, 284, 120292.

Yong, W.F. and Zhang, H. (2021). Recent advances in polymer blend membranes for gas separation and pervaporation. *Progress in Materials Science*, 116, 100713.

Zhang, H., Jiao, Y., Zhao, Q., Li, C., Cui, P., Wang, Y., Zheng, S., Li, X., Zhu, Z., Gao, J. (2022). Sustainable separation of ternary azeotropic mixtures based on enhanced extractive distillation/pervaporation structure and multi-objective optimization. *Separation and Purification Technology*, 298, 121685.

Zheng, P., Li, C., Wang, N., Li, J., An, Q. (2019). The potential of pervaporation for biofuel recovery from fermentation: An energy consumption point of view. *Chinese Journal of Chemical Engineering*, 27(6), 1296–1306.

Zhou, F. (2005). *Novel Pervaporation for Separating Acetic Acid and Water Mixtures Using Hollow Fiber Membranes*. Georgia Institute of Technology, Atlanta.

List of Authors

Leela Manohar AESHALA
Department of Chemical Engineering
National Institute of Technology
Warangal
India

Ebrahim ALLAHKARAMI
Department of Extraction and Processing
Institute of Strategic Elements
Iranian Research Organization for Science and Technology (IROST)
Tehran
Iran

Esmaeil ALLAHKARAMI
Faculty of Petroleum
Gas and Petrochemical Engineering
Persian Gulf University
Bushehr
and
Persian Gulf Star Oil Company
Bandar Abbas
Iran

Raj Kumar ARYA
Department of Chemical Engineering
Dr. B. R. Ambedkar National Institute of Technology
Jalandhar
India

Chitresh Kumar BHARGAVA
IITB-Monash Research Academy
Indian Institute of Technology
Mumbai
India

Karthikeyan C.
Department of Chemical Engineering
Indian Institute of Technology
Madras
Tamil Nadu
India

Avinash CHANDRA
Department of Chemical Engineering
Thapar Institute of Engineering & Technology
Patiala
India

Juli DAS
School of Biosciences
and Technology
Vellore Institute of Technology
Tamil Nadu
India

Dipaloy DATTA
Department of Chemical Engineering
Malaviya National Institute of
Technology Jaipur (MNIT)
India

J. Paulo DAVIM
Department of Mechanical
Engineering
Universidade de Aveiro
Portugal

Diptanu DE
School of Biosciences
and Technology
Vellore Institute of Technology
Tamil Nadu
India

Neha DHIMAN
Department of Chemistry
Goswami Ganesh Dutta Sanatan
Dharma College
Chandigarh
India

Sangeeta GARG
Department of Chemical Engineering
Ambedkar National Institute
of Technology
Punjab
India

Abhishek GUPTA
Department of Chemical Engineering
Faculty of Engineering
and Technology
Marwadi University
Gujarat
India

Girish GUPTA
Dr. Shanti Swarup Bhatnagar
University Institute of Chemical
Engineering and Technology
Panjab University
Chandigarh
India

G.C. JEEVITHA
School of Biosciences and
Technology
Vellore Institute of Technology
Tamil Nadu
India

Kedar JOSHI
Department of Chemical Engineering
Faculty of Engineering
and Technology
Marwadi University
Gujarat
India

MUTHAMIZHI K.
Department of Chemical Engineering
National Institute of Technology
Trichy
India

Abhishek KUMAR
Department of Chemical Engineering
National Institute of Technology
Hamirpur
India

Rahul KUMAR
College of Technology
Sardar Vallabhbhai Patel University
of Agriculture & Technology
Meerut
India

Aanchal MITTAL
School of Biosciences and
Technology
Vellore Institute of Technology
Tamil Nadu
India

A.B. MOHAMMED BASITH
Hochschule Offenburg
Germany

Ananya MUNNANGI
Department of Chemical Engineering
Indian Institute of Technology
Madras
Tamil Nadu
India

Tapas PALAI
Department of Chemical Engineering
National Institute of Technology
Hamirpur
India

Vyomesh M. PARSANA
Chemical Engineering Department,
V.V.P. Engineering College
Gujarat Technological University
India

K.S.N.V. PRASAD
Department of Chemical Engineering
BV Raju Institute of Technology
Telangana
India

B.K. PUROHIT
Department of Chemical Engineering
BV Raju Institute of Technology
Telangana
India

A.V. RAGHAVENDRA RAO
Department of Chemical Engineering
BV Raju Institute of Technology
Telangana
India

V. Sravani SAMEERA
Department of Chemical Engineering
BV Raju Institute of Technology
Telangana
India

Pramita SEN
Department of Chemical Engineering
Heritage Institute of Technology
Kolkata
India

G. SRINIVAS
Department of Chemical Engineering
BV Raju Institute of Technology
Telangana
India

Thangallapalli SRINIVAS
Department of Chemical Engineering
BV Raju Institute of Technology
Telangana
India

Rompicherla SRIVIDYA
Department of Chemical Engineering
BV Raju Institute of Technology
Telangana
India

Dharmesh SUR
Department of Chemical Engineering
Faculty of Engineering
and Technology
Marwadi University
Gujarat
India

K. TAMILSELVI
Department of Chemical Engineering
Engineering College
Anjalai Ammal Mahalingam
Tamil Nadu
India

Kshitij TEWARI
Department of Chemical and
Biomedical Engineering
West Virginia University
Morgantown
USA

Amit K. THAKUR
Energy Cluster
UPES
Dehradun
India

Devyani THAPLIYAL
Department of Chemical Engineering
National Institute of Technology
Punjab
India

George D. VERROS
Department of Chemistry
Aristotle University of Thessaloniki
Greece

Index

A, B

C, D

V, W

Summary of Volume 2

Chapter 2. Reactive Chromatography: A Concept of Multifunctional Reactors
Praveen Kumar GHODKE, Sudip DAS and Rohidas BHOI

Chapter 3. Mathematical Modeling of a Batch Reactor and a Non-Isothermal CSTR with Their Respective Simulation Using MATLAB and ASPEN PLUS

Karthikeyan C., Praveen Kumar V., Preetha V. and Faheem ARAKKAL

Part 2. Material Properties and Advanced Applications

Chapter 4. Properties of Materials and Selection Criteria

Dharmesh SUR, Abhishek GUPTA, Swati DUBEY and Avanish KUMAR

Other titles from

iSTE

in

Chemical Engineering

2023

LAURENT André
Towards Process Safety 4.0 in the Factory of the Future

2020

BELAADI Salah
Thermodynamic Processes 1: Systems without Physical State Change
Thermodynamic Processes 2: State and Energy Change Systems

DAL PONT Jean-Pierre, DEBACQ Marie
Process Industries 1: Sustainability, Managerial and Scientific Fundamentals
Process Industries 2: Digitalization, a New Key Driver for Industrial Management

SCHAER Éric, ANDRÉ Jean-Claude
Process Engineering Renewal 1: Background and Training
Process Engineering Renewal 2: Research
Process Engineering Renewal 3: Prospects

2019

HARMAND Jérôme, LOBRY Claude, RAPAPORT Alain, SARI Tewfik
Optimal Control in Bioprocesses: Pontryagin's Maximum Principle in Practice
(Chemostat and Bioprocesses Set – Volume 3)

2018

LOBRY Claude
The Consumer–Resource Relationship: Mathematical Modeling
(Chemostat and Bioprocesses Set – Volume 2)

2017

HARMAND Jérôme, LOBRY Claude, RAPAPORT Alain, SARI Tewfik
The Chemostat: Mathematical Theory of Microorganims Cultures
(Chemostat and Bioprocesses Set – Volume 1)

2016

SOUSTELLE Michel
Phase Transformations
(Chemical Thermodynamics Set – Volume 5)
Ionic and Electrochemical Equilibria
(Chemical Thermodynamics Set – Volume 6)
Thermodynamics of Surfaces and Capillary Systems
(Chemical Thermodynamics Set – Volume 7)

2015

SOUSTELLE Michel
Phase Modeling Tools: Application to Gases
(Chemical Thermodynamics Set – Volume 1)
Modeling of Liquid Phases
(Chemical Thermodynamics Set – Volume 2)
Thermodynamic Modeling of Solid Phases
(Chemical Thermodynamics Set – Volume 3)
Chemical Equilibria
(Chemical Thermodynamics Set – Volume 4)

2014

DAL PONT Jean-Pierre, AZZARO-PANTEL Catherine
New Approaches to the Process Industries: The Manufacturing Plant of the Future

HAMAIDE Thierry, DETERRE Rémi, FELLER Jean-François
Environmental Impact of Polymers

2012

DAL PONT Jean-Pierre
Process Engineering and Industrial Management

2011

MORY Mathieu
Fluid Mechanics for Chemical Engineering

SOUSTELLE Michel
An Introduction to Chemical Kinetics

2010

SOUSTELLE Michel
Handbook of Heterogenous Kinetics

Printed and bound by CPI Group (UK) Ltd, Croydon, CR0 4YY
16/09/2026
14963089-0006